T0345171

Fundamentals of Mathematical Statistics

Fundamentals of Mathematical Statistics is meant for a standard one-semester advanced undergraduate or graduate-level course in Mathematical Statistics. It covers all the key topics—statistical models, linear normal models, exponential families, estimation, asymptotics of maximum likelihood, significance testing, and models for tables of counts. It assumes a good background in mathematical analysis, linear algebra, and probability but includes an appendix with basic results from these areas. Throughout the text, there are numerous examples and graduated exercises that illustrate the topics covered, rendering the book suitable for teaching or self-study.

Features

- A concise yet rigorous introduction to a one-semester course in Mathematical Statistics
- Covers all the key topics
- Assumes a solid background in Mathematics and Probability
- Numerous examples illustrate the topics
- Many exercises enhance understanding of the material and enable course use

This textbook will be a perfect fit for an advanced course in Mathematical Statistics or Statistical Theory. The concise and lucid approach means it could also serve as a good alternative, or supplement, to existing texts.

Steffen Lauritzen is Emeritus Professor of Statistics at the University of Copenhagen and the University of Oxford as well as Honorary Professor at Aalborg University. He is most well known for his work on graphical models, in particular represented in a monograph from 1996 with that title, but he has published in a wide range of topics. He has received numerous awards and honours, including the Guy Medal in Silver from the Royal Statistical Society, where he also is an Honorary Fellow. He was elected to the Royal Danish Academy of Sciences and Letters in 2008 and became a Fellow of the Royal Society in 2011.

CHAPMAN & HALL/CRC
Texts in Statistical Science Series

Recently Published Titles

Sampling
Design and Analysis, Third Edition
Sharon L. Lohr

Theory of Statistical Inference
Anthony Almudevar

Probability, Statistics, and Data
A Fresh Approach Using R
Darrin Speegle and Brain Claire

Bayesian Modeling and Computation in Python
Osvaldo A. Martin, Raviv Kumar and Junpeng Lao

Bayes Rules!
An Introduction to Applied Bayesian Modeling
Alicia Johnson, Miles Ott and Mine Dogucu

Stochastic Processes with R
An Introduction
Olga Korosteleva

Introduction to Design and Analysis of Scientific Studies
Nathan Taback

Practical Time Series Analysis for Data Science
Wayne A. Woodward, Bivin Philip Sadler and Stephen Robertson

Statistical Theory
A Concise Introduction, Second Edition
Felix Abramovich and Ya'acov Ritov

Applied Linear Regression for Longitudinal Data
With an Emphasis on Missing Observations
Frans E.S. Tan and Shahab Jolani

Fundamentals of Mathematical Statistics
Steffen Lauritzen

For more information about this series, please visit: https://www.routledge.com/Chapman--HallCRC-Texts-in-Statistical-Science/book-series/CHTEXSTASCI

Fundamentals of Mathematical Statistics

Steffen Lauritzen

University of Copenhagen, Denmark

University of Oxford, United Kingdom

CRC Press is an imprint of the
Taylor & Francis Group, an **informa** business
A CHAPMAN & HALL BOOK

First edition published 2023

by CRC Press
6000 Broken Sound Parkway NW, Suite 300, Boca Raton, FL 33487-2742

and by CRC Press
4 Park Square, Milton Park, Abingdon, Oxon, OX14 4RN

CRC Press is an imprint of Taylor & Francis Group, LLC

Library of Congress Cataloging-in-Publication Data

Names: Lauritzen, Steffen L., author.
Title: Fundamentals of mathematical statistics / Steffen Lauritzen.
Description: First edition. | Boca Raton, FL : C&H/CRC Press, 2023. | Series: Chapman and Hall/CRC texts in statistical science | Includes bibliographical references and index.
Identifiers: LCCN 2022037917 (print) | LCCN 2022037918 (ebook) | ISBN 9781032223827 (hbk) | ISBN 9781032223834 (pbk) | ISBN 9781003272359 (ebk)
Subjects: LCSH: Mathematical statistics.
Classification: LCC QA276 .L324 2023 (print) | LCC QA276 (ebook) | DDC 519.5--dc23/eng20221209
LC record available at https://lccn.loc.gov/2022037917
LC ebook record available at https://lccn.loc.gov/2022037918

ISBN: 978-1-032-22382-7 (hbk)
ISBN: 978-1-032-22383-4 (pbk)
ISBN: 978-1-003-27235-9 (ebk)

DOI: 10.1201/9781003272359

Publisher's note: This book has been prepared from camera-ready copy provided by the authors.

Contents

Preface

This book has been prepared for use in the second-year course in Mathematical Statistics at the University of Copenhagen, as a part of the curriculum for a B.Sc. in Mathematics. It gives an exposition of fundamental elements of Mathematical Statistics and parametric inference, chiefly based on the method of maximum likelihood and focusing on exponential families, which in particular enables establishing the asymptotic properties of the method of maximum likelihood with some rigour.

The general structure and contents have been inspired by Hansen (2012) who follows on from more than 50 years of teaching tradition in Copenhagen, notably based on lecture notes by Steen Andersson and Søren Tolver Jensen. These again build on the first comprehensive course in Mathematical Statistics created by Hans Brøns in the 1960s, inspired by Hald (1952) and refined over the years by many colleagues.

A key concept in this tradition is the focus on the *statistical model* and its properties, rather than the design of various ad hoc methods; the model identifies a framework used to reason about the problems considered, and the methods are derived from general principles, here mostly the method of maximum likelihood. An important feature has been to give a geometric treatment of the multivariate normal distribution and associated linear normal models on a finite-dimensional vector space.

I have also been inspired by material prepared by Helle Sørensen who has been teaching the course before me. Draft editions of the book have been used for teaching this course by Niels Richard Hansen, Anders Tolver, and myself, and I am grateful for detailed and constructive comments received from Anders Tolver and Niels Richard Hansen at various stages in the process of writing this text.

The book assumes that a course in Probability based on Measure Theory is taught in parallel or prior to this course and that the students have a good background in Mathematical Analysis and Linear Algebra. Some important results from these areas are briefly described in appendices.

It is useful and probably necessary that the students previously have been acquantied with general statistical thinking and simple statistical methods at least up to the level of simple linear regression in the normal distribution.

Copenhagen, July 2022. Steffen Lauritzen

List of Figures

List of Figures

List of Tables

Chapter 1

Statistical Models

1.1 Models and parametrizations

The basic object in mathematical statistics is a *statistical model* defined formally as follows:

Definition 1.1. A *statistical model* consists of a measure space $(\mathcal{X}, \mathbb{E})$ and a family $\mathcal{P}$ of probability measures on $(\mathcal{X}, \mathbb{E})$. The space $(\mathcal{X}, \mathbb{E})$ is the *representation space* of the model.

The interpretation of a statistical model is that *data* x is an observed outcome $X = x$ of a random variable X with values in $\mathcal{X}$ and distribution given by an *unknown* probability measure $P \in \mathcal{P}$. The task of the statistician is to say something sensible about P, based on the observation x. This process is known as *statistical inference*. In the machine learning literature, it is more common to refer to this task as that of *learning* P from data x.

This formulation of the basic statistical problem and associated inference task was introduced by *Sir Ronald Aylmer Fisher* (1890–1962) in Fisher (1922), and we shall sometimes refer to such a statistical model as a *Fisherian statistical model*. An alternative approach to statistical inference is based on a *Bayesian statistical model*, but we shall only briefly touch on this approach in Section 6.6.

The reader should be aware that the notion of a statistical model as used here is quite different from the notion of a model as used by most scientists. A scientist would rather think of the specific P as the model, rather than the family $\mathcal{P}$. This is a source of considerable confusion, also among statisticians who sometimes use the term 'true model' for the specific P. We shall avoid this term in the subsequent developments.

The role of a statistical model is primarily to specify the framework within which our reasoning about the problem takes place. Sometimes the model has the purpose of giving a good description of reality, but other times—and quite often—it is rather used as a basis for saying something interesting about reality, for example that certain aspects of the observations do not conform with the model used, thereby providing a basis for further understanding of the reality behind the observations.

It is convenient that the family $\mathcal{P}$ is *parametrized*, i.e. that we have a surjective map $\nu : \Theta \mapsto \mathcal{P}$ such that

$$\mathcal{P} = \{\nu(\theta) \mid \theta \in \Theta\} = \{P_\theta \mid \theta \in \Theta\}$$

DOI: 10.1201/9781003272359-1

where we traditionally write P_θ instead of $\nu(\theta)$. Such a map is a *parametrization* of the statistical model, the space Θ is the *parameter space*, and θ is a *parameter*. Sometimes the parametrization is just a technical device to label the measures in $\mathcal{P}$, but the parameter θ might be a quantity of its own interest in the scientific domain where data x have been collected to determine θ.

As we shall discuss in detail later, statistical inference can take many shapes. Methods of *estimation* are concerned with systematic attempts of guessing the value of θ that is behind the observation x (see Chapter 4); *set estimation* generates a subset $C(x) \subseteq \Theta$ as a function of x within which the unknown parameter θ may be; this is discussed in Chapter 6. Methods for *hypothesis testing*, described in Chapter 7, are concerned with assessing whether the observation x conforms with a value $\theta \in \Theta_0$, where Θ_0 is a pre-specified subset of parameter values in Θ. Clearly, these activities are intimately related.

Generally the parameter space Θ can be any set, but in many cases Θ is a well-behaved subset of $\mathbb{R}^k$ with non-empty interior, in which case it is common to say that the family or model is *parametric* and has *dimension* k. Also, we would mostly assume that the map $\theta \mapsto P_\theta$ is injective, so that $\theta_1 \neq \theta_2 \implies P_{\theta_1} \neq P_{\theta_2}$ and often we shall assume or exploit that the parametrization is a 'nice' function of θ in some specified way.

If there is no simple parametrization with $\Theta \subseteq \mathbb{R}^k$, we say the model is *non-parametric*. An example of a non-parametric model could be the family $\mathcal{P}_2$ of probability measures on $(\mathcal{X}, \mathbb{E}) = (\mathbb{R}, \mathbb{B})$ with finite second moments, i.e.

$$P \in \mathcal{P}_2 \iff \int_{-\infty}^{\infty} x^2 \, P(dx) < \infty.$$

This book has its focus on parametric statistical models although we from time to time give examples of non-parametric ones.

1.1.1 *Examples of statistical models*

Here we shall briefly mention a number of common and useful statistical models. We shall focus on their mathematical specification rather than their application to specific problems, and many of them will reappear from time to time in later sections and chapters.

Example 1.2. [The simple Poisson model] In this model we consider a random variable X taking values in $\mathcal{X} = \mathbb{N}_0 = \{0, 1, 2, \ldots\}$, and the *family of probability measures* is $\mathcal{P} = \{P_\lambda \mid \lambda \in \Lambda\}$, where P_λ is the Poisson distribution

$$P_\lambda(X = x) = \frac{\lambda^x}{x!} e^{-\lambda}, \quad x \in \mathbb{N}_0. \tag{1.1}$$

We must also specify the possible values of λ and use $\Lambda = \mathbb{R}_+ = (0, \infty)$ as *parameter space*. This model has *representation space* $(\mathbb{N}_0, \mathbb{P}(\mathbb{N}_0))$, where $\mathbb{P}(A)$ denotes the *power set* of A, i.e. the set of all subsets of A.

A variant of this model appears when we have independent and identically distributed observations $X_1, \ldots, X_n$ from such a model. This would lead to a model with *representation space* $(\mathbb{N}_0^n, \mathbb{P}(\mathbb{N}_0^n))$, *parameter space* $\Lambda = \mathbb{R}_+$, and the *family of probability measures* $\mathcal{P} = \{P_\lambda^{\otimes n} \mid \lambda \in \Lambda\}$.

Note that the parameter space Λ is the same for the two variants of the simple Poisson model, whereas the representation space and family of probability measures is obtained by an n-fold product of the entities for the models corresponding to a single observation.

The simple Poisson model is often used to describe the number of events happening in a time period or in a specific region; a notable example is the number of α-particles emitted in a time interval from radioactive materials where this model is known to give a very accurate description of the phenomenon.

In principle we could modify the Poisson models above by adding $\lambda = 0$ to the parameter space, allowing for the possibility that $X = 0$ with probability one. However, this will typically add complications to other aspects of the analysis of the model so this is not normally done. □

We shall often consider the situation where we have independent and identically distributed observations $X_1 = x_1, \dots, X_n = x_n$ from a given distribution P. In such cases we also say that we have a *sample* $x = (x_1, \dots, x_n)$ or $X = (X_1, \dots, X_n)$ of size n from P, depending on whether or not we are emphasizing the randomness of the values. Our next example is related to the Poisson model:

Example 1.3. [The exponential distribution] Consider a sample $(X_1, \dots, X_n)$ from an exponential distribution where X_i have density

$$f_\mu(x) = \frac{e^{-x/\mu}}{\mu}, \quad x > 0$$

with respect to Lebesgue measure on $\mathbb{R}_+$. The *parameter space* is $M = \mathbb{R}_+$, the *representation space* is $(\mathbb{R}_+^n, \mathbb{B}(\mathbb{R}_+^n))$, where $\mathbb{B}(\mathbb{R}_+^n)$ are the Borel sets of $\mathbb{R}_+^n$, and the *family of probability distributions* is $\mathcal{P} = \{P_\mu^{\otimes n}, \mu \in M\}$, where P_μ is the distribution with density f_μ as above.

The exponential distribution is closely associated with the Poisson distribution as P_μ is the distribution of the waiting time to an event, where the number of events in a time interval of length t is Poisson distributed with parameter $\lambda = t/\mu$. The quantity $1/\mu$ is also known as the *rate* of the exponential distribution. □

The next model is the basic model for tossing a coin repeatedly.

Example 1.4. [The simple Bernoulli model] Here we consider n independent and identically distributed random variables with

$$P_\mu(X_i = 1) = 1 - P_\mu(X_i = 0) = \mu,$$

where $0 < \mu < 1$. The *representation space* is $(\{0,1\}^n, \mathbb{P}(\{0,1\}^n))$, the *parameter space* is the open interval $M = (0,1)$, and the *family of probability measures* is $\mathcal{P} = \{P_\mu^{\otimes n} \mid \mu \in M\}$. As in the Poisson model, we have in principle the possibility of adding the endpoints of the interval $(0,1)$ to the parameter space, but this is usually not done. □

A classic is the simple normal model, corresponding to measuring a quantity $\xi \in \mathbb{R}$ with variance $\sigma^2 \in \mathbb{R}_+$.

Example 1.5. [Simple normal model] We consider independent and identically distributed random variables $X_1, \ldots, X_n$ with $X_i \sim \mathcal{N}(\xi, \sigma^2)$; i.e. the distribution of X_i has density

$$f_\omega(x) = \frac{1}{\sqrt{2\pi\sigma^2}} e^{-\frac{(x-\xi)^2}{2\sigma^2}}$$

with respect to standard Lebesgue measure on $\mathbb{R}$, and $\omega = (\xi, \sigma^2) \in \mathbb{R} \times \mathbb{R}_+ = \Omega$. Here the *representation space* is $(\mathbb{R}^n, \mathbb{B}(\mathbb{R})^n)$, and the *parameter space* is Ω. Submodels of this model appear, for example, when either ξ or σ^2 is assumed to be fixed and known. □

The next class of models is useful for describing positive random quantities.

Example 1.6. [Gamma models] We consider $X_1, \ldots, X_n$ independent and identically distributed with X_i having density

$$f_{\alpha,\beta}(x) = \frac{x^{\alpha-1}}{\beta^\alpha \Gamma(\alpha)} e^{-x/\beta} \tag{1.2}$$

with respect to standard Lebesgue measure on $\mathbb{R}_+$. We let $\theta = (\alpha, \beta)$ and the *parameter space* be $\Theta = \mathbb{R}_+^2$; the *representation space* is $(\mathbb{R}_+^n, \mathbb{B}(\mathbb{R}_+^n))$; and the *family* is $\mathcal{P} = \{P_\theta^{\otimes n} \mid \theta \in \Theta\}$, where $P_\theta = P_{(\alpha,\beta)}$ has density as (1.2).

We note that this gamma model has a number of interesting submodels, obtained by restricting the parameter space appropriately:

a) The *exponential distribution model* discussed in Example 1.3, obtained by modifying the parameter space to $\Theta_1 = \{(\alpha, \beta) \in \Theta \mid \alpha = 1\}$.

b) The gamma model with *fixed shape parameter* obtained by modifying the parameter space to $\Theta_2 = \{(\alpha, \beta) \in \Theta \mid \alpha = \alpha_0\}$ for fixed $\alpha_0 \in \mathbb{R}_+$.

c) The gamma model with fixed scale parameter obtained by modifying the *parameter space* to $\Theta_3 = \{(\alpha, \beta) \in \Theta \mid \beta = \beta_0\}$ for fixed $\beta_0 \in \mathbb{R}_+$.

d) The gamma model with *fixed mean* obtained by modifying the parameter space to $\Theta_4 = \{(\alpha, \beta) \in \Theta \mid \alpha\beta = \mu_0\}$ for some fixed $\mu_0 \in \mathbb{R}_+$.

Clearly this list is far from exhaustive. □

Although this book focuses on parametric models, we shall give an example of a non-parametric model, here distributions with decreasing densities.

Example 1.7. [Decreasing density] This model is often used for modelling waiting times until the next of a series of recurrent events occurs. We assume that $X_1, \ldots, X_n$ are independent and identically distributed positive random variable with X_i having a non-increasing density f with respect to Lebesgue measure on $\mathbb{R}_+$, i.e. the density satisfies

$$\int_0^\infty f(x)\,dx = 1, \quad x < y \implies f(x) \geq f(y). \tag{1.3}$$

The *representation space* is again $(\mathbb{R}^n, \mathbb{B}(\mathbb{R}^n))$, the *family of probability measures* consists of those with density in the *parameter space* $\mathcal{F}$, where

$$\mathcal{F} = \{f : \mathbb{R}_+ \mapsto [0, \infty) \mid f \text{ satisfies } (1.3)\}.$$

We note again the slightly weird terminology that this is a *non-parametric model* even though the parameter space is huge. It simply indicates that there is no simple parametrization of the model with a finite-dimensional parameter space. □

The next two models are useful for illustrating various results that shall appear later.

Example 1.8. [The simple uniform model] Here we consider a random variable X which is uniformly distributed on the interval $(0, \theta)$ where $\theta \in \Theta = \mathbb{R}_+$. That is, the density of X is

$$f_\theta(x) = \frac{1}{\theta}\mathbf{1}_{(0,\theta)}(x)$$

with respect to Lebesgue measure on $\mathbb{R}_+$. The *representation space* is $(\mathbb{R}_+, \mathbb{B}(\mathbb{R}_+))$, the *parameter space* $\Theta = \mathbb{R}_+$, and the *family* is $\mathcal{P} = \{P_\theta \mid \theta \in \Theta\}$, where P_θ has density f_θ as above. We refrain from an explicit description of the variant with n repeated observations $X_1, \dots, X_n$. □

And finally, we conclude with yet another model that mostly appears for illustration of some of the theoretical concepts we shall consider later.

Example 1.9. [The Cauchy model] This model is, for example, used to describe the distribution of hitting points along the x-axis from a source placed at $\theta = (\alpha, \beta)$ that emits particles uniformly in all directions. Here X has density

$$f_{\alpha,\beta}(x) = \frac{1}{\pi}\frac{\beta}{((x-\alpha)^2 + \beta^2)}, \quad x \in \mathcal{R}$$

with respect to standard Lebesgue measure on $\mathbb{R}$. The *representation space* is $(\mathbb{R}, \mathbb{B}(\mathbb{R}))$ and the *parameter space* is $\Theta = \mathbb{R} \times \mathbb{R}_+$. □

In the following we shall see other examples of statistical models, and other variants of the models listed above. We shall then not always give all the formal details in terms of identifying representation space, family, and parameter space, as these mostly will be obvious in the given context.

1.1.2 Reparametrization

In many cases it is convenient to work with several different parametrizations of a statistical model when the associated family of probability can be represented in more than one way, i.e.

$$\mathcal{P} = \{\nu_\theta \mid \theta \in \Theta\} = \{\xi_\lambda \mid \lambda \in \Lambda\}.$$

This can be for a number of different reasons; maybe one of the parametrizations is particularly natural for interpretation in the scientific context where the model is to be used, maybe the parameter has a simple description in terms of the properties of the model, or a particular parametrization may be more convenient for the mathematical manipulations used when analysing the model.

We would normally assume that both parametrizations are bijective implying that there must be a bijective function ϕ such that $\lambda = \phi(\theta)$, and we shall then say that

ϕ is a *reparametrization* of the model. In this case, we have that

$$\xi_\lambda = \nu_{\phi^{-1}(\lambda)}, \quad \nu_\theta = \xi_{\phi(\theta)}.$$

It is important to know whether methods used for statistical inference are *equivariant* in the sense that, based on data x, methods yield the same inference for ν_θ and ξ_λ if λ and θ are related as above. We shall comment on that from time to time. Below we shall give a list of examples of relevant reparametrizations of some of the models described in Section 1.1.1.

Example 1.10. [Poisson model] In Example 1.2 we parametrized the Poisson family through its mean λ

$$E_\lambda(X) = \sum_{x=0}^{\infty} \frac{x\lambda^x}{x!} e^{-\lambda} = \lambda;$$

but an alternative parameter could be the *null fraction* $\eta = P_\lambda(X = 0) = e^{-\lambda}$ with the family of distributions represented as

$$P_\eta(X = x) = \eta \frac{(-\log \eta)^x}{x!}.$$

Here the function $\phi : \mathbb{R}_+ \mapsto (0, 1)$ given as $\phi(\lambda) = e^{-\lambda}$ is bijective, so this constitutes a valid reparametrization of the model with a new parameter space $H = (0, 1)$. □

Example 1.11. [Exponential distribution] In Example 1.3 we parametrized the family through its mean μ:

$$\mathbf{E}_\mu(X) = \int_0^\infty \frac{xe^{-x/\mu}}{\mu}\, dx = \mu$$

but it is sometimes more practical to parametrize the family using the *rate* $\lambda = 1/\mu$ and then write the density as

$$g_\lambda(x) = \lambda e^{-\lambda x}, \quad x > 0.$$

Since the map $\phi : \mathbb{R}_+ \mapsto \mathbb{R}_+$ given as $\phi(\mu) = 1/\mu$ is bijective, this constitutes a valid reparametrization of the model with a new parameter space $\Lambda = \mathbb{R}_+$. □

Example 1.12. [Gamma models] The gamma models have a large variety of interesting parametrizations, and we shall see several of those in later chapters. In Example 1.6 we have used the shape α and scale β, but as for the exponential model we would sometimes replace the scale with the rate $\lambda = 1/\beta$; as earlier, the map $\phi : \mathbb{R}_+^2 \mapsto \mathbb{R}_+^2$ defined as $\phi(\alpha, \beta) = (\alpha, 1/\beta)$ is bijective, so this yields an alternative parametrization.

Yet another possibility is to use the mean $\mu = \alpha\beta$ and variance $\sigma^2 = \alpha\beta^2$ as parameters, and the reparametrization function is $\psi : \mathbb{R}_+^2 \mapsto \mathbb{R}_+^2$ defined as $\psi(\alpha, \beta) = (\alpha\beta, \alpha\beta^2)$, which again is bijective. □

Example 1.13. [Decreasing density] For a more subtle example, we consider the non-parametric model in Example 1.7 with decreasing densities. Instead of using the density as 'parameter', we could, for example, use the cumulative distribution function

$$F(x) = P(X \leq x), \quad x \geq 0.$$

The parameter space would then be

$$\mathcal{F} = \{F : [0, \infty) \to [0, 1] \mid F \text{ is concave and right continuous}\}$$

as there would be a bijection ϕ from the set of (equivalence classes of) decreasing densities and $\mathcal{F}$, since a density is equivalent to a decreasing function if and only if the corresponding distribution function is concave.

A yet more subtle parametrization uses a classical result of A. Ya. Khinchin, saying that *a random variable X with values in $\mathbb{R}_+$ has a concave distribution function if and only if X has the same distribution as $X = YU$, where U and Y are independent, U is uniformly distributed on* $(0, 1)$, *and Y has an arbitrary distribution on* $[0, \infty)$. This relation defines a natural bijection between $\mathcal{F}$ and the set $\mathcal{G}$ of all distribution functions on $\mathbb{R}_+$ via the representation for $G \in \mathcal{G}$:

$$F_G(x) = \int_0^\infty \frac{\min(x, y)}{y} G(dy),$$

where we have let $0/0 = 1$; hence also $\mathcal{G}$ may be used as a parameter space for this model. □

1.1.3 Parameter functions

We consider a statistical model with representation space $(\mathcal{X}, \mathbb{E})$ and associated family $\mathcal{P}$. A *parameter function* is formally a mapping of the form $\phi : \mathcal{P} \mapsto \Lambda$ where Λ is some set, often a subset of $\mathbb{R}^k$.

If $\nu : \Theta \mapsto \mathcal{P}$ is a parametrization of $\mathcal{P}$, we may instead think of the parameter function as a map $\phi' : \Theta \mapsto \Lambda$ by composing it with ν, i.e. $\phi' = \phi \circ \nu$. Similarly, if ν is an injective parametrization of $\mathcal{P}$, any function $\phi : \Theta \mapsto \Lambda$ can be considered a parameter function by composing it with ν^{-1} as $\phi \circ \nu^{-1}$. In the following, we shall—in the case of injective parametrizations—interchangeably regard a parameter function as defined on Θ or $\mathcal{P}$ without further ado.

We might not necessarily be interested in the full parameter θ, but sometimes only specific aspects of it. This is in particular the case when Θ is high-dimensional, where we for example could be interested in parameter functions as

a) A *specific coordinate* of the parameter: $\phi(\theta) = \theta_1$ if $\Theta \subseteq \mathbb{R}^k$.

b) The *mean* of the distribution: $\phi(\theta) = \mathbf{E}_\theta(X)$.

c) The *variance* of the distribution: $\phi(\theta) = \mathbf{V}_\theta(X)$.

d) The *median* of the distribution: $P_\theta\{X \leq \phi(\theta)\} = 1/2$.

e) The *coefficient of variation* of the distribution:

$$\phi(\theta) = \mathbf{C}_\theta\{X\} = \sqrt{\mathbf{V}_\theta\{X\}}/\mathbf{E}_\theta\{X\}$$

but there are many other possibilities.

1.1.4 Nuisance parameters and parameters of interest

If the parameter space has the form $\Theta = \Lambda \times \Xi$ we may wish to focus on the component $\phi(\theta) = \lambda$, whereas the component $\xi \in \Xi$ just is something we must take into account. We say that λ is a *parameter of interest* and ξ a *nuisance parameter* reflecting that we consider ξ as a disturbing factor.

1.2 Likelihood, score, and information

In this section we shall exclusively consider a subclass of statistical models where the associated family of probability measures is *dominated*. Formally we define

Definition 1.14. A family $\mathcal{P}$ of probability measures on $(\mathcal{X}, \mathbb{E})$ is said to be *μ-dominated* if μ is a σ-finite measure on $(\mathcal{X}, \mathbb{E})$ so that every $P \in \mathcal{P}$ has a density with respect to μ.

In other words, a family is dominated if for all $P \in \mathcal{P}$, there exists a non-negative real-valued function f such that it holds for all $A \in \mathbb{E}$ that

$$P(A) = \int_A f(x)\,\mu(dx).$$

All the numbered examples in Section 1.1.1 are dominated, and this will be the case for almost all models considered in this book. An example of a non-dominated family is the family $\mathcal{P}_2$ of probability measures on $(\mathcal{X}, \mathbb{E}) = (\mathbb{R}, \mathbb{B})$ with finite second moments. These include measures with both discrete and continuous support and therefore does not admit a dominating measure. If a dominated family $\mathcal{P}$ is parametrized, we may write

$$P_\theta(A) = \int_A f_\theta(x)\,\mu(dx).$$

The important fact is that μ is the same for all values of θ in this expression. Although each f_θ is only defined up to a μ null-set, we shall in the following assume that we once and for all have chosen a specific family $\mathcal{F} = \{f_\theta \mid \theta \in \Theta\}$ of densities so we without ambiguity may identify the parameter space Θ, the family of probability measures $\mathcal{P}$, and the family of densities $\mathcal{F}$.

1.2.1 The likelihood function

The notion of likelihood function was introduced by R. A. Fisher, and despite its simplicity, it is arguably the most fundamental concept in statistical theory.

1.2.1.1 Formal definition

We consider an injectively parametrized statistical model on $(\mathcal{X}, \mathbb{E})$ with a μ-dominated family and associated family of densities $\mathcal{F} = \{f_\theta \mid \theta \in \Theta\}$.

Definition 1.15. For every $x \in \mathcal{X}$, the *likelihood function* $L_x : \Theta \mapsto [0, \infty)$ is the function given as

$$L_x(\theta) = f_\theta(x)$$

and the *log-likelihood function* $\ell_x : \Theta \mapsto [-\infty, \infty)$ is the logarithm of the likelihood function:

$$\ell_x(\theta) = \ell(x, \theta) = \log L_x(\theta).$$

In other words, if we consider $f_x(\theta)$ as a function of two variables, a *density* appears if we consider θ fixed and x varying, whereas the *likelihood* appears by considering x fixed and θ varying. Given how fundamental the idea of likelihood is, its rationale is surprisingly simple; if

$$L_x(\theta_1) > L_x(\theta_2)$$

or, equivalently,

$$f_{\theta_1}(x) > f_{\theta_2}(x),$$

then it is more *likely* that the data x have been generated by P_{θ_1} than by P_{θ_2}. In other words, high values of $L_x(\theta)$ or $\ell_x(\theta)$ means that data x support P_θ as the generating distribution.

It is important to realize that the likelihood function is only well-defined up to a multiplicative constant. When we defined the likelihood function above, we chose a base measure μ arbitrarily, and the absolute value of the likelihood function depends highly on that choice.

Suppose that we choose another dominating measure $\tilde{\mu} = g \cdot \mu$. Then the two likelihood functions L and $\tilde{L}$ relative to μ and $\tilde{\mu}$ will satisfy the relation

$$L_x(\theta) = g(x)\tilde{L}_x(\theta),$$

because the density of P_θ with respect to $\tilde{\mu}$ will be f_θ/g, where f_θ is the density of P_θ with respect to μ. Since the data x are fixed, the two likelihood functions are proportional but not identical. Indeed, one may show that any two versions of the likelihood function are always proportional in the following sense:

Theorem 1.16. *Let $\mathcal{P} = \{P_\theta \mid \theta \in \Theta\}$ be a parametrized statistical model with representation space $(\mathcal{X}, \mathbb{E})$ and let μ and $\tilde{\mu}$ be dominating measures for $\mathcal{P}$. Then the densities and associated likelihood functions L and $\tilde{L}$ may be chosen such that there is a measurable function $h : \mathcal{X} \mapsto \mathbb{R}_+$ satisfying*

$$L_x(\theta) = h(x)\tilde{L}_x(\theta), \quad \textit{for all } \theta \in \Theta \textit{ and all } x \in \mathcal{X}.$$

Proof. The proof is somewhat technical and omitted. The interested reader may find further details in Hansen (2012, p. 77 ff). □

In other words, the likelihood function is only well-defined up to a positive multiplicative constant (depending on x only) and only *ratios* of likelihood functions makes solid sense since it then holds that

$$\frac{\tilde{L}_x(\theta_1)}{\tilde{L}_x(\theta_2)} = \frac{h(x)\tilde{L}_x(\theta_1)}{h(x)\tilde{L}_x(\theta_2)} = \frac{L_x(\theta_1)}{L_x(\theta_2)}.$$

Similarly, only *differences* of log-likelihood functions are meaningful. Therefore, it is common to ignore such constant terms when making likelihood calculations.

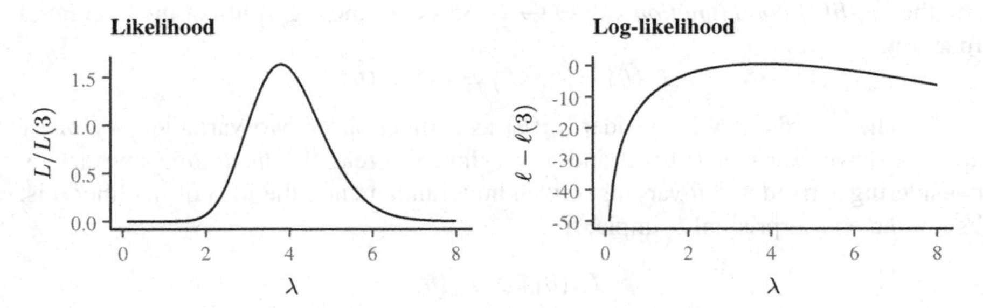

Figure 1.1 – The Poisson likelihood and log-likelihood functions for five observations $(3, 3, 6, 1, 6)$, normalized with the value of the function at $\lambda = 3$.

Example 1.17. [Poisson likelihood] Consider a simple Poisson model and an observation $X = x$. The likelihood and log-likelihood functions become

$$L_x(\lambda) = \lambda^x e^{-\lambda}, \quad \ell_x(\lambda) = x \log \lambda - \lambda.$$

We have ignored the factor $1/x!$ which enters as a multiplicative constant in L_x and as an additive constant $-\log x!$ in ℓ_x. If we are considering the variant of the Poisson model with $X_1, \ldots, X_n$ being independent and identically distributed, we get from observations $x_1, \ldots, x_n$ that

$$\ell_x(\lambda) = \log \lambda \sum_{i=1}^{n} x_i - n\lambda$$

where we again have ignored terms that do not depend on λ. Figure 1.1 displays the likelihood and log-likelihood functions for a sample of five observations $(x_1, \ldots, x_5) = (3, 3, 6, 1, 6)$, i.e. with $n = 5$, normalized with the value of the likelihood at $\lambda = 3$. □

The likelihood and log-likelihood in the Poisson example are well behaved and this is mostly the case in situations within the scope of this book. However, for completeness, we shall also consider an example where the likelihood is not so well behaved.

Example 1.18. [Uniform likelihood] Consider a sample $X = (X_1, \ldots, X_n)$ from the family of uniform distributions on the interval $(0, \theta)$ for $\theta \in \Theta = \mathbb{R}_+$ as in Example 1.8. We get for the likelihood function

$$L_x(\theta) = f_\theta(x) = \prod_{i=1}^{n} \frac{1}{\theta} \mathbf{1}_{(0,\theta)}(x_i) = \theta^{-n} \mathbf{1}_{(y_n,\infty)}(\theta)$$

where $y_n = \max(x_1, \ldots, x_n)$ is the largest observation. The likelihood function is plotted in Figure 1.2 for 10 simulated observations from a uniform distribution on the interval $(0, 2.5)$. The likelihood function is sharply peaked at the largest observation, which necessarily must be smaller than the true value. □

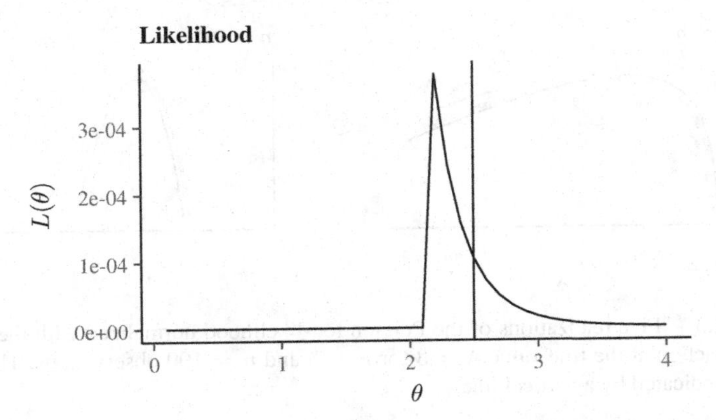

Figure 1.2 – The likelihood function in the uniform model for 10 simulated observations. The true value $\theta = 2.5$ of the parameter is indicated by a vertical line.

1.2.1.2 *Equivariance of the likelihood function*

Consider a bijective reparametrization $\phi : \Theta \mapsto \Lambda$ with $\lambda = \phi(\theta)$. The family of densities is then represented in the new parametrization as

$$\{g_\lambda \mid \lambda \in \Lambda\} = \{f_\theta \mid \theta \in \Theta\}$$

so the log-likelihood function $\ell_x(\lambda) = \log g_\lambda(x)$ in the new parametrization for $\lambda = \phi(\theta)$ will be

$$\ell_x(\lambda) = \ell_x(\phi(\theta)) = \tilde{\ell}_x(\theta),$$

where we have let $\tilde{\ell}$ denote the likelihood function in the θ-parametrization. In other words, we would have

$$\ell_x(\lambda_1) > \ell_x(\lambda_2) \iff \tilde{\ell}_x(\theta_1) > \tilde{\ell}_x(\theta_2) \text{ whenever } \phi(\theta_i) = \lambda_i, i = 1, 2.$$

So low or high values of $\tilde{\ell}_x(\theta)$ match low or high values of $\ell_x(\lambda)$ if $\lambda = \phi(\theta)$, reflecting that the *likelihood function is equivariant* under bijective reparametrizations.

1.2.1.3 *Likelihood as a random variable*

Although the likelihood appears by considering the density for a fixed x as a function of the unknown parameter θ, different likelihood functions appear for different outcomes x of the random variable X associated with the model under investigation.

Thus the likelihood function itself, and its logarithm, can be considered as a *random function* $L_X(\cdot)$ and $\ell_X(\cdot)$ and we shall do so without paying specific attention to the σ-algebra making the corresponding map measurable; when a specific choice of the family $\mathcal{F}$ of densities has been made, the map and notion are defined without ambiguity.

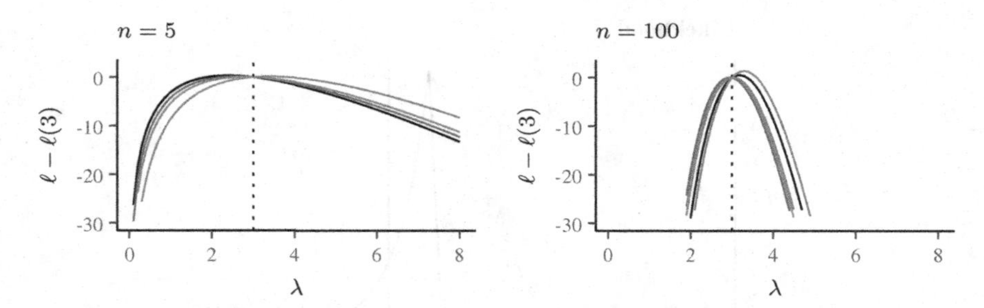

Figure 1.3 – Five realizations of the Poisson log-likelihood normalized with the value of the function at the true value $\lambda = 3$ for $n = 5$ and $n = 100$ observations. The true value is indicated by a vertical line.

An example of the variability of these random likelihood functions for the simple Poisson model is displayed in Figure 1.3. We note that the variability of the likelihood function is much larger in the diagram to the left, corresponding to $n = 5$ than in the diagram to the right, where $n = 100$. We also note that the log-likelihood function is more peaked for $n = 100$ and the peak occurs close to the true value of the parameter. In addition, the shape of the log-likelihood function is not far from a parabola when $n = 100$. We shall later—in Chapter 5—see that this is not a coincidence; rather a typical behaviour of the log-likelihood function. These fact suggest that the value of θ at which the log-likelihood peaks could be a good guess for the true value of that parameter. Indeed this shall later—in Section 4.3—be formalized into the idea of *maximum likelihood estimation.*

1.2.2 Score and information

Consider now a μ-dominated family of probability measures on $(\mathcal{X}, \mathbb{E})$ with associated family of densities $\mathcal{F} = \{f_\theta \mid \theta \in \Theta\}$ where $\Theta \subseteq \mathbb{R}^k$ is open and recall that a function $g : \mathbb{R}^k \to \mathbb{R}^m$ is said to be *smooth* if g is infinitely often differentiable, i.e. if $g \in C^\infty$. If Θ is an open subset of $\mathbb{R}^k$ and the likelihood function is smooth for μ-almost all x, we say that the family is *smooth.*

Remark 1.19. *We emphasize that almost all results in this book would hold if 'smooth' is interpreted as twice continuously differentiable (C^2) rather than infinitely often differentiable (C^∞). But for simplicity we restrict to C^∞-functions to avoid having to keep track of the degree of differentiability.*

We now introduce two fundamental quantities.

Definition 1.20. The *score function* for a smooth family is

$$S_x(\theta) = S(x, \theta) = D\ell_x(\theta) = \begin{pmatrix} \frac{\partial \ell_x(\theta)}{\partial \theta_1} & \cdots & \frac{\partial \ell_x(\theta)}{\partial \theta_k} \end{pmatrix} = \nabla \ell_x(\theta)^\top$$

and the *information function* is

$$I_x(\theta) = I(x,\theta) = -D^2\ell_x(\theta) = -\begin{pmatrix} \frac{\partial^2 \ell_x(\theta)}{\partial\theta_1^2} & \frac{\partial^2 \ell_x(\theta)}{\partial\theta_1\partial\theta_2} & \cdots & \frac{\partial^2 \ell_x(\theta)}{\partial\theta_1\partial\theta_k} \\ \frac{\partial^2 \ell_x(\theta)}{\partial\theta_2\partial\theta_1} & \frac{\partial^2 \ell_x(\theta)}{\partial\theta_2^2} & \cdots & \frac{\partial^2 \ell_x(\theta)}{\partial\theta_2\partial\theta_k} \\ \vdots & \vdots & \ddots & \vdots \\ \frac{\partial^2 \ell_x(\theta)}{\partial\theta_k\partial\theta_1} & \frac{\partial^2 \ell_x(\theta)}{\partial\theta_k\partial\theta_2} & \cdots & \frac{\partial^2 \ell_x(\theta)}{\partial\theta_k^2} \end{pmatrix}.$$

In other words, the score function is the derivative and the information function is the negative Hessian of the log-likelihood function. The arbitrary constant associated with the log-likelihood function disappears in the differentiation process.

The idea behind the score function is that it measures how sensitive the log-likelihood function is to changes in the parameter θ, the logic being that if it does not change much with the parameter θ, there is little information about θ. The formal measure of information is then obtained by a second differentiation. Also, the score function is zero at any local maximum of the log-likelihood function, so a zero score at θ', say, might indicate that the true value is close to θ'. As we shall show below in Theorem 1.23, the expectation of the information function is equal to the variance of the score function under mild regularity conditions. Also, since the log-likelihood function ℓ_x often approximately has the shape of a parabola—see Fig. 1.3—the first and second derivative of ℓ_x roughly identifies the shape and location of the log-likelihood function.

Example 1.21. [Normal score and information] It is illuminating to understand the score and information function by considering a normal model with known variance σ^2, i.e. the family of densities

$$f_\theta(x) = \frac{1}{\sqrt{2\pi\sigma^2}}e^{-\frac{(x-\theta)^2}{2\sigma^2}}, \quad \theta \in \mathbb{R}.$$

The log-likelihood function becomes

$$\ell_x(\theta) = -\frac{(x-\theta)^2}{2\sigma^2},$$

where we have ignored additive constants that do not depend on θ. This yields the score function

$$S(x,\theta) = \ell'_x(\theta) = \frac{x-\theta}{\sigma^2},$$

and the information function is obtained by changing sign and differentiating once more

$$I(x,\theta) = -\ell''_x(\theta) = \frac{1}{\sigma^2} = \mathbf{V}_\theta\{S(X,\theta)\} = \mathbf{E}_\theta\{I(X,\theta)\}.$$

Thus the information is simply the inverse variance, or *concentration*, supporting the intuition that the information function captures the accuracy available to determine θ. We shall later, in Chapter 4 and Chapter 5, see that this interpretation can be extended to more general situations: the Fisher information yields the inherent precision with which a parameter may be determined from data.

The identities in the last line are known as *Bartlett's identities* and established for general models under suitable smoothness and regularity conditions in Theorem 1.23 below. □

We consider a μ-dominated parametrized family $\mathcal{P} = \{P_\theta, \theta \in \Theta\}$ of probability measures on $(\mathcal{X}, \mathbb{E})$. We define

Definition 1.22. A smooth family $\mathcal{F} = \{f_\theta \mid \theta \in \Theta\}$ of densities with respect to μ is said to be *stable* if every $\theta \in \Theta$ has an open neighbourhood U_θ and there are μ-integrable functions g_θ and h_θ such that for all $x \in \mathcal{X}$, $i, j = 1, \dots k$, and all $\eta \in U_\theta$:

$$\left|\frac{\partial}{\partial \eta_i} L_x(\eta)\right| \leq g_\theta(x), \quad \left|\frac{\partial^2}{\partial \eta_i \partial \eta_j} L_x(\eta)\right| \leq h_\theta(x). \tag{1.4}$$

In words, the derivative $DL_x(\theta)$ and Hessian $D^2 L_x(\theta)$ of the likelihood function are uniformly locally bounded in a neighbourhood of θ by integrable functions.

The condition (1.4) ensures that differentiation with respect to θ of integrals of these functions commutes with integration so differentiation can be performed inside the integration sign, see for example Schilling (2017, Theorem 12.5).

As we may consider the log-likelihood function $\ell_X(\theta)$ as a random object, so we may also consider the score $S(X, \theta)$ and information $I(X, \theta)$. We have the following important identities for their simple moments:

Theorem 1.23 (Bartlett's identities). *Consider a statistical model on $(\mathcal{X}, \mathbb{E})$ with a μ-dominated family $\{P_\theta \mid \theta \in \Theta\}$ where the associated family of densities is smooth and stable. It then holds that*

$$\mathbf{E}_\theta\{S(X, \theta)\} = 0, \quad \mathbf{V}_\theta\{S(X, \theta)^\top\} = \mathbf{E}_\theta\{S(X, \theta)^\top S(X, \theta)\} = \mathbf{E}_\theta\{I(X, \theta)\}.$$

Proof. Since the family is smooth and stable, we may differentiate under the integral sign. As

$$\int_{\mathcal{X}} f_\theta(x)\, \mu(dx) = 1,$$

we get by differentiation

$$\begin{aligned} 0 &= \frac{\partial}{\partial \theta_i}\left(\int_{\mathcal{X}} f_\theta(x)\, \mu(dx)\right) = \int_{\mathcal{X}} \frac{\partial}{\partial \theta_i} f_\theta(x)\, \mu(dx) \\ &= \int_{\mathcal{X}} \frac{\frac{\partial}{\partial \theta_i} f_\theta(x)}{f_\theta(x)} f_\theta(x)\, \mu(dx) = \int_{\mathcal{X}} \frac{\partial}{\partial \theta_i} \ell(\theta, x) f_\theta(x)\, \mu(dx) = \mathbf{E}_\theta\{S(X, \theta)_i\} \end{aligned}$$

and the first identify is established.

Differentiating a second time we find

$$0 = \frac{\partial}{\partial \theta_j}\left(\int_{\mathcal{X}} \frac{\partial}{\partial \theta_i} f_\theta(x)\, \mu(dx)\right) = \int_{\mathcal{X}} \frac{\partial^2}{\partial \theta_i \partial \theta_j} f_\theta(x)\, \mu(dx)$$

so also

$$\mathbf{E}_\theta\left\{\frac{\frac{\partial^2 f_\theta(X)}{\partial \theta_i \partial \theta_j}}{f_\theta(X)}\right\} = \int_{\mathcal{X}} \frac{\partial^2}{\partial \theta_i \partial \theta_j} f_\theta(x)\, \mu(dx) = 0$$

and since

$$\frac{\partial^2}{\partial\theta_i\partial\theta_j}\ell(\theta,x) = \frac{\frac{\partial^2}{\partial\theta_i\partial\theta_j}f_\theta(x)}{f_\theta(x)} - \frac{\frac{\partial}{\partial\theta_i}f_\theta(x)}{f_\theta(x)}\cdot\frac{\frac{\partial}{\partial\theta_j}f_\theta(x)}{f_\theta(x)}$$
$$= \frac{\frac{\partial^2}{\partial\theta_i\partial\theta_j}f_\theta(x)}{f_\theta(x)} - S(x,\theta)_i S(x,\theta)_j,$$

we further obtain, using that $\mathbf{E}_\theta\{S(X,\theta)\} = 0$,

$$\mathbf{E}_\theta\left\{-\frac{\partial^2}{\partial\theta_i\partial\theta_j}\ell(\theta,X)\right\} = -\mathbf{E}_\theta\left\{\frac{\frac{\partial^2 f_\theta(X)}{\partial\theta_i\partial\theta_j}}{f_\theta(X)}\right\} + \mathbf{V}_\theta\{S(X,\theta)\}_{ij}$$
$$= \mathbf{V}_\theta\{S(X,\theta)\}_{ij}$$

as desired. □

For a smooth and stable family, the quantity

$$i(\theta) = \mathbf{V}_\theta\{S(X,\theta)^\top\} = \mathbf{E}_\theta\{I(X,\theta)\}$$

is known as the *Fisher information* (matrix) or *expected information* and it plays a central role in the following developments and in many other areas of mathematical statistics.

We need another important quantity associated with smooth statistical families and define:

Definition 1.24. [Quadratic score] Consider a statistical model on $(\mathcal{X},\mathbb{E})$ with a μ-dominated family $\{P_\theta \mid \theta \in \Theta\}$ where the associated family of densities is smooth and stable and the Fisher information matrix is positive definite. The *quadratic score* is the quantity

$$Q(X,\theta) = S(X,\theta)i(\theta)^{-1}S(X,\theta)^\top$$

where $i(\theta)$ is the Fisher information and $S(X,\theta)$ the score statistic.

Thus the quadratic score measures the length of the score statistics with respect to the inner product determined by its inverse variance, i.e. the inverse Fisher information. We then have

Theorem 1.25. *Consider a statistical model on $(\mathcal{X},\mathbb{E})$ with a μ-dominated family $\{P_\theta \mid \theta \in \Theta\}$ with $\Theta \subseteq \mathbb{R}^k$ an open set. If the associated family of densities is smooth and stable and $i(\theta)$ is positive definite, it holds that*

$$\mathbf{E}_\theta\{Q(X,\theta)\} = k.$$

Proof. We have

$$\begin{aligned}
\mathbf{E}_\theta\{Q(X,\theta)\} &= \mathbf{E}_\theta\left\{S(X,\theta)i(\theta)^{-1}S(X,\theta)^\top\right\}\\
&= \mathbf{E}_\theta\left\{\operatorname{tr}(S(X,\theta)i(\theta)^{-1}S(X,\theta)^\top)\right\}\\
&= \mathbf{E}_\theta\left\{\operatorname{tr}(i(\theta)^{-1}S(X,\theta)^\top S(X,\theta))\right\}\\
&= \operatorname{tr}\left(\mathbf{E}_\theta\left\{i(\theta)^{-1}S(X,\theta)^\top S(X,\theta)\right\}\right)\\
&= \operatorname{tr}\left(i(\theta)^{-1}\mathbf{E}_\theta\left\{S(X,\theta)^\top S(X,\theta)\right\}\right)\\
&= \operatorname{tr}\left\{i(\theta)^{-1}i(\theta)\right\} = \operatorname{tr}(I_k) = k,
\end{aligned}$$

as desired. We used Bartlett's identities from Theorem 1.23 for the penultimate equation and properties of the trace of a matrix in several of the steps above; see Section A.5 for the latter. □

Example 1.26. [Poisson score and information] In the simple Poisson model, we have previously (see Example 1.17) calculated the log-likelihood function to be

$$\ell_x(\lambda) = x \log \lambda - \lambda$$

which is clearly smooth, so the family of densities for the Poisson model is smooth. By differentiation we further get

$$S(x, \lambda) = \frac{x}{\lambda} - 1, \quad I(x, \lambda) = \frac{x}{\lambda^2}.$$

We shall in the next section argue that the family is also stable. This implies that Bartlett's identities hold, which in this case can also be verified directly:

$$\mathbf{E}\{S(X, \lambda)\} = \frac{\mathbf{E}(X)}{\lambda} - 1 = \frac{\lambda}{\lambda} - 1 = 1 - 1 = 0$$

and we may calculate the Fisher information in two different ways:

$$i(\lambda) = \mathbf{V}\{S(X, \lambda)\} = \frac{1}{\lambda^2}\mathbf{V}(X) = \frac{1}{\lambda}; \quad i(\lambda) = \mathbf{E}\{I(X, \lambda)\} = \frac{\mathbf{E}(X)}{\lambda^2} = \frac{1}{\lambda}$$

where we have used that a Poisson distribution has $\mathbf{E}_\lambda(X) = \mathbf{V}_\lambda(X) = \lambda$, so the identities are satisfied. The quadratic score becomes

$$Q(x, \lambda) = \left(\frac{x}{\lambda} - 1\right)^2 \lambda = \frac{(x - \lambda)^2}{\lambda}$$

and we may easily verify that $\mathbf{E}_\lambda\{Q(X, \lambda)\} = \mathbf{V}_\lambda(X)/\lambda = 1$, as Theorem 1.25 says. □

Since the quadratic score has expectation equal to the model dimension and the score itself has mean zero at the true parameter, large values of the quadratic score at a parameter value will point to that parameter value not being true. In other words, the quadratic score $Q(x, \theta)$ behaves in a similar way to the negative of the log-likelihood function $-\ell(x, \theta)$ and indeed may be seen as a quadratic approximation to $-\ell(x, \theta)$.

Example 1.27. [Score and information in the gamma distribution] For a two-dimensional example, consider the gamma family in Example 1.6 with densities

$$f_{\alpha,\beta}(x) = \frac{x^{\alpha-1}}{\beta^\alpha \Gamma(\alpha)} e^{-x/\beta}, \quad x > 0, \quad (\alpha, \beta) \in \mathbb{R}^2.$$

The log-likelihood function becomes

$$\ell_x(\alpha, \beta) = (\alpha - 1) \log x - \alpha \log \beta - \log \Gamma(\alpha) - \frac{x}{\beta}.$$

Also this family is smooth and stable, as we shall see in Chapter 3, so Bartlett's identities hold. The score function is obtained by differentiation

$$S(x,\alpha,\beta) = \left(\log x - \log\beta - \Psi(\alpha), -\frac{\alpha}{\beta} + \frac{x}{\beta^2}\right)$$

where $\Psi(\alpha) = D\log\Gamma(\alpha)$ is known as the *digamma function*.

As $\mathbf{E}_{(\alpha,\beta)}(X) = \alpha\beta$ we see directly that the second component has mean zero. From the fact that also the first component must have mean zero, we deduce that

$$\mathbf{E}_{(\alpha,\beta)}(\log X) = \log\beta + \Psi(\alpha).$$

Changing signs and differentiating a second time yields the information function

$$I(x,\alpha,\beta) = \begin{pmatrix} \Psi_1(\alpha) & \frac{1}{\beta} \\ \frac{1}{\beta} & \frac{2x-\alpha\beta}{\beta^3} \end{pmatrix}.$$

Here $\Psi_1(\alpha) = \Psi'(\alpha) = D^2\log\Gamma(\alpha)$ is known as the *trigamma function*. Taking expectations yields the Fisher information matrix

$$i(\alpha,\beta) = \begin{pmatrix} \Psi_1(\alpha) & \frac{1}{\beta} \\ \frac{1}{\beta} & \frac{2\mathbf{E}_{\alpha,\beta}(X)-\alpha\beta}{\beta^3} \end{pmatrix} = \begin{pmatrix} \Psi_1(\alpha) & \frac{1}{\beta} \\ \frac{1}{\beta} & \frac{\alpha}{\beta^2} \end{pmatrix}.$$

Note that from Bartlett's identities, we have that

$$i(\alpha,\beta) = \mathbf{E}_{\alpha,\beta}\{I(X,\alpha,\beta)\} = \mathbf{V}_{\alpha,\beta}\{S(X,\alpha,\beta)^\top\} = \mathbf{V}_{\alpha,\beta}\{(\log X, X/\beta^2)^\top\}$$

which conforms with the fact that

$$\mathbf{V}_{\alpha,\beta}(X/\beta^2) = \alpha\beta^2/\beta^4 = \alpha/\beta^2.$$

As we shall see later, there are also other ways of deriving these results. □

Example 1.28. [Uniform not smooth] The uniform distribution is not smooth, as the likelihood function is not differentiable at $\theta = \max(x_1,\ldots,x_n)$, see Figure 1.2. So score and information is not well-defined for this type of model. □

1.2.3 Reparametrization and repetition

It is important to understand how likelihood, score, and information behave when we reparametrize a model. We first establish that the properties of smoothness and stability are preserved by *diffeomorphisms*, i.e. reparametrizations that are bijective and smooth in both directions.

Lemma 1.29. *Consider a family of densities $\mathcal{F} = \{f_\theta \mid \theta\in\Theta\}$ with $\Theta\subseteq\mathbb{R}^k$ open, and a smooth and bijective map $\phi:\Theta\mapsto\Lambda$ with $\det(D\phi(\theta))\neq 0$. Then ϕ is a diffeomorphism and $\mathcal{F}$ is smooth and stable if and only if the reparametrized family $\mathcal{F} = \{g_\lambda \mid \lambda\in\Lambda\}$ is smooth and stable.*

Proof. The inverse function theorem (Theorem A.20) ensures that ϕ is a diffeomorphism. The statements concerning smoothness follows as this property is closed under composition of functions. We let $\tilde{L}_x(\theta)$ and $L_x(\lambda)$ denote the likelihood functions expressed in the two parametrizations. Using composite differentiation (Rudin, 1976, Theorem 9.15) yields

$$\frac{\partial \tilde{L}_x(\theta)}{\partial \theta_i} = \sum_u \frac{\partial L_x(\lambda)}{\partial \lambda_u} \frac{\partial \phi_u(\theta)}{\partial \theta_i}$$

or, in matrix form

$$D\tilde{L}_x(\theta) = DL_x(\phi(\theta))D\phi(\theta),$$

where $D\phi(\theta)$ is the Jacobian of ϕ. So the left-hand side is locally bounded by integrable functions if $DL_x(\phi(\theta))$ is.

Differentiating a second time yields

$$\frac{\partial^2 \tilde{L}_x(\theta)}{\partial \theta_i \partial \theta_j} = \sum_{uv} \frac{\partial^2 L_x(\lambda)}{\partial \lambda_u \partial \lambda_v} \frac{\partial \phi_u(\theta)}{\partial \theta_i} \frac{\partial \phi_v(\theta)}{\partial \theta_i} + \sum_u \frac{\partial L_x(\lambda)}{\partial \lambda_u} \frac{\partial^2 \phi_u(\theta)}{\partial \theta_i \partial \theta_j}$$

which again is locally bounded by integrable functions if the derivatives on the right hand side are. To see the converse we just reverse the roles of θ and λ which can be done since the inverse function theorem ensures that ϕ^{-1} also satisfies the conditions of the lemma. □

Lemma 1.29 ensures that for many purposes, we do not have to worry about which parametrization we use for a given model, as long as reparametrizations are diffeomorphisms. However, we must track how the score and information change when parameters change, and this is summarized in the theorem below.

Theorem 1.30. *Consider a family of densities $\mathcal{F} = \{f_\theta \mid \theta \in \Theta\}$ that is smooth and stable. Let $\phi : \Theta \mapsto \Lambda$ be a diffeomorphism as in Lemma 1.29. Further, let S, I, i, and Q denote the score, information function, Fisher information, and quadratic score with respect to λ; and $\tilde{S}$, $\tilde{I}$, $\tilde{\imath}$, and $\tilde{Q}$ the corresponding quantities with respect to θ. It then holds that*

$$\tilde{S}(x,\theta) = S(x,\lambda)D\phi(\theta) \tag{1.5}$$

$$\tilde{I}(x,\theta) = D\phi(\theta)^\top I(x,\lambda)D\phi(\theta) - \sum_u S(x,\lambda)_u D^2\phi_u(\theta) \tag{1.6}$$

$$\tilde{\imath}(\theta) = D\phi(\theta)^\top i(\phi(\theta))D\phi(\theta) = D\phi(\theta)^\top i(\lambda)D\phi(\theta), \tag{1.7}$$

$$\tilde{Q}(x,\theta) = Q(x,\lambda), \tag{1.8}$$

where $D\phi$ is the Jacobian of ϕ and $D^2\phi_u$ is the Hessian of ϕ_u.

Proof. Lemma 1.29 yields that all quantities are well defined in the reparametrized model. Composite differentiation applied to the score function yields

$$\tilde{S}(x,\theta) = D\tilde{\ell}_x(\theta) = D\ell_x(\phi(\theta))D\phi(\theta) = \sum_u \frac{\partial \ell_x(\lambda)}{\partial \lambda_u} \frac{\partial \phi_u(\theta)}{\partial \theta_i} = S(x,\lambda)D\phi(\theta)$$

which establishes (1.5). For the information function, we get by further differentiation

$$\tilde{I}(x,\theta)_{ij} = \sum_{uv} -\frac{\partial^2 \ell_x(\lambda)}{\partial\lambda_u \partial\lambda_v}\frac{\partial\phi_u(\theta)}{\partial\theta_i}\frac{\partial\phi_v(\theta)}{\partial\theta_j} + \sum_u -\frac{\partial\ell_x(\lambda)}{\partial\lambda_u}\frac{\partial^2\phi_u(\theta)}{\partial\theta_i\partial\theta_j}$$

which in a more compact form is (1.6). Then (1.7) can be derived from (1.5) as

$$\begin{aligned}\tilde{i}(\theta) &= \mathbf{V}_\theta\{\tilde{S}(X,\theta)\} = \mathbf{E}_\theta\{\tilde{S}(X,\theta)^\top \tilde{S}(X,\theta)\} \\ &= D\phi(\theta)^\top \mathbf{E}_\lambda\{S(X,\lambda)^\top S(X,\lambda)\} D\phi(\theta) = D\phi(\theta)^\top i(\lambda) D\phi(\theta)\end{aligned}$$

or alternatively from (1.6) since

$$\mathbf{E}_\lambda\left\{\sum_u S(X,\lambda)_u D^2\phi_u(\theta)\right\} = \sum_u \mathbf{E}_\lambda\left\{S(X,\lambda)_u\right\} D^2\phi_u(\theta) = 0$$

so

$$\tilde{i}(\theta) = \mathbf{E}_\lambda\{\tilde{I}(X,\theta)\} = \mathbf{E}_\lambda\{D\phi(\theta)^\top I(X,\lambda) D\phi(\theta)\} = D\phi(\theta)^\top i(\lambda) D\phi(\theta).$$

Finally, we get

$$\begin{aligned}\tilde{Q}(x,\theta) &= \tilde{S}(x,\theta)\tilde{i}(\theta)^{-1}\tilde{S}(x,\theta)^\top \\ &= S(x,\lambda)D\phi(\theta)\left(D\phi(\theta)^\top i(\lambda) D\phi(\theta)\right)^{-1} D\phi(\theta)^\top S(x,\lambda)^\top \\ &= S(x,\lambda) i(\lambda)^{-1} S(x,\lambda)^\top.\end{aligned}$$

This completes the proof. □

Note in particular that *the quadratic score is equivariant* as was also true for the likelihood and log-likelihood function.

Example 1.31. [Score for reparametrized Poisson] Consider again the simple Poisson family, but let us parametrize the family with $\theta = \log\lambda$ instead of the mean λ. In Example 1.26, we calculated the score, information function, and Fisher information for λ to be

$$S(x,\lambda) = \frac{x}{\lambda} - 1, \quad I(x,\lambda) = \frac{x}{\lambda^2}, \quad i(\lambda) = \frac{1}{\lambda}.$$

To use the identities in Theorem 1.30, we note that $D\phi(\theta) = e^\theta$ so we get

$$\tilde{S}(x,\theta) = S(x,\lambda)e^\theta = \left(\frac{x}{\lambda} - 1\right)e^\theta = x - e^\theta$$

since $\lambda = e^\theta$. Further,

$$\tilde{I}(x,\theta) = e^{2\theta}\frac{x}{\lambda^2} - S(x,\lambda)e^\theta = x - x + e^\theta = e^\theta$$

and

$$\tilde{Q}(x,\theta) = \frac{(x - e^\theta)^2}{e^\theta} = Q(x,\lambda).$$

Clearly we get the same results by calculating the quantities directly in the θ-parametrization. Then the Poisson density becomes

$$f_\theta(x) = \frac{e^{\theta x}}{x!} e^{-e^\theta}$$

so we get by differentiation (ignoring $x!$ as usual)

$$\ell_x(\theta) = x\theta - e^\theta; \quad S(x,\theta) = x - e^\theta; \quad I(x,\theta) = e^\theta = i(\theta).$$

The quantities are often most easily found by direct calculation as we did here at the end, but Theorem 1.30 is important for theoretical considerations. □

Example 1.32. [Exponential distribution—rate or mean] The model for the simple exponential distribution may either be parametrized by the rate or the mean. Parametrizing with the rate yields

$$f_\lambda(x) = \lambda e^{-\lambda x}, \quad x > 0$$

with associated log-likelihood, score, information, and quadratic score being

$$\ell_x(\lambda) = \log \lambda - \lambda x, \quad S(x,\theta) = \frac{1}{\lambda} - x,$$
$$I(x,\lambda) = i(\lambda) = \frac{1}{\lambda^2}, \quad Q(x,\lambda) = (\lambda x - 1)^2.$$

The mean of an exponential distribution with rate λ is $\mathbf{E}_\lambda(X) = 1/\lambda$, conforming with the property that $\mathbf{E}_\lambda\{S(X,\lambda)\} = 0$ and $\mathbf{E}_\lambda\{Q(X,\lambda)\} = 1$. Now parametrizing with the mean $\theta = 1/\lambda$ yields

$$f_\theta(x) = \frac{1}{\theta} e^{-x/\theta}, \quad x > 0$$

with associated log-likelihood, score, and information being

$$\tilde{\ell}_x(\theta) = -\log\theta - x/\theta, \quad \tilde{S}(x,\theta) = -\frac{1}{\theta} + \frac{x}{\theta^2}, \quad \tilde{I}(x,\theta) = \frac{2x}{\theta^3} - \frac{1}{\theta^2}$$

yielding the Fisher information

$$\tilde{i}(\theta) = \mathbf{E}_\theta\left(I(X,\theta)\right) = \frac{2\theta}{\theta^3} - \frac{1}{\theta^2} = \frac{1}{\theta^2}$$

and quadratic score

$$\tilde{Q}(x,\theta) = \theta^2\left(\frac{x}{\theta^2} - \frac{1}{\theta}\right)^2 = \left(\frac{x}{\theta} - 1\right)^2 = (\lambda x - 1)^2 = Q(x,\lambda).$$

We note that this conforms well with Theorem 1.30 since with $\phi(\theta) = 1/\theta$ we have $D\phi(\theta) = -1/\theta^2$, $D^2\phi(\theta) = 2/\theta^3$ and thus, for example, using (1.6) we get

$$\tilde{I}(x,\theta) = \frac{1}{\theta^4}\frac{1}{\lambda^2} - \left(\frac{1}{\lambda} - x\right)\frac{2}{\theta^3} = \frac{1}{\theta^2} - (\theta - x)\frac{2}{\theta^3} = \frac{2x}{\theta^3} - \frac{1}{\theta^2}$$

as we also found above. Similarly (1.7) yields

$$\tilde{i}(\theta) = i(\lambda)\frac{1}{\theta^4} = \frac{1}{\lambda^2}\frac{1}{\theta^4} = \frac{1}{\theta^2}.$$

Note that also in this case it seems simpler to calculate the quantities directly than using the formulae in Theorem 1.30. □

In the special case where the reparametrization is affine, the relations involving score and information simplify.

Corollary 1.33. *Consider a family of densities $\mathcal{F} = \{f_\theta \mid \theta \in \Theta\}$ that is smooth and stable. Let $\phi : \Theta \mapsto \Lambda$ be an affine and bijective reparametrization, i.e. $\lambda = \phi(\theta) = a + B\theta$. Let further S, I, and i denote the score, information function, and Fisher information with respect to λ; and $\tilde{S}$, $\tilde{I}$, and $\tilde{i}$ the corresponding quantities with respect to θ. Then it holds that*

$$\tilde{S}(x,\theta) = S(x,\lambda)B \tag{1.9}$$

$$\tilde{I}(x,\theta) = B^\top I(\lambda, x)B \tag{1.10}$$

$$\tilde{i}(\theta) = B^\top i(\phi(\theta))B = B^\top i(\lambda)B. \tag{1.11}$$

Proof. This follows immediately from Theorem 1.30 since $D\phi(\theta) = B$ and $D^2\phi(\theta) = 0$. □

Also here is important to understand the behaviour of score and information when we have repeated observations. So consider a sample $X = (X_1, \ldots, X_n)$ of independent and identically distributed random variables with distributions from a parametrized family $\mathcal{P} = \{P_\theta \mid \theta \in \Theta\}$.

Theorem 1.34. *Assume that $\mathcal{P}$ is smooth and stable. Then so is $\mathcal{P}^{\otimes n}$. The score and information functions S_n, I_n, and i_n for $\mathcal{P}^{\otimes n}$ are related to the corresponding functions S, I, and i for $\mathcal{P}$ as*

$$S_n(X,\theta) = \sum_{i=1}^n S(X_i,\theta), \quad I_n(X,\theta) = \sum_{i=1}^n I(X_i,\theta), \quad i_n(\theta) = ni(\theta)$$

and the quadratic score becomes

$$Q_n(X,\theta) = S_n(X,\theta)i_n(\theta)^{-1}S_n(X,\theta)^\top = n\bar{S}_n(X,\theta)i(\theta)^{-1}\bar{S}_n(X,\theta)^\top \tag{1.12}$$

where $\bar{S}_n(X,\theta) = S_n(X,\theta)/n$ is the average score.

Proof. This is immediate from the definition of the concepts involved. □

As a consequence we have the following:

Corollary 1.35. *Assume that $\mathcal{P} = \{P_\theta \mid \theta \in \Theta\}$ is smooth and stable with $\Theta \subseteq \mathbb{R}^k$ and assume the Fisher information $i(\theta)$ is positive definite. Then the quadratic score $Q_n(X,\theta)$ associated with $\mathcal{P}^{\otimes n}$ for n repeated samples converges in distribution as*

$$Q_n(X,\theta) \xrightarrow{\mathcal{D}}_{n\to\infty} \chi^2(k)$$

where $\chi^2(k)$ denotes the χ^2-distribution with k degrees of freedom.

Proof. From the Central Limit Theorem (Theorem A.16), it follows that

$$\sqrt{n}\bar{S}_n(X,\theta) \xrightarrow{\mathcal{D}}_{n\to\infty} \mathcal{N}_k(0, i(\theta)).$$

The result now follows from Theorem A.11 and (1.12). □

1.3 Exercises

Exercise 1.1. Consider the simple normal model in Example 1.5 with the modification that the mean ξ is restricted to $\xi \in \mathbb{R}_+$ and reparametrize the model in terms of the mean and coefficient of variation

$$\mathbf{C}(X) = \sqrt{\mathbf{V}(X)}/\mathbf{E}(X).$$

Exercise 1.2. A drawing pin is thrown several times at random until it has landed five times flat on its head, and the total number of throws needed to achieve this is recorded. Describe a suitable statistical model for this experiment, performed with the purpose of determining the probability that the pin lands on its head. *Hint: use the negative binomial distribution.*

Exercise 1.3. If $Y = \log X$ and $Y \sim \mathcal{N}(\xi, \sigma^2)$, X is said to have a *log-normal* distribution. Identify the mean and coefficient of variation in this distribution and parametrize the family with these quantities.

Exercise 1.4. Consider the simple normal model in Example 1.5 and determine the score, information function, Fisher information, and quadratic score in the parametrization using the mean ξ and standard deviation σ.

Exercise 1.5. Consider the simple Poisson model, parametrized with the null fraction as in Example 1.10 and determine the score, information function, Fisher information, and quadratic score in this parametrization.

Exercise 1.6. Consider the simple normal model as in Exercise 1.1 and determine the score, information function, Fisher information, and quadratic score when parametrized with mean and coefficient of variation.

Exercise 1.7. Consider the family of log-normal distributions as in Exercise 1.3 and determine the score, information function, Fisher information, and quadratic score when parametrized with the mean and coefficient of variation.

Exercise 1.8. Consider the family of negative binomial distributions as in Exercise 1.2 and determine the score, information function, Fisher information, and quadratic score in a suitable parametrization.

Exercise 1.9. Consider the family of distributions with densities

$$f_\theta(x) = a(\theta)x^{\theta-1}\mathbf{1}_{(0,1)}(x), \quad \theta \in \Theta = \mathbb{R}_+$$

with respect to standard Lebesgue measure on $\mathbb{R}$, where $a(\theta)$ is a normalizing constant.

a) Show that $a(\theta) = \theta$;

b) Show that $Y = -\log X$, where X has the density above, follows an exponential distribution with mean $\mathbf{E}_\theta(Y) = 1/\theta$;

c) Find the score, information function, Fisher information, and quadratic score;

d) Verify Bartlett's identities directly by calculating relevant quantities and showing that they are equal.

Exercise 1.10. Consider the Cauchy model in Example 1.9 for the case when the scale parameter $\beta = 1$ is known.

a) Determine the score and information function;

b) Investigate by simulation whether Bartlett's identities hold in this case.

Chapter 2

Linear Normal Models

Linear normal models are the basis of a huge body of statistical methods and we shall briefly describe the basic elements of these. In particular we give a short treatment of the normal distribution on a Euclidean vector space, but abstain from a full treatment of the statistical issues of linear models, as this would require and deserve a text on its own.

The most important results for the developments in the following chapters are the decomposition theorems: Theorem 2.24 and the extended form in Theorem 2.27. The proof of these theorems can be quite cumbersome in some expositions. Here a geometric interpretation and proof is given which hopefully is both simple and illuminating. To achieve this properly, it is most convenient to discuss the normal distribution as a distribution on a finite-dimensional vector space with an inner product, in other words a Euclidean space. This will be done in Section 2.2. Before we move into this generality, we first recall some standard results for the multivariate normal distribution on $\mathbb{R}^d$ in the next section.

2.1 The multivariate normal distribution

Recall from, e.g. Jacod and Protter (2004, Ch. 16) that $X = (X_1, \ldots, X_d)^\top$ follows a multivariate normal distribution on $\mathbb{R}^d$ with *mean* $\xi \in \mathbb{R}^d$ and *covariance matrix* $\Sigma \in \mathbb{R}^{d\times d}$ if it holds for any $\lambda = (\lambda_1, \ldots, \lambda_d)^\top \in \mathbb{R}^d$ that the corresponding linear combination is univariate normal with mean $\lambda^\top\xi$ and variance $\lambda^\top\Sigma\lambda$:

$$\lambda^\top X \sim \mathcal{N}(\lambda^\top\xi, \lambda^\top\Sigma\lambda).$$

Here and elsewhere we let $\mathcal{N}(a, 0) = \delta_a$ denote Dirac measure at a, i.e. the distribution with all mass at a:

$$\delta_a(B) = \mathbf{1}_B(a), \qquad \text{for all } B \in \mathbb{B}(\mathbb{R}).$$

Then $\Sigma = \{\sigma_{ij}\}$ is a positive semi-definite (hence symmetric) matrix, $\mathbf{E}(X_i) = \xi_i$, and $\mathbf{V}(X_i, X_j) = \sigma_{ij}$. We then write $X \sim \mathcal{N}_d(\xi, \Sigma)$.

We first show that the distribution is well-defined and recall that any distribution of a random variable X with values in $\mathbb{R}^k$ is uniquely determined by its *characteristic function* (Jacod and Protter, 2004, Theorem 14.1)

$$\varphi_X(u) = \mathbf{E}(e^{iu^\top X}).$$

DOI: 10.1201/9781003272359-2

Proposition 2.1. *Let Σ be any positive semidefinite $d \times d$ matrix and $\xi \in \mathbb{R}^m$. Then the normal distribution $\mathcal{N}_d(\xi, \Sigma)$ exists and has characteristic function*

$$\varphi_X(u) = \mathbf{E}(e^{iu^\top X}) = \exp\{iu^\top \xi - u^\top \Sigma u/2\}. \tag{2.1}$$

Proof. Let $Z_1, \dots Z_d$ be independent and identically normally distributed as $\mathcal{N}(0,1)$ and let $Z = (Z_1, \dots Z_d)^\top$. The characteristic function φ_Z of Z is then

$$\varphi_Z(u) = \prod_{i=1}^{d} e^{-u_i^2/2} = \exp\{-\|u\|^2/2\}.$$

Since Σ is positive semidefinite, the spectral theorem (Horn and Johnson, 2013, Corollary 2.5.11) implies that there is an orthogonal matrix U and a diagonal matrix Λ such that $\Sigma = U\Lambda U^\top$. The diagonal elements of Λ are the non-negative eigenvalues $\lambda_1 \geq \dots \geq \lambda_d$ of Σ; some of these may be identical, and some may be zero if Σ is only positive semidefinite. We then let $\sqrt{\Lambda}$ be the diagonal matrix with $\sqrt{\lambda_i}$ in the diagonal, let $A = U\sqrt{\Lambda}U^\top$, and $X = \xi + AZ$. Note that since U is orthogonal we have $U^\top U = I_d$ and thus

$$AA^\top = U\sqrt{\Lambda}U^\top U\sqrt{\Lambda}U^\top = U\Lambda U^\top = \Sigma.$$

It now holds that any linear combination of the coordinates of X corresponds to an affine combination of elements of Z as

$$\lambda^\top X = \lambda^\top \xi + (\lambda^\top A)Z$$

and thus

$$\lambda^\top X \sim \mathcal{N}(\lambda\xi, \|\lambda^\top A\|^2) = \mathcal{N}(\lambda\xi, \lambda^\top AA^\top \lambda) = \mathcal{N}(\lambda\xi, \lambda^\top \Sigma\lambda)$$

so $X \sim \mathcal{N}_d(\xi, \Sigma)$, as desired. We get for the characteristic function

$$\begin{aligned}
\varphi_X(u) &= \mathbf{E}(e^{iu^\top X}) = \mathbf{E}(e^{iu^\top(\xi+AZ)}) = e^{iu^\top \xi}\mathbf{E}(e^{iu^\top AZ}) \\
&= e^{iu^\top \xi}\varphi_Z(A^\top u) = e^{iu^\top \xi}\exp\{-\|A^\top u\|^2/2\} \\
&= e^{iu^\top \xi}\exp\{-u^\top AA^\top u/2\} = \exp\{iu^\top \xi - u^\top \Sigma u/2\}.
\end{aligned}$$

The result follows. □

If Σ is positive definite, Σ is also invertible, and the inverse $K = \Sigma^{-1}$ is the *concentration matrix* of the normal distribution. We then say that the normal distribution is *regular*; then the distribution of X has a density with respect to *standard Lebesgue measure* on $\mathbb{R}^d$, i.e. the Lebesgue measure giving measure 1 to the unit cube

$$\{u \in \mathbb{R}^d \mid 0 \leq u_i \leq 1, i = 1, \dots d\}.$$

Proposition 2.2. *Let $\xi \in \mathbb{R}^d$, $\Sigma \in \mathbb{R}^{d\times d}$ be a positive definite matrix, and $K = \Sigma^{-1}$. Then the normal distribution $\mathcal{N}_d(\xi, \Sigma)$ has density*

$$\begin{aligned} f_{(\xi,\Sigma)}(x) &= (2\pi)^{-d/2}(\det \Sigma)^{-1/2} \exp\{-(x-\xi)^\top \Sigma^{-1}(x-\xi)/2\} \\ &= (2\pi)^{-d/2}(\det K)^{1/2} \exp\{-(x-\xi)^\top K(x-\xi)/2\} \end{aligned}$$

with respect to standard Lebesgue measure on $\mathbb{R}^d$.

Proof. See Corollary 16.2 of Jacod and Protter (2004) although their formula (16.5) has a typographical error. □

Example 2.3. Consider the case $d = 2$ and let

$$\xi = \begin{pmatrix} 1 \\ 2 \end{pmatrix}, \quad \Sigma_\rho = \begin{pmatrix} 1 & 2\rho \\ 2\rho & 4 \end{pmatrix}.$$

If $-1 < \rho < 1$, this is a regular normal distribution since $\det \Sigma_\rho = 4(1-\rho^2) > 0$. The concentration matrix becomes

$$K_\rho = \Sigma_\rho^{-1} = \frac{1}{4(1-\rho^2)} \begin{pmatrix} 4 & -2\rho \\ -2\rho & 1 \end{pmatrix}$$

and the distribution therefore has density

$$\begin{aligned} f_{\xi,\Sigma_\rho}(x) &= \frac{1}{4\pi\sqrt{1-\rho^2}} \exp\{-(x-\xi)^\top K_\rho(x-\xi)/2\} \\ &= \frac{1}{4\pi\sqrt{1-\rho^2}} \exp\{-Q_\rho(x-\xi)/2\} \qquad (2.2) \\ &= \frac{1}{4\pi\sqrt{1-\rho^2}} \exp\left\{-\frac{(x_1-1)^2}{2(1-\rho^2)} + \frac{\rho(x_1-1)(x_2-2)}{2(1-\rho^2)} - \frac{(x_2-2)^2}{8(1-\rho^2)}\right\} \end{aligned}$$

with respect to Lebesgue measure on $\mathbb{R}^2$. Here ρ is the correlation between X_1 and X_2. The contours of the density are ellipses centred at $\xi = (1, 2)$, with major axes and rotation depending on the correlation, see Figure 2.1. □

Affine images of normal distributions are again normal. More precisely

Proposition 2.4. *Let $X \sim \mathcal{N}_d(\xi, \Sigma)$, $a \in \mathbb{R}^m$, and let $B \in \mathbb{R}^{m\times d}$ be an $m \times d$ matrix. If $Y = a + BX$ then $Y \sim \mathcal{N}_m(a + B\xi, B\Sigma B^\top)$.*

Proof. Clearly Y takes values in $\mathbb{R}^m$. The characteristic function of Y is

$$\begin{aligned} \varphi_Y(u) &= \mathbf{E}(e^{iu^\top Y}) = \mathbf{E}(e^{iu^\top(a+BX)}) = e^{iu^\top a}\mathbf{E}(e^{iu^\top BX}) \\ &= e^{iu^\top a}\varphi_X(B^\top u) = e^{iu^\top a}\exp\{iu^\top B\xi - u^\top B\Sigma B^\top u/2\} \\ &= \exp\{iu^\top \mu - u^\top \Psi u/2\}, \end{aligned}$$

where $\mu = a + B\xi$ and $\Psi = B\Sigma B^\top$; the result follows. □

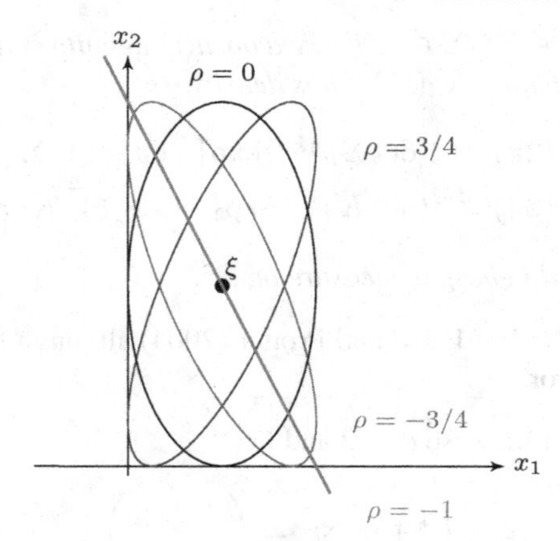

Figure 2.1 – Contour ellipses for the density of normal distributions with mean equal to $\xi = (1, 2)^\top$, variances $\mathbf{V}(X_1) = 1$ and $\mathbf{V}(X_2) = 4$, and varying correlations $\rho = -3/4, 0, 3/4$. Each ellipse indicates when the quadratic form Q_ρ in the exponent of (2.2) takes the value 1. When $\rho = -1$, the distribution degenerates to be supported on the affine subspace $x_2 = -2x_1 + 4$, as indicated by the thick line in the diagram; see Example 2.8 for details.

Example 2.5. As a special case of this, assume the random vector X partitioned into components X_1 and X_2, where $X_1 \in \mathbb{R}^r$ and $X_2 \in \mathbb{R}^s$ with $r + s = d$. Its mean vector and covariance matrix can then be partitioned accordingly into blocks as

$$\xi = \begin{pmatrix} \xi_1 \\ \xi_2 \end{pmatrix} \quad \text{and} \quad \Sigma = \begin{pmatrix} \Sigma_{11} & \Sigma_{12} \\ \Sigma_{21} & \Sigma_{22} \end{pmatrix}$$

such that Σ_{11} has dimensions $r \times r$ and so on. Let X be distributed as $\mathcal{N}_d(\xi, \Sigma)$, where X, ξ and Σ are partitioned as above. Then the marginal distribution of X_1 is also normal:

$$X_1 \sim \mathcal{N}_r(\xi_1, \Sigma_{11}).$$

This follows by letting $a = 0$ and $B = (I_r : 0_{r\times s})$ in Proposition 2.4. □

Independence of components in the multivariate normal distribution may easily be checked via the covariance matrix. More precisely, we have:

Proposition 2.6. *Assume the random vector X is partitioned into components X_1 and X_2, where $X_1 \in \mathbb{R}^r$ and $X_2 \in \mathbb{R}^s$ with $r + s = d$ and its mean vector and covariance matrix partitioned accordingly as in Example 2.5. Then X_1 and X_2 are independent if and only if $\Sigma_{12} = 0$.*

Proof. We get for the characteristic function

$$\begin{aligned} \varphi_X(u) &= \exp\{-iu^\top\xi - u^\top\Sigma u/2\} \\ &= e^{iu_1^\top\xi_1 + iu_2^\top\xi_2}\exp\{-u_1^\top\Sigma_{11}u_1/2 - u_2^\top\Sigma_{22}u_2/2 - u_1^\top\Sigma_{12}u_2\} \\ &= \varphi_{X_1}(u_1)\varphi_{X_2}(u_2)\exp\{-u_1^\top\Sigma_{12}u_2\} \end{aligned}$$

and hence $\varphi_X(u) = \varphi_{X_1}(u_1)\varphi_{X_2}(u_2)$ if and only if $\Sigma_{12} = 0$ whence the result follows. □

Combining Proposition 2.1 and Proposition 2.4, we obtain a structure theorem for multivariate normal distributions.

Theorem 2.7. *Let X be a random vector with values in $\mathbb{R}^d$, $\xi \in \mathbb{R}^d$, and let $\Sigma \in \mathbb{R}^{d\times d}$ be positive semidefinite. Then $X \sim \mathcal{N}_d(\xi, \Sigma)$ if and only if there is a matrix $A \in \mathbb{R}^{d\times d}$ with $AA^\top = \Sigma$ and $X \overset{\mathcal{D}}{=} \xi + AZ$ where $Z \sim \mathcal{N}_d(0, I_d)$. Here I_d is the $d \times d$ identity matrix.*

Proof. If $AA^\top = \Sigma$ and $Y = \xi + AZ$ Proposition 2.4 yields that

$$Y \sim \mathcal{N}_d(\xi, AA^\top) = \mathcal{N}_d(\xi, \Sigma).$$

The converse is Proposition 2.1. □

The distribution of $\xi + AZ$ is supported on the affine subspace $\xi + \mathrm{rg}(A)$ where $\mathrm{rg}(A) = A(\mathbb{R}^d)$ is the *range* of A, i.e. the image of the linear map with matrix A. If Σ is regular, $\mathrm{rg}(A) = \mathbb{R}^d$ and hence $\xi + \mathrm{rg}(A) = \xi + \mathbb{R}^d = \mathbb{R}^d$. If Σ is only semi-definite, we say that the normal distribution is *singular* and it does not have a density with respect to Lebesgue measure on $\mathbb{R}^d$, but it is rather supported on a proper affine subspace of $\mathbb{R}^d$.

Example 2.8. Consider the family of bivariate normal distributions in Example 2.3 but assume this time that $\rho = -1$ so the covariance matrix is

$$\Sigma_{-1} = \begin{pmatrix} 1 & -2 \\ -2 & 4 \end{pmatrix}.$$

This is singular since $\det \Sigma_{-1} = 0$. Indeed Σ_{-1} has eigenvalues 5 and 0 with the vector $u = (1, -2)^\top$ an eigenvector for the eigenvalue 5 and $v = (2, 1)^\top$ for the eigenvalue 0. Thus we have $\Sigma_{-1} = AA^\top$, where

$$A = A^\top = \frac{1}{\sqrt{5}} \begin{pmatrix} 1 & -2 \\ -2 & 4 \end{pmatrix} = \frac{1}{\sqrt{5}} \Sigma_{-1}$$

and thus

$$X \overset{\mathcal{D}}{=} \begin{pmatrix} 1 \\ 2 \end{pmatrix} + \frac{1}{\sqrt{5}} \begin{pmatrix} 1 & -2 \\ -2 & 4 \end{pmatrix} \begin{pmatrix} Z_1 \\ Z_2 \end{pmatrix},$$

where $Z = (Z_1, Z_2)^\top \sim \mathcal{N}(0, I_2)$. We conclude that this distribution is supported on the affine subspace of points satisfying the relation $x_2 - 2 = -2(x_1 - 1)$ or, equivalently, the line with equation $x_2 = -2x_1 + 4$; see Figure 2.1 for an illustration. An alternative representation of the distribution is as

$$X \overset{\mathcal{D}}{=} \xi + BW,$$

where $W \sim \mathcal{N}(0, 1)$ but $B = (1, -2)$ since then also $\Sigma_{-1} = BB^\top$, so the representation in this form is far from unique. □

Finally we recall the important $\chi^2(k)$-distribution:

Definition 2.9. The *$\chi^2(k)$-distribution*, where k are the *degrees of freedom* is the distribution of

$$Z_1^2 + \cdots + Z_k^2$$

where $Z_1, \ldots, Z_k$ are independent and identically distributed as $\mathcal{N}(0,1)$.

The $\chi^2(k)$-distribution is also a Gamma distribution $\Gamma(k/2, 2)$ with shape parameter $k/2$ and scale parameter 2, i.e. it has density

$$f_k(x) = \frac{x^{k/2-1}}{2^k \Gamma(k/2)} e^{-x/2}, \quad x > 0$$

with respect to Lebesgue measure on $\mathbb{R}_+$; its characteristic function is

$$\phi_{\chi^2(k)}(t) = \mathbf{E}(e^{it(Z_1^2+\cdots+Z_k^2)}) = \mathbf{E}(e^{itZ_1^2})^k = \frac{1}{(1-2it)^{k/2}}$$

and it has mean k and variance $2k$. See, for example Jacod and Protter (2004, Example 15.6) for some of these facts.

2.2 The normal distribution on a vector space

2.2.1 *Random vectors in V*

Let $(V, \langle\cdot,\cdot\rangle)$ be a Euclidean space and let $\mathbb{B}(V)$ be the Borel sets of V i.e. the σ-algebra generated by open sets. As for $\mathbb{R}^d$, a probability distribution on $(V, \mathbb{B}(V))$ is uniquely determined by its *characteristic function*

$$\varphi_X(u) = \mathbf{E}(e^{i\langle u, X\rangle}).$$

In particular, the normal distribution with *mean* $\xi \in V$ and *covariance* Σ is the distribution with characteristic function

$$\varphi_X(u) = \exp\{i\langle u, \xi\rangle - \langle u, \Sigma u\rangle/2\}.$$

Here $\Sigma \in \mathcal{L}(V, V)$ is a positive semidefinite and self-adjoint linear map so that for all $u, v \in V$ we have

$$\langle u, \Sigma u\rangle \geq 0 \text{ and } \langle u, \Sigma v\rangle = \langle \Sigma u, v\rangle.$$

In other words, X has a normal distribution on V if and only if any linear form $\langle u, X\rangle$ follows a normal distribution on $\mathbb{R}$ as $Y_u = \langle u, X\rangle \sim \mathcal{N}(\langle u, \xi\rangle, \langle u, \Sigma u\rangle)$. This follows since the characteristic function of Y_u is

$$\begin{aligned} \varphi_{Y_u}(t) &= \mathbf{E}(e^{itY_u}) = \mathbf{E}(e^{it\langle u, X\rangle}) = \mathbf{E}(e^{i\langle tu, X\rangle}) \\ &= \exp\{i\langle tu, \xi\rangle - \langle tu, \Sigma tu\rangle/2\} = \exp\{it\langle u, \xi\rangle - t^2\langle u, \Sigma u\rangle/2\}. \end{aligned}$$

Then the *mean* $\xi \in V$ is determined by the equation

$$\mathbf{E}(\langle u, X\rangle) = \langle u, \xi\rangle \text{ for all } u \in V.$$

The map Σ is the *covariance operator* of the distribution or, briefly, the *covariance*. This reflects the fact that for all linear forms determined by $u, v \in V$ we have

$$\mathbf{V}(\langle u, X\rangle, \langle v, X\rangle) = \langle u, \Sigma v\rangle \tag{2.3}$$

which follows by using Proposition A.2 and the relation

$$\mathbf{V}(\langle u, X\rangle) = \langle u, \Sigma u\rangle.$$

For a given orthonormal basis, we define the *mean vector* and *covariance matrix* of a random variable on V as the mean vector and covariance matrix of its coordinates with respect to this basis, i.e. for the normal distribution above the vector and matrix with entries

$$\xi_i = \mathbf{E}(\langle e_i, X\rangle), \quad \sigma_{ij} = \mathbf{V}(\langle e_i, X\rangle, \langle e_j, X\rangle) = \langle e_i, \Sigma e_j\rangle$$

and we shall use Σ to denote both the covariance operator and its matrix with respect to a given basis. We have the following analogue of Proposition 2.4.

Proposition 2.10. *Let $(V, \langle\cdot,\cdot\rangle_V)$ and $(W, \langle\cdot,\cdot\rangle_W)$ be Euclidean spaces and assume $X \sim \mathcal{N}_V(\xi, \Sigma)$. For $a \in W$ and $B \in \mathcal{L}(V, W)$ it holds that*

$$Y = a + BX \sim \mathcal{N}_W(a + B\xi, B\Sigma B^*).$$

Proof. The characteristic function of Y is

$$\begin{aligned}
\varphi_Y(u) &= \mathbf{E}(e^{i\langle u,Y\rangle_W}) = e^{i\langle u,a\rangle_W}\mathbf{E}(e^{i\langle u,BX\rangle}) = e^{i\langle u,a\rangle_W}\mathbf{E}(e^{i\langle B^*u,X\rangle_V}) \\
&= e^{i\langle u,a\rangle_W}\varphi_X(B^*u) = e^{i\langle u,a\rangle_W}\exp\{i\langle B^*u,\xi\rangle_V - \langle B^*u, \Sigma B^*u\rangle_V/2\} \\
&= e^{i\langle u,a\rangle_W}\exp\{i\langle u, B\xi\rangle_W - \langle u, B\Sigma B^*u\rangle_W/2\} \\
&= \exp\{i\langle u,\mu\rangle_W - \langle u, \Psi u\rangle_W/2\},
\end{aligned}$$

where $\mu = a + B\xi$ and $\Psi = B\Sigma B^*$; the result follows. ☐

Further we have a structure theorem, similar to Theorem 2.7:

Theorem 2.11. *Let X be a random vector with values in the Euclidean space $(V, \langle\cdot,\cdot\rangle)$, $\xi \in V$, and let $\Sigma \in \mathcal{L}(V, V)$ a positive semidefinite and self-adjoint linear map. Then $X \sim \mathcal{N}_V(\xi, \Sigma)$ if and only if there is a linear map $A \in \mathcal{L}(V, V)$ with $AA^* = \Sigma$. Then X has the same distribution as $Y = \xi + AZ$ where $Z \sim \mathcal{N}_V(0, I)$ and I is the identity map on V.*

Proof. Let $e_1, \dots, e_d$ be an orthonormal basis for V and $Y = (Y_1, \dots, Y_d)^\top$ the coordinates of X with respect to this basis. Then $Y \sim \mathcal{N}_d(\mu, \Gamma)$ where μ are the coordinates of ξ and $\Gamma_{ij} = \langle e_i, \Sigma e_j\rangle$ the representation of Σ in this basis. Theorem 2.7 yields that

$$Y \stackrel{\mathcal{D}}{=} \mu + BW,$$

where $\Gamma = BB^\top$ and $W \sim \mathcal{N}_d(0, I_d)$. Now let Z be the random variable in V with coordinates W and A the linear map represented by B in the chosen basis, i.e.

$$Z = \sum_{i=1}^{d} W_i e_i, \quad Au = \sum_{i=1}^{d}\sum_{j=1}^{d} B_{ij}\langle u, e_j\rangle u_i.$$

Then $Z \sim \mathcal{N}_V(0, I)$, $\Sigma = AA^*$ and $X \stackrel{\mathcal{D}}{=} \xi + AZ$, as required. □

The distribution of the random variable Y in Theorem 2.11 is supported on the affine subspace $\xi + \mathrm{rg}(A)$ with $\mathrm{rg}(A)$ being the image of the map A. Although the covariance Σ may be represented as $\Sigma = AA^*$ in a multitude of ways, see Example 2.8, the space $\mathrm{rg}(A)$ does not depend on this representation since Lemma A.6 yields that $\mathrm{rg}(\Sigma) = \mathrm{rg}(A)$. We thus have:

Corollary 2.12. *The normal distribution $\mathcal{N}_V(\xi, \Sigma)$ is supported on the affine subspace $\xi + \mathrm{rg}(\Sigma)$ of V.*

We next define a *concentration operator* or just *concentration* of the normal distribution $\mathcal{N}_V(\xi, \Sigma)$ as any positive self-adjoint linear map $K \in \mathcal{L}(V, V)$ satisfying for all $u \in \mathrm{rg}(\Sigma)$ and all $v \in V$ that

$$\langle Ku, \Sigma v\rangle = \langle u, v\rangle. \tag{2.4}$$

Proposition 2.13. *A positive self-adjoint map K is a concentration operator for Σ if and only if it satisfies the relation*

$$\Sigma K \Sigma = \Sigma. \tag{2.5}$$

Proof. If K is self-adjoint and satisfies (2.4) we have for all $u, v \in V$

$$\langle \Sigma K \Sigma u, v\rangle = \langle K\Sigma u, \Sigma v\rangle = \langle \Sigma u, v\rangle$$

implying that $\Sigma K \Sigma = \Sigma$.

Conversely, if $\Sigma K \Sigma = \Sigma$ we have for all $u = \Sigma w$ and $v \in V$ that

$$\langle Ku, \Sigma v\rangle = \langle K\Sigma w, \Sigma v\rangle = \langle \Sigma K \Sigma w, v\rangle = \langle \Sigma w, v\rangle = \langle u, v\rangle$$

which shows (2.4). □

A positive and self-adjoint map K that satisfies (2.5) is also known as a *generalized inverse* of Σ, sometimes written as $K = \Sigma^-$.

Example 2.14. If Σ is singular, a generalized inverse is far from unique. For example, if we have

$$\Sigma = \begin{pmatrix} 3 & 0 \\ 0 & 0 \end{pmatrix},$$

then any positive definite matrix of the form

$$K_{a,b} = \begin{pmatrix} 1/3 & a \\ a & b \end{pmatrix} \tag{2.6}$$

is a concentration matrix for Σ. □

Proposition 2.15. *If Σ is invertible, its inverse $K = \Sigma^{-1}$ is the unique concentration operator for Σ.*

Proof. If Σ is invertible we may let $K = \Sigma^{-1}$ which then clearly satisfies $\Sigma K \Sigma = \Sigma$. And conversely, if we pre- and post-multiply the relation $\Sigma K_1 \Sigma$ with $K = \Sigma^{-1}$ we conclude $K_1 = K$. □

A concentration operator defines an inner product $\langle \cdot, \cdot \rangle_K$ on V by the relation

$$\langle u, v \rangle_K = \langle Ku, v \rangle.$$

Even if Σ is singular and the concentration therefore is not unique, the restriction of the associated inner product onto the range of Σ is unique.

Proposition 2.16. *Let K_1 and K_2 be concentration operators for Σ. It then holds for all $u, v \in \mathrm{rg}(\Sigma)$ that*

$$\langle u, v \rangle_{K_1} = \langle K_1 u, v \rangle = \langle K_2 u, v \rangle = \langle u, v \rangle_{K_2}.$$

Proof. By Proposition A.2, it is sufficient to show that the inner products agree on the diagonal. So let $u = \Sigma v \in \mathrm{rg}(\Sigma)$. Then for $i = 1, 2$ we have

$$\langle K_i u, u \rangle = \langle K_i \Sigma v, \Sigma v \rangle = \langle \Sigma K_i \Sigma v, v \rangle = \langle \Sigma v, v \rangle,$$

showing that the inner products coincide. □

Example 2.17. As a simple illustration of Proposition 2.13, we may consider the singular covariance

$$\Sigma = \begin{pmatrix} 3 & 0 \\ 0 & 0 \end{pmatrix}$$

as in Example 2.14 above. Here we have

$$\mathrm{rg}(\Sigma) = \left\{ u = \begin{pmatrix} \alpha \\ 0 \end{pmatrix} : \alpha \in \mathbb{R} \right\}$$

Then any matrix of the form (2.6) will satisfy for $u, v \in \mathrm{rg}(\Sigma)$

$$\langle u, v \rangle_K = \begin{pmatrix} \alpha & 0 \end{pmatrix} K_{a,b} \begin{pmatrix} \beta \\ 0 \end{pmatrix} = \frac{\alpha\beta}{3}$$

reflecting that this does not depend on a nor b. □

In the special case where Σ is invertible, the corresponding normal distribution has density with respect to Lebesgue measure on V:

Proposition 2.18. *Let $\xi \in V$ and $\Sigma \in \mathcal{L}(V, V)$ be positive, invertible and self-adjoint. Let now $K = \Sigma^{-1}$. Then the normal distribution $\mathcal{N}_V(\xi, \Sigma)$ has density*

$$f_{(\xi,\Sigma)}(x) = \frac{\det K^{1/2}}{(2\pi)^{\dim V/2}} \exp\left\{-\|x - \xi\|_K^2/2\right\} \tag{2.7}$$

with respect to standard Lebesgue measure on V. Here $\det K$ *is the determinant of the concentration matrix with entries $k_{ij} = \langle Ke_i, e_j\rangle$ where $e_1, \ldots, e_{\dim V}$ is any basis for V that is orthonormal with respect to the inner product $\langle\cdot,\cdot\rangle$.*

Proof. This follows from Proposition 2.2 since the standard Lebesque measure on V is obtained by transforming the standard Lebesque measure on $\mathbb{R}^d$ onto V via the coordinates with respect to any orthonormal basis; see Appendix A.1. □

If Σ is only semi-definite, we say that the normal distribution is *singular* and it does not have a density with respect to Lebesgue measure on V; rather $Y = X - \xi$ has density with respect to Lebesque measure on the subspace $L = \mathrm{rg}(\Sigma)$ which itself is a Euclidean space. For the sake of completeness, we mention that this density is very similar to that above:

$$f_{(\xi,\Sigma)}(y) = \frac{\det_{\mathrm{L}} K^{1/2}}{(2\pi)^{\dim L/2}} \exp\left\{-\|y\|_K^2/2\right\}$$

where now $\det_{\mathrm{L}} K$ is the determinant of the restriction of the concentration matrix K onto L with respect to an orthonormal basis $e_1, \ldots, e_{\dim L}$ for L, or, in other words,

$$\det{}_{\mathrm{L}} K = \prod_{i=1}^{\dim L} \lambda_i^{-1},$$

where $\lambda_1 \geq \cdots \geq \lambda_d$ are the ordered eigenvalues of Σ. We refrain from giving a formal proof of this, as this density plays no role in subsequent developments.

2.2.2 *Projections with respect to the concentration*

Let now $L \subseteq V$ be a linear subspace of V and assume that Σ is regular so the inner product $\langle\cdot,\cdot\rangle_K$ is well-defined on V. We shall refer to $\langle\cdot,\cdot\rangle_K$ as the *concentration inner product* whereas we refer to $\langle\cdot,\cdot\rangle$ as the *base inner product*. The (orthogonal) projection Π_L onto L of an element $u \in V$ with respect to the concentration inner product is then determined as the unique point $\Pi_L u \in L$ satisfying for all $w \in L$

$$\langle u - \Pi_L u, w\rangle_K = 0 \tag{2.8}$$

or, equivalently

$$\langle u - \Pi_L u, Kw\rangle = 0.$$

Note that for simplicity we here and elsewhere omit the qualifier 'orthogonal' and simply say 'projection' to mean 'orthogonal projection' when there is no ambiguity. We have

Theorem 2.19. *A linear map $P \in \mathcal{L}(V, V)$ is a projection with respect to $\langle \cdot, \cdot \rangle_K$ if and only if P satisfies $P^2 = P$ and $P^* K = KP$, where P^* is the adjoint of P with respect to the base inner product.*

Proof. Theorem A.3 gives that we just have to check whether P is self-adjoint with respect to $\langle \cdot, \cdot \rangle_K$. But we have for all $u, v \in V$ that

$$\langle u, Pv \rangle_K = \langle Ku, Pv \rangle = \langle P^* Ku, v \rangle \quad \text{and} \quad \langle Pu, v \rangle_K = \langle KPu, v \rangle$$

whence the result follows. □

Remark 2.20. *The condition for self-adjointness with respect to $\langle \cdot, \cdot \rangle_K$ may be expressed in terms of the covariance Σ as*

$$\Sigma P^* = P \Sigma$$

since we may pre- and post-multiply the relation $P^* K = KP$ with Σ.

Common ways of specifying a linear space are as follows as an image of a linear map, typically from $\mathbb{R}^k$, i.e. as $L = \text{rg}(A) = A(\mathbb{R}^k)$, where $A \in \mathcal{L}(\mathbb{R}^k, V)$, or: As the kernel of a linear map, i.e. as $\ker(B) = \{v \in V \mid B(v) = 0\}$ for $B \in \mathcal{L}(V, \mathbb{R}^m)$. It is then useful to express the projections with respect to the concentration inner product in terms of K, Σ, and the maps A and B. Indeed we have the following, where we have replaced $\mathbb{R}^k$ with general Euclidean spaces U and W.

Proposition 2.21. *Let $(U, \langle \cdot, \cdot \rangle_U)$, $(V, \langle \cdot, \cdot \rangle_V)$, and $(W, \langle \cdot, \cdot \rangle_W)$ be Euclidean spaces. Suppose $\Sigma \in \mathcal{L}(V, V)$ is a positive definite and self-adjoint covariance, $L = \text{rg}(A)$ or $L = \ker(B)$, where $A \in \mathcal{L}(U, V)$ is injective and $B \in \mathcal{L}(W, V)$ surjective. Then $A^* KA$ and BKB^* are both invertible and the projection $\Pi_L \in \mathcal{L}(V, V)$ onto L with respect to the concentration inner product $\langle \cdot, \cdot \rangle_K$ is given as*

$$\Pi_L = A(A^* KA)^{-1} A^* K \text{ or, equivalently, } \Pi_L = I - B^*(BKB^*)^{-1} BK,$$

where I is the identity on V and $K = \Sigma^{-1}$.

Proof. Let $Q = A^* KA : U \to U$ and assume $u \in \ker(Q)$. Then we have

$$0 = \langle u, Qu \rangle_U = \langle u, A^* KAu \rangle_U = \langle Au, KAu \rangle_V = ||Au||_K^2$$

whence we conclude that $Au = 0$ and therefore $u = 0$ since A is assumed injective; hence we conclude that $Q = A^* KA$ is injective. The similar result for B follows by replacing A with B^* which is injective by Proposition A.4. Next, let $H = A(A^* KA)^{-1} A^* K$. Then

$$H^2 = A(A^* KA)^{-1} A^* K \, A(A^* KA)^{-1} A^* K = A(A^* KA)^{-1} A^* K = H$$

so H is idempotent. Further, it satisfies $H^* K = KH$ since

$$H^* K = (A(A^* KA)^{-1} A^* K)^* K = KA(A^* KA)^{-1} A^* K = KH.$$

Finally we show that $\mathrm{rg}(H) = \mathrm{rg}(A)$. The expression for H yields directly that $\mathrm{rg}(H) \subseteq \mathrm{rg}(A)$. For the reverse inclusion, we let Au denote a generic element of $\mathrm{rg}(A)$ and get

$$HAu = A(A^*KA)^{-1}A^*KAu = Au$$

and hence $Au \in \mathrm{rg}(H)$, showing that $\mathrm{rg}(A) \subseteq \mathrm{rg}(H)$, as required.

The statements in terms of B follow by first replacing A with B^* in the expression for H above, implying that $B^*(BKB^*)^{-1}BK$ is the projection onto $\mathrm{rg}(B^*)$, then using Proposition A.4 to conclude that $\mathrm{rg}(B^*) = \ker(B)^\perp = L^\perp$ and thus $I - B^*(BKB^*)^{-1}BK$ is the projection onto L. □

2.2.3 *Derived distributions*

Here we briefly describe some important distributional results associated with the normal distribution. We first need a lemma.

Lemma 2.22. *Assume that $X \sim \mathcal{N}_V(\xi, \Sigma)$ on $(V, \langle\cdot,\cdot\rangle)$ and let $A \in \mathcal{L}(V, W_1)$ and $B \in \mathcal{L}(V, W_2)$ where $(W_1, \langle\cdot,\cdot\rangle_1)$ and $(W_2, \langle\cdot,\cdot\rangle_2)$ are Euclidean spaces. Then $Y = AX$ and $Z = BX$ are independent if and only if $A\Sigma B^* = 0$.*

Proof. By Proposition 2.10 the distribution of (Y, Z) is normal on $W_1 \oplus W_2$ with mean $(A\xi, B\xi)$ and covariance Σ_{AB} given as

$$\langle (w_1, w_2), \Sigma_{AB}(w_1, w_2)\rangle_{12} = \langle w_1, A\Sigma A^* w_1\rangle + 2\langle w_1, A\Sigma B^* w_2\rangle + \langle w_2, B\Sigma B^* w_2\rangle.$$

We thus get for the characteristic function

$$\begin{aligned}
\varphi_{Y,Z}(w_1, w_2) &= \mathbf{E}\left(e^{i(\langle w_1, Y\rangle_1 + \langle w_2, Z\rangle_2)}\right) \\
&= e^{i(\langle w_1, A\xi\rangle_1 + \langle w_2, B\xi\rangle_2)} \\
&\quad \times e^{-\langle w_1, A\Sigma A^* w_1\rangle/2 - \langle w_1, A\Sigma B^* w_2\rangle - \langle w_2, B\Sigma B^* w_2\rangle/2} \\
&= \varphi_Y(w_1)\varphi_Z(w_2)e^{-\langle w_1, A\Sigma B^* w_2\rangle}.
\end{aligned}$$

Thus we have that $\varphi_{Y,Z}(w_1, w_2) = \varphi_Y(w_1)\varphi_Z(w_2)$ if and only if it holds that $A\Sigma B^* = 0$, and the result follows. □

Note in particular the following corollary which is a minor modification of Proposition 2.6.

Corollary 2.23. *Let $(V, \langle\cdot,\cdot\rangle)$ be a Euclidean space and $X \sim \mathcal{N}_V(\xi, \Sigma)$. Then disjoint sets $Y_A = (Y_1, \ldots Y_m)^\top$ and $Y_B = (Y_{m+1}, \ldots, Y_{m+k})^\top$ of coordinates $Y_i = \langle e_i, X\rangle$ for X with respect to an orthonormal basis $e = (e_1, \ldots, e_d)$ for V are independent if and only if*

$$\sigma_{ij} = 0 \quad \textit{for all } i = 1, \ldots, m,\ j = m+1, \ldots, m+k,$$

where $\sigma_{ij} = \langle e_i, \Sigma e_j\rangle$ are the entries of the covariance matrix for X in this basis.

Proof. Let A be the linear map that sends X into the coordinates Y_A and B the map that sends X to the coordinates Y_B. Then, in the basis considered, the matrix for $A\Sigma B^*$ is given as

$$(A\Sigma B^*)_{ij} = \langle e_i, A\Sigma B^* e_j \rangle = \langle e_i, \Sigma e_j \rangle = \sigma_{ij}$$

and the result follows from Lemma 2.22. □

We are now ready to show the following important result, known as the *decomposition theorem* for the normal distribution.

Theorem 2.24. *Let* $X \sim \mathcal{N}_V(\xi, \Sigma)$ *on* $(V, \langle\cdot,\cdot\rangle)$ *where* $\dim V = d$ *and assume that* Σ *is invertible with concentration* $K = \Sigma^{-1}$*. Further, let* $L \subseteq V$ *be a subspace of dimension* $\dim L = m$ *and let* Π_L *denote the projection onto* L *with respect to the concentration inner product* $\langle\cdot,\cdot\rangle_K$*. Let* I *denote the identity on* V*. Then the following hold:*

(a) *The projections* $\Pi_L X$ *and* $X - \Pi_L X$ *are independent and normally distributed;*

(b) *The covariances of* $\Pi_L X$ *and* $X - \Pi_L X$ *are* $\Pi_L \Sigma$ *and* $(I - \Pi_L)\Sigma$*;*

(c) K *is a concentration operator for both of* $\Pi_L X$ *and* $X - \Pi_L X$*;*

(d) *If* $\xi \in L$*, then* $\|X - \Pi_L X\|_K^2 \sim \chi^2(k)$*, where* $k = \dim V - \dim L = d - m$.

Proof. From Remark 2.20 we have that $\Pi_L \Sigma = \Sigma \Pi_L^*$ implying that

$$(I - \Pi_L)\Sigma\Pi_L^* = (I - \Pi_L)\Pi_L\Sigma = (\Pi_L^2 - \Pi_L)\Sigma = (\Pi_L - \Pi_L)\Sigma = 0$$

where we have used that Π_L is idempotent. Lemma 2.22 now yields that $\Pi_L X$ and $X - \Pi_L X = (I - \Pi_L)X$ are independent and since Π_L and $I - \Pi_L$ are linear, they are also normally distributed. This establishes (a).

The covariance for $\Pi_L X$ is

$$\mathbf{V}(\Pi_L X) = \Pi_L \Sigma \Pi_L^* = \Pi_L^2 \Sigma = \Pi_L \Sigma$$

and by analogy we get $\mathbf{V}(X - \Pi_L X) = (I - \Pi_L)\Sigma$ which yields (b).

But then K satisfies for $u \in L = \mathrm{rg}(\Pi_L \Sigma)$

$$\langle Ku, \Pi_L \Sigma v \rangle = \langle \Pi_L^* K u, \Sigma v \rangle = \langle K \Pi_L u, \Sigma v \rangle = \langle Ku, \Sigma v \rangle = \langle u, v \rangle$$

showing that K is a concentration for $\Pi_L \Sigma$. The calculations for $X - \Pi_L X$ are again analogous; hence (c) is established.

If $\xi \in L$ we have the mean of $X - \Pi_L X$ is $\xi - \Pi_L \xi = 0$ and hence it holds that the residual $X - \Pi_L X$ is distributed as $\mathcal{N}_V(0, (I - \Pi_L)\Sigma)$. Now choose a basis $e_1, \ldots, e_d$ for V that is orthogonal for the concentration inner product and where $e_1, \ldots, e_m$ form a basis for L and thus $e_{m+1}, \ldots, e_d$ a basis for $L^\perp$. Then

$$\|X - \Pi_L X\|_K^2 = \sum_{i=1}^{k} Z_{m+i}^2 = \sum_{i=m+1}^{d} Z_i^2$$

where $Z_i = \langle X, e_i \rangle_K$ are the coordinates in this basis. But we have for the covariances of Z_i

$$\begin{aligned} \mathbf{V}(Z_i, Z_j) &= \mathbf{V}(\langle Ke_i, X \rangle, \langle Ke_j, X \rangle) \\ &= \langle Ke_i, \Sigma K e_j \rangle = \langle Ke_i, e_j \rangle = \langle e_i, e_j \rangle_K. \end{aligned}$$

Since $\langle e_i, e_j \rangle_K = 0$ for $i \neq j$ and $\langle e_i, e_i \rangle_K = 1$, we get that $Z_{m+1}, \ldots, Z_d$ are independent and standard normally distributed whence Definition 2.9 yields that $\|X - \Pi_L X\|_K^2 \sim \chi^2(k)$, where $k = d - m = \dim V - \dim L$, as required for (d). This completes the proof. □

Remark 2.25. *Note in particular the conclusion in Theorem 2.24(c), saying that although the covariance for the projection $\Pi_L X$ is $\Pi_L \Sigma$ and thus differs from Σ, the concentration K is still a valid concentration; here one should bear in mind that only the restriction of K to L matters, cf. Proposition 2.16.*

We illustrate the use of the decomposition theorem in a very simple example that is both classical and in some sense generic.

Example 2.26. Consider $V = \mathbb{R}^n$ with standard inner product $\langle u, v \rangle = u^\top v$ and the normal distribution $\mathcal{N}_V(\xi, \sigma^2 I_n)$. Assume $\xi \in L$ where

$$L = \{\xi \in \mathbb{R}^n \mid \xi_1 = \xi_2 = \cdots = \xi_n = \mu,\ \mu \in \mathbb{R}\}$$

is the one-dimensional subspace of $\mathbb{R}^n$ where all coordinates are identical, i.e. the space spanned by the constant vector $c = (1, \ldots, 1)^\top$. From (A.2), the orthogonal projection $\Pi_L X$ of X onto L is determined as

$$\Pi_L X = \frac{\langle X, c \rangle}{\|c\|^2} c = \frac{X^\top c}{c^\top c} c = \frac{\sum_{i=1}^n X_i}{n} c = \bar{X} c = (\bar{X}, \ldots, \bar{X})^\top.$$

Similarly, the residual $X - \Pi_L X$ is

$$X - \Pi_L X = (X_1 - \bar{X}, \ldots, X_n - \bar{X})^\top$$

and Theorem 2.24(a) now yields that $\bar{X}$ and the residual $(X_1 - \bar{X}, \ldots, X_n - \bar{X})^\top$ are independent.

The matrix for Π_L is E/n where E is the matrix with all elements equal to 1:

$$E = \begin{pmatrix} 1 & 1 & \cdots & 1 \\ 1 & 1 & \cdots & 1 \\ \vdots & \vdots & \vdots & \vdots \\ 1 & 1 & \cdots & 1 \end{pmatrix}.$$

Hence from Theorem 2.24(b) we deduce that the covariance matrices for $\Pi_L X$ and $X - \Pi_L X$ are

$$\mathbf{V}(\Pi_L X) = \frac{\sigma^2}{n} E, \quad \mathbf{V}(X - \Pi_L X) = \sigma^2 \left(I - \frac{1}{n} E \right).$$

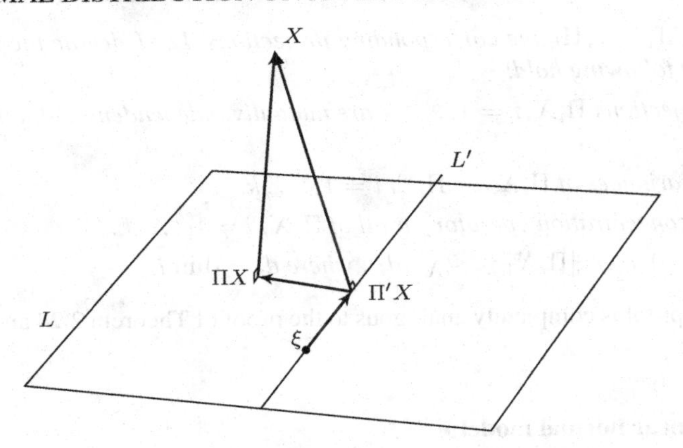

Figure 2.2 – Visualization of Theorem 2.27. If $X \sim \mathcal{N}_V(\xi, \Sigma)$ with $\xi \in L' \subseteq L \subseteq V$, $\Pi_1 X = X - \Pi X$, $\Pi_2 X = \Pi X - \Pi' X$, and $\Pi_3(X) = \Pi' X$ are all independent if projections are orthogonal with respect to $\langle\cdot,\cdot\rangle_K$. Further, since $\xi \in L'$, the squared norms $\|X - \Pi X\|_K^2$ and $\|\Pi X - \Pi' X\|_K^2$ are χ^2-distributed with degrees of freedom equal to $\dim V - \dim L$ and $\dim L - \dim L'$ respectively.

Note that both of these covariance matrices are singular, reflecting that $\Pi_L X$ is supported on the one-dimensional subspace L, and $X - \Pi_L X$ is supported on the orthogonal complement $L^\perp$. Still $K = \sigma^{-2} I_n$ is a concentration matrix for both of these distributions, as shown in Theorem 2.24(c). Using this fact, we get from Theorem 2.24(d) that

$$\|X - \Pi_L X\|_K^2 = \frac{1}{\sigma^2} \sum_{i=1}^{n} (X_i - \bar{X})^2$$

follows a $\chi^2(n-1)$ distribution since $\dim V = n$ and $\dim L = 1$. □

There is a natural generalization of the decomposition theorem to multiple subspaces. A geometric illustration of this decomposition theorem is given in Fig. 2.2. An *orthogonal decomposition* of V with respect to an inner product on V is a system $\Pi_i, i = 1, \ldots k$ of k projections onto mutually orthogonal spaces $L_i, i = 1, \ldots, k$ so that

$$\Pi_i \Pi_j = 0 \text{ for } i \neq j, \quad I = \sum_{i=1}^{k} \Pi_i$$

or, in other words, a decomposition of V as

$$V = L_1 \oplus \cdots \oplus L_k$$

into mutually orthogonal subspaces. We then have

Theorem 2.27. *Let $X \sim \mathcal{N}_V(\xi, \Sigma)$ on $(V, \langle\cdot,\cdot\rangle)$ where $\dim V = d$ and assume that Σ is invertible with concentration $K = \Sigma^{-1}$. Further, let $L_1, \ldots, L_k$ denote an orthogonal decomposition of V with respect to the concentration inner product*

$\langle \cdot, \cdot \rangle_K$, and $\Pi_1, \ldots, \Pi_k$ the corresponding projections. Let I denote the identity on V. Then the following hold:

(a) *The projections $\Pi_i X, i = 1, \ldots, k$ are mutually independent and normally distributed;*

(b) *The covariances of $\Pi_i X$ are $\Pi_i \Sigma$, $i = 1, \ldots, k$.*

(c) *K is a concentration operator for all of $\Pi_i X$, $i = 1, \ldots, k$.*

(d) *If $\Pi_i \xi = 0$, then $\|\Pi_i X\|_K^2 \sim \chi^2(d_i)$, where $d_i = \dim L_i$.*

Proof. The proof is completely analogous to the proof of Theorem 2.24 and therefore omitted. □

2.3 The linear normal model

2.3.1 Basic structure

The linear normal model is obtained by assuming X to be normally distributed on a d-dimensional Euclidean vectorspace $(V, \langle \cdot, \cdot \rangle)$ with mean ξ in a linear subspace $L \subseteq V$ and covariance $\sigma^2 \Sigma$, where $\Sigma \in \mathcal{L}(V, V)$ is a fixed and known positive definite and self-adjoint map. Thus X has density

$$f_{(\xi,\sigma^2)}(x) = \frac{(\det K)^{1/2}}{(2\pi\sigma^2)^{d/2}} e^{\frac{-\|x-\xi\|_K^2}{2\sigma^2}}$$

with respect to standard Lebesgue measure λ_V on V; here $K = \Sigma^{-1}$ is the concentration of the distribution and the squared norm in the exponent is using the concentration inner product determined by K.

This model has *representation space* $(V, \mathbb{B}(V))$, *parameter space* $L \times \mathbb{R}_+$, and the *family of probability measures* is

$$\mathcal{P} = \{\mathcal{N}_V(\xi, \sigma^2\Sigma) \mid (\xi, \sigma^2) \in L \times \mathbb{R}_+\}.$$

We may without loss of generality reparametrize the model by using $\langle \cdot, \cdot \rangle_K$ as the base inner product on V, leading to $\Sigma = K = I$ and the density

$$f_{(\xi,\sigma^2)}(x) = \frac{1}{(2\pi\sigma^2)^{d/2}} e^{\frac{-\|x-\xi\|^2}{2\sigma^2}}$$

with respect to 'standard Lebesgue measure on V', the norm $\|\cdot\|$ now referring to the new base inner product $\langle \cdot, \cdot \rangle_I$. We shall later, in Chapter 3, see that the family is also smooth and stable so that Bartlett's identities hold for the score and information; see also Section 2.3.2 below.

Special instances of this model include linear regression, models for comparing means, analysis of variance, and many others. And from time to time we shall consider the submodel where also σ^2 is fixed and known.

Example 2.28. [Single normal sample] Consider $X = (X_1, \ldots, X_n)^\top$ where $X_1, \ldots, X_n$ are identically distributed as $\mathcal{N}(\mu, \sigma^2)$. This may be represented as a linear normal model by considering X as an element of $V = \mathbb{R}^n$ with standard inner product $\langle u, v\rangle = u^\top v$ and mean $\xi = (\mu, \dots, \mu)^\top$ an element of the one-dimensional subspace

$$L = \{\xi \in \mathbb{R}^n \mid \xi_1 = \xi_2 = \cdots = \xi_n = \mu,\ \mu \in \mathbb{R}\}$$

as in Example 2.26 above. Then the model can either be parametrized by the general parametrization $(\xi, \sigma^2) \in L \times \mathbb{R}_+$, or by $(\mu, \sigma^2) \in \mathbb{R} \times \mathbb{R}_+$ using the correspondence $\mu \leftrightarrow \xi = (\mu, \dots, \mu)^\top$. □

Example 2.29. [Simple linear regression] Another special instance of the linear model is that of simple linear regression. We consider $X_1, \dots, X_d$ to be independent and normally distributed as

$$X_i \sim \mathcal{N}(\alpha + \beta t_i, \sigma^2), \quad i = 1, \dots, d,$$

where $t = (1, \dots, t_d)$ are known real numbers and $\alpha, \beta \in \mathbb{R}$ and $\sigma^2 \in \mathbb{R}_+$ are unknown. This is a general linear model with $V = \mathbb{R}^d$, standard inner product, and mean $\xi \in L$, where L is the two-dimensional subspace of $\mathbb{R}^d$ determined as

$$L = \{\xi \in \mathbb{R}^d \mid \xi_i = \alpha + \beta t_i,\ (\alpha, \beta) \in \mathbb{R}^2\} = A(\mathbb{R}^2).$$

Here A is the *design matrix*,

$$A^\top = \begin{pmatrix} 1 & \cdots & 1 \\ t_1 & \cdots & t_d \end{pmatrix}.$$

We may make a *reparametrization* by parametrizing L by the intercept α and slope β of the regression line. The *parameter space* then becomes $\mathbb{R}^2 \times \mathbb{R}_+$. Note that A has full rank 2 unless all t_i are identical.

If A has full rank, the vectors $u_1 = (1, \dots, 1)^\top$ and $u_2 = (t_1, \dots, t_d)^\top$ form a basis for L or, alternatively the orthogonal set v_1, v_2 where $u_1 = v_1$ and

$$v_2^\top = (t_1 - \bar{t}, \dots, t_d - \bar{t})^\top$$

with $\bar{t} = (t_1 + \cdots + t_d)/d$. It now follows from (A.2) that we have

$$\Pi_L X = \bar{X} v_1 + \frac{\sum_{i=1}^d X_i (t_i - \bar{t})}{\sum_{i=1}^d (t_i - \bar{t})^2} v_2$$

with the ith coordinate

$$(\Pi_L X)_i = \bar{X}\left(1 - \frac{\bar{t}}{\sum_i (t_i - \bar{t})^2}\right) + \frac{\sum_i X_i t_i}{\sum_i (t_i - \bar{t})^2} t_i.$$

We may also express this in matrix form using Proposition 2.21 as

$$\Pi_L X = A(A^\top A)^{-1} A^\top X = A \begin{pmatrix} d & \sum_i t_i \\ \sum_i t_i & \sum_i t_i^2 \end{pmatrix}^{-1} A^\top X$$

leading to the singular covariance matrix for $\Pi_L X$

$$\begin{aligned} \mathbf{V}(\Pi_L X) &= \Pi_L \Sigma = \sigma^2 \Pi_L = \sigma^2 A \begin{pmatrix} d & \sum_i t_i \\ \sum_i t_i & \sum_i t_i^2 \end{pmatrix}^{-1} A^\top \\ &= \frac{\sigma^2}{d\sum_i t_i^2 - (\sum_i t_i)^2} A \begin{pmatrix} \sum_i t_i^2 & -\sum_i t_i \\ -\sum_i t_i & d \end{pmatrix} A^\top \\ &= \frac{\sigma^2}{\sum_i (t_i - \bar{t})^2} A \begin{pmatrix} \bar{t^2} & -\bar{t} \\ -\bar{t} & 1 \end{pmatrix} A^\top, \end{aligned}$$

and similarly for the residual $X - \Pi_L X$.

Note that the uvth element c_{uv} of this singular covariance matrix C is given by the expression

$$c_{uv} = \frac{\sigma^2((\bar{t^2} - (\bar{t})^2) + (t_u - \bar{t})(t_v - \bar{t}))}{\sum_i (t_i - \bar{t})^2} = \frac{\sigma^2}{d}\left(1 + \frac{(t_u - \bar{t})(t_v - \bar{t})}{(\bar{t^2} - (\bar{t})^2)}\right)$$

and despite this apparently complicated expression, it is still true that $K = \sigma^{-2} I_n$ is valid as a concentration matrix for $\Pi_L X$, see Remark 2.25. Indeed, we have the alternative expression for C as

$$C = \frac{\sigma^2}{\|v_1\|^2} v_1 v_1^\top + \frac{\sigma^2}{\|v_2\|^2} v_2 v_2^\top,$$

where v_1 and v_2 are the orthogonal basis for L as found above. □

In a linear model we are typically interested in the mean $\xi \in L$, whereas the variance σ^2 appears as a nuisance parameter; we need to take σ^2 into account as it determines the precision with which the mean may be determined. But other times our focus could be on the variance rather than the mean, as in the example below.

Example 2.30. [Double measurements] Suppose we are interested in determining the precision of a measuring instrument. We may then collect n units and measure each unit twice, resulting in $2n$ independent observations

$$X_i, Y_i \sim \mathcal{N}(\xi_i, \sigma^2),\ i = 1, \ldots, n$$

where $\xi_i \in \mathbb{R}$ is characteristic for the ith unit and σ^2 is the precision of the instrument. This is a linear normal model on $V = \mathbb{R}^n \times \mathbb{R}^n$ with the usual inner product and mean $\xi \in L$, where

$$L = \{(u, v) \in \mathbb{R}^n \times \mathbb{R}^n \mid u_i = y_i, i = 1 \ldots, n\}$$

Here, the parameter of interest is $\phi(\xi, \sigma^2) = \sigma^2$, whereas $\xi = (\xi_1, \ldots, \xi_n)$ is only a necessary disturbance, and thus a nuisance parameter. □

2.3.2 *Likelihood, score, and information*

Consider now a linear normal model given as $X \sim \mathcal{N}_V(\xi, \sigma^2\Sigma)$ with Σ known and $\xi \in L$, and let us for simplicity assume that also σ^2 is known. To calculate the log-likelihood, score, and information, we shall assume that $V = \mathbb{R}^d$ with standard inner product and the linear space L is determined by a design matrix $A \in \mathbb{R}^{d\times k}$ where A has full rank $k < d$ as in Example 2.29 so that $\xi = A\theta$ with $\theta \in \Theta = \mathbb{R}^k$, i.e. the linear subspace is simply $L = A(\mathbb{R}^k)$ and the parameter space is Θ.

We get for the log-likelihood function (ignoring constant terms)

$$\begin{aligned}\ell_x(\theta) &= -\frac{\|x - A\theta\|_K^2}{2\sigma^2} = -\frac{(x - A\theta)^\top K(x - A\theta)}{2\sigma^2}\\ &= -\frac{x^\top x}{2\sigma^2} + \frac{x^\top KA\theta}{\sigma^2} - \frac{\theta^\top A^\top KA\theta}{2\sigma^2}\end{aligned}$$

and differentiate to find the score function

$$S(x,\theta) = \frac{x^\top KA}{\sigma^2} - \frac{\theta^\top A^\top KA}{\sigma^2} = \frac{(x - A\theta)^\top KA}{\sigma^2} = \frac{1}{\sigma^2}\langle x - A\theta, A\rangle_K. \quad (2.9)$$

Further, when we change sign and differentiate further, the information function becomes

$$I(x,\theta) = \frac{1}{\sigma^2}A^\top KA = i(\theta) \quad (2.10)$$

which is constant in x and therefore equal to the Fisher information. In this case, the Fisher information is also constant in θ. Also, note that the Fisher information is inversely proportional to the variance factor σ^2.

As mentioned, we shall show in Chapter 3 that linear normal models are regular exponential families and therefore smooth and stable, so Bartlett's identities hold. But here it can also be seen directly: inspecting the expression (2.9) for the score yields

$$\mathbf{E}\{S(X,\theta)\} = \mathbf{E}\left\{\frac{(X - A\theta)^\top KA}{\sigma^2}\right\} = 0$$

and

$$\begin{aligned}\mathbf{E}\{S(X,\theta)^\top S(X,\theta)\} &= \frac{1}{\sigma^4}\mathbf{E}\left\{A^\top K(X - A\theta)(X - A\theta)^\top KA\right\}\\ &= \frac{\sigma^2}{\sigma^4}A^\top K\Sigma KA = \frac{1}{\sigma^2}A^\top KA = i(\theta).\end{aligned}$$

The quadratic score becomes

$$\begin{aligned}Q(X,\theta) &= S(X,\theta)i(\theta)^{-1}S(X,\theta)^\top\\ &= (x - A\theta)^\top KA(A^\top KA)^{-1}A^\top K(x - A\theta)/\sigma^2\\ &= (x - A\theta)^\top KH(x - A\theta)/\sigma^2\\ &= \langle X - A\theta, H(X - A\theta)\rangle_K/\sigma^2 = \|\Pi_L(X - A\theta)\|_{K/\sigma^2}^2.\end{aligned}$$

since the somewhat lengthy expression

$$H = A(A^\top KA)^{-1}A^\top K = A(A^\top (K/\sigma^2)A)^{-1}A^\top (K/\sigma^2)$$

is the matrix for the orthogonal projection onto L with respect to $\langle \cdot, \cdot \rangle_{K/\sigma^2}$, also known as the *hat-matrix*; see Proposition 2.21.

As we have $\mathbf{E}(X - A\theta) = 0$, Theorem 2.27 implies that the quadratic score $Q(X, \theta)$ is exactly $\chi^2(k)$ distributed in the linear model with fixed covariance. This should be compared to Theorem 1.25 and Corollary 1.35.

2.4 Exercises

Exercise 2.1. Let $X \sim \mathcal{N}_3(\xi, \Sigma)$ where

$$\xi = \begin{pmatrix} 1 \\ 3 \\ 4 \end{pmatrix}, \quad \Sigma = \begin{pmatrix} 6 & 3 & 0 \\ 3 & 8 & 0 \\ 0 & 0 & 1 \end{pmatrix}.$$

a) Find the marginal distributions of $(X_1, X_2)^\top$ and $(X_1, X_3)^\top$.

b) Find the distribution of $(X_2, X_3, X_1)^\top$.

c) Find the distribution of $Y = X_1 - X_2 + X_3$.

Exercise 2.2. Let $X \sim \mathcal{N}_3(\xi, \Sigma)$ where

$$\xi = \begin{pmatrix} 1 \\ -1 \\ 2 \end{pmatrix}, \quad \Sigma = \begin{pmatrix} 4 & 3 & 1 \\ 3 & 8 & 1 \\ 1 & 1 & 2 \end{pmatrix}.$$

a) Argue that X has a density with respect to Lebesgue measure on $\mathbb{R}^3$.

b) Find the distribution of

$$Y = \begin{pmatrix} X_1 - X_2 + 2X_3 \\ 2X_1 + X_3 \end{pmatrix}.$$

c) Find the concentration matrix for the distribution of Y.

Exercise 2.3. Let $X \sim \mathcal{N}_2(\xi, \Sigma)$ where

$$\xi = \begin{pmatrix} 1 \\ 0 \end{pmatrix}, \quad \Sigma = \begin{pmatrix} 2 & 2 \\ 2 & 4 \end{pmatrix}$$

and define $Y \in \mathbb{R}^2$ by $Y_1 = X_2 - X_1$ and $Y_2 = X_1 - 1$.

a) Argue that X has a density with respect to Lebesgue measure on $\mathbb{R}^2$.

b) Find the concentration matrix K of X.

c) Write an expression for the density, and sketch the level curves.

d) What is the distribution of Y?

e) Are Y_1 and Y_2 independent?

f) Give an expression for the projection onto $L = \{x \in \mathbb{R}^2 : x_1 = 2x_2\}$ with respect to the concentration inner product $\langle \cdot, \cdot \rangle_K$.

Exercise 2.4. Let $X = (X_1, X_2)^\top$ follow a normal distribution on $\mathbb{R}^2$ with expectation 0 and covariance matrix

$$\Sigma(a) = \begin{pmatrix} a & a/2 \\ a/2 & a \end{pmatrix},$$

where $a \in \mathbb{R}$.

a) For which values of a is $\Sigma(a)$ a valid covariance matrix?

b) For which values of a is the distribution regular?

c) Find the distribution of $Y = (Y_1, Y_2)^\top = (X_1 + X_2, X_1 - X_2)^\top$ and show that Y_1 and Y_2 are independent.

d) Find the distribution of $Z = Y_1^2/3Y_2^2$.

Exercise 2.5. Let $X \sim \mathcal{N}_3(0, I_3)$ and define $Y = a + BX$, where

$$a = \begin{pmatrix} 1 \\ 0 \\ -1 \end{pmatrix}, \quad B = \begin{pmatrix} 1 & -2 & 1 \\ 1 & 1 & 1 \\ 2 & -1 & 2 \end{pmatrix}.$$

a) What is the distribution of Y?

b) Which of the pairs (Y_1, Y_2), (Y_1, Y_3), and (Y_2, Y_3) are independent?

c) Show that the distribution of Y is singular and identify the support of the distribution.

Exercise 2.6. Let $X \sim \mathcal{N}_2(0, \Sigma)$ where

$$\Sigma = \begin{pmatrix} 1 & 2 \\ 2 & 4 \end{pmatrix}.$$

a) Show that the distribution of X is singular.

b) Identify the support of the distribution.

c) Find all valid concentration operators K for X.

Exercise 2.7. Let $X = (X_1, X_2, X_3)^\top$ follow a normal distribution on $\mathbb{R}^3$ with expectation 0 and covariance matrix

$$\Sigma(a) = \begin{pmatrix} 1 & a & a \\ a & 4 & 0 \\ a & 0 & a \end{pmatrix},$$

where $a \in \mathbb{R}$.

a) For which values of a is $\Sigma(a)$ a valid covariance matrix?

b) What is the distribution of $Y = 2X_1 - X_3$?

c) Show that the distribution of X is singular if $a = 2(\sqrt{2} - 1)$.

d) Find the support of the distribution of X for the case $a = 2(\sqrt{2} - 1)$.

e) Determine the set of valid concentration operators for $\Sigma(a)$ when $a = 2(\sqrt{2} - 1)$.

Exercise 2.8. Let $X \sim \mathcal{N}_3(\xi, \sigma^2 I_3)$ where

$$\xi = \begin{pmatrix} a+b \\ a \\ b \end{pmatrix}.$$

Further, let L be the linear subspace of $\mathbb{R}^3$ determined as

$$L = \operatorname{span}\left\{ \begin{pmatrix} 1 \\ 1 \\ 0 \end{pmatrix}, \begin{pmatrix} 1 \\ 0 \\ 1 \end{pmatrix} \right\}.$$

a) Determine $\Pi_L X$ and find its distribution.

b) Find the distribution of $\|X - \Pi_L X\|^2$, where $\|\cdot\|$ denotes the usual Euclidean norm on $\mathbb{R}^3$.

Exercise 2.9. Let $X \sim \mathcal{N}_3(0, \Sigma_\rho)$ where $-1 < \rho < 1$ and

$$\Sigma_\rho = \begin{pmatrix} 1 & \rho & \rho^2 \\ \rho & 1 & \rho \\ \rho^2 & \rho & 1 \end{pmatrix}.$$

a) Show that the specification above defines a regular normal distribution on $\mathbb{R}^3$.

b) Show that the concentration matrix K_ρ is given as

$$K_\rho = \frac{1}{1-\rho^2} \begin{pmatrix} 1 & -\rho & 0 \\ -\rho & 1+\rho^2 & -\rho \\ 0 & -\rho & 1 \end{pmatrix}.$$

c) Let now L be the subspace spanned by the constant vector $e = (1,1,1)^\top$. Determine the projection $\Pi_\rho X$ onto L with respect to the inner product $\langle \cdot, \cdot \rangle_{K_\rho}$.

d) Determine the distribution of $\Pi_\rho X$ and $X - \Pi_\rho X$.

Chapter 3

Exponential Families

3.1 Regular exponential families

There is a large and important class of statistical models that have a common mathematical structure which make their analysis particularly simple. These are models where the associated family $\mathcal{P}$ of probability measures is a so-called *exponential family*. Such models are also known as *exponential models*. We define:

Definition 3.1. An *exponential family* of probability distributions on $(\mathcal{X}, \mathbb{E})$ is a parametrized family $\mathcal{P} = \{P_\theta \mid \theta \in \Theta\}$ of the form

$$P_\theta(A) = \int_A \frac{1}{c(\theta)} e^{\langle \theta, t(x) \rangle} \, d\mu(x), \tag{3.1}$$

where V is a k-dimensional Euclidean vector space with inner product $\langle \cdot, \cdot \rangle$, $\Theta \subseteq V$ is the space of *canonical parameters*, $t : \mathcal{X} \to V$ is the *canonical statistic*, and μ is a σ-finite *base measure* of the family.

Remark 3.2. *We note in particular that an exponential family of measures is dominated by the base measure μ and the form of the density implies that the support of the measures P_θ is the same for all $\theta \in \Theta$.*

We shall in the following assume that the exponential family $\mathcal{P}$ is *minimally represented* and *regular* which means that

minimal There is no $(\lambda, c) \in \mathbf{V} \setminus \{0\} \times \mathbb{R}$ with $\langle \lambda, t(x) \rangle = c$ a.e. μ.

regular The parameter space Θ is an open and convex subset of V.

Then $k = \dim V$ is the *dimension* of the exponential family. Since $P_\theta(\mathcal{X}) = 1$, we must have for all $\theta \in \Theta$:

$$c(\theta) = \int_{\mathcal{X}} e^{\langle \theta, t(x) \rangle} \, d\mu(x) < \infty \tag{3.2}$$

so $c(\theta)$ is a finite *normalizing constant*. The family is said to be *full* if it is also as large as possible:

$$\Theta = \left\{ \theta \in V \;\middle|\; c(\theta) = \int_{\mathcal{X}} e^{\langle \theta, t(x) \rangle} \, d\mu(x) < \infty \right\}.$$

DOI: 10.1201/9781003272359-3

In much standard literature, the term 'regular' incorporates that the family is also full. However, in this text we shall not be concerned with fullness. It is practical to introduce the function

$$\psi(\theta) = \log c(\theta)$$

which is known as the *cumulant function* of the exponential family and we may then alternatively write the density in (3.1) as

$$f_\theta(x) = e^{\langle \theta, t(x) \rangle - \psi(\theta)} \tag{3.3}$$

and the associated log-likelihood function as

$$\ell_x(\theta) = \langle \theta, t(x) \rangle - \psi(\theta).$$

In the following we often assume that an orthonormal basis for V has been chosen so that we without loss of generality can identify V with $\mathbb{R}^k$ and write the inner product in terms of coordinates as

$$\langle \theta, t(x) \rangle = \theta^\top t(x).$$

3.2 Examples of exponential families

Many of the statistical models we have seen in the previous chapters are indeed exponential families. In each case, however, a little reformulation is needed to see that this is the case.

Example 3.3. [Bernoulli model as exponential model] Consider the Bernoulli model in Example 1.4 with densities for $\mu \in (0, 1)$ given as

$$f_\mu(x) = \mu^x (1 - \mu)^{(1-x)}, \quad x \in \{0, 1\}$$

with respect to counting measure on $\mathcal{X} = \{0, 1\}$. To identify this as an exponential family, we introduce the parameter

$$\theta = \log \frac{\mu}{1 - \mu}$$

representing the *log-odds* of the distribution and note that we then have

$$\mu = \frac{e^\theta}{1 + e^\theta}, \quad 1 - \mu = \frac{1}{1 + e^\theta}$$

so we may rewrite the density in terms of this parameter as

$$f_\theta(x) = \frac{e^{\theta x}}{1 + e^\theta}, \quad x \in \{0, 1\}.$$

When the parameter μ varies in the unit interval, the log-odds θ vary in all of $\mathbb{R}$. Thus we may identify the Bernoulli family as an exponential family with the following characteristics:

base measure counting measure on $\{0, 1\}$;
canonical parameter $\theta = \log(\mu/(1-\mu)) \in \Theta = \mathbb{R}$;
canonical statistic $t(x) = x$;
normalizing constant $c(\theta) = 1 + e^{\theta}$;
cumulant function $\psi(\theta) = \log c(\theta) = \log(1 + e^{\theta})$.

The family is regular since $\Theta = \mathbb{R}$ is an open subset of $\mathbb{R}$ and if $\lambda \cdot 1 = \lambda \cdot 0$, we must have $\lambda = 0$ so it is also minimally represented and the dimension of the family is one.

Note that we could also have represented the Bernoulli model with the canonical parameter $\tilde{\theta} = (\tilde{\theta}_1, \tilde{\theta}_2)^\top \in \mathbb{R}^2$ with

$$\tilde{\theta}_1 = \log \mu, \quad \tilde{\theta}_2 = \log(1-\mu)$$

and

$$\tilde{t}(x) = \begin{pmatrix} \tilde{t}_1(x) \\ \tilde{t}_2(x) \end{pmatrix} = \begin{pmatrix} x \\ 1-x \end{pmatrix}$$

since then

$$f_{\tilde{\theta}}(x) = \mu^x (1-\mu)^{1-x} = \exp\{\tilde{\theta}^\top \tilde{t}(x)\}, \quad x \in \{0, 1\},$$

making the family appear to be two-dimensional. However, in this form, the family is over-parametrized and the representation is not minimal since $\tilde{t}_1(x) + \tilde{t}_2(x) = 1$ for all $x \in \{0, 1\}$. □

Example 3.4. [Poisson model as an exponential model] The simple Poisson family in Example 1.2 is an exponential family. To see this we reparametrize to $\theta = \log \lambda$ where λ is the mean of the distribution and write

$$f_\theta(x) = \frac{e^{\theta x}}{x!} e^{-e^{\theta}} = \frac{e^{\theta x}}{e^{e^{\theta}}} \frac{1}{x!},$$

identifying the Poisson model as a one-dimensional exponential model with
base measure $\mu = \frac{1}{x!} \cdot m$, where m is counting measure on $\mathbb{N}_0$;
canonical parameter $\theta \in \Theta = \mathbb{R}_+$;
canonical statistic $t(x) = x$;
normalizing constant $c(\theta) = e^{e^{\theta}}$;
cumulant function $\psi(\theta) = \log c(\theta) = e^{\theta}$.

The model is minimally represented as λX is constant if and only if $\lambda = 0$ and it is regular since $\Theta = \mathbb{R}_+$ is an open and convex subset of $\mathbb{R}$.

Note that if we consider the variant with $X_1, \ldots, X_n$ independent and identically Poisson distributed, we have the density

$$f_\theta(x_1, \ldots, x_n) = \prod_{i=1}^{n} \frac{e^{\theta x_i}}{x_i!} e^{-e^{\theta}} = \frac{e^{\theta \sum_i x_i}}{e^{ne^{\theta}}} \prod_{i=1}^{n} \frac{1}{x!},$$

so this is again an exponential family with the same parameter space, but base measure $\mu^{\otimes n}$, canonical statistic $t(x_1, \ldots, x_n) = \sum_i x_i$, and cumulant function $\psi_n(\theta) = n\psi(\theta)$. As we shall see below in Theorem 3.11, this is a general phenomenon for exponential families and a major reason for their importance. □

Example 3.5. The simple normal model in Example 1.5 is a regular exponential model. To see this, we first rewrite the density for a single observation as

$$f_\omega(x) = \frac{1}{\sqrt{2\pi\sigma^2}} e^{-\frac{(x-\xi)^2}{2\sigma^2}} = \frac{1}{\sqrt{2\pi\sigma^2}} e^{-\frac{x^2}{2\sigma^2} + \frac{x\xi}{\sigma^2} - \frac{\xi^2}{2\sigma^2}} = \frac{1}{e^{\frac{\xi^2}{2\sigma^2}} \sqrt{2\pi\sigma^2}} e^{\frac{x\xi}{\sigma^2} - \frac{x^2}{2\sigma^2}}.$$

We next reparametrize and introduce the parameter $\theta = (\theta_1, \theta_2)$

$$\theta_1 = \frac{\xi}{\sigma^2}, \; \theta_2 = \frac{1}{\sigma^2} \quad \text{with} \quad \xi = \theta_1/\theta_2, \; \sigma^2 = 1/\theta_2$$

and let

$$c(\theta) = e^{\frac{\xi^2}{2\sigma^2}} \sqrt{2\pi\sigma^2} = e^{\frac{\theta_1^2}{2\theta_2}} \sqrt{\frac{2\pi}{\theta_2}}.$$

With these quantities the density may be rewritten as

$$f_\theta(x) = \frac{e^{\theta_1 x + \theta_2(-x^2/2)}}{c(\theta)}.$$

When ω varies in $\Omega = \mathbb{R} \times \mathbb{R}_+$, θ varies in all of $\Theta = \mathbb{R} \times \mathbb{R}_+$ which is an open and convex set. We thus have a two-dimensional regular exponential family with

base measure standard Lebesgue measure on $\mathbb{R}$;

canonical parameter $\theta = (\theta_1, \theta_2)^\top = (\xi\sigma^{-2}, \sigma^{-2})^\top \in \Theta$;

canonical statistic $t(x) = (x, -x^2/2)^\top$;

normalizing constant $c(\theta) = \exp(\theta_1^2/(2\theta_2))\sqrt{2\pi/\theta_2}$;

cumulant function $\psi(\theta) = \log c(\theta) = \theta_1^2/(2\theta_2) + \frac{1}{2}\log(2\pi) - \frac{1}{2}\log\theta_2$.

The family is minimally represented for if we assume that $\lambda^\top t(X)$ is almost everywhere constant, we have

$$\lambda_1 x - \frac{1}{2}\lambda_2 x^2 + c = 0 \quad \text{almost everywhere,}$$

but this is a polynomium of degree at most two so must be identically zero.

As seen from the above, identifying the simple normal model as an exponential model is a bit involved and may not in itself be so helpful; but it will be important when considering related models. We refrain from given the details here of the extension to the case of n observations. □

Example 3.6. [Linear normal model] Consider the linear normal model as discussed in Section 2.3. This corresponds to the family of densities

$$f_{(\xi,\sigma^2)}(x) = \frac{1}{(2\pi\sigma^2)^{d/2}} e^{\frac{-\|x-\xi\|^2}{2\sigma^2}}$$

with respect to standard Lebesgue measure λ_V on $(V, \langle\cdot,\cdot\rangle)$, where $(\xi, \sigma^2) \in L \times \mathbb{R}_+$ for L being a linear subspace of V.

To see this is an exponential family, we rewrite the quadratic term in the exponent:

$$\begin{aligned} -\frac{\|x-\xi\|^2}{2\sigma^2} &= -\frac{\|x\|^2}{2\sigma^2} - \frac{\|\xi\|^2}{2\sigma^2} + \frac{\langle \xi, x\rangle}{\sigma^2} \\ &= \left\langle \frac{\xi}{\sigma^2}, \Pi_L(x) \right\rangle - \frac{\|x\|^2}{2\sigma^2} - \frac{\|\xi\|^2}{2\sigma^2} \end{aligned}$$

where Π_L is the orthogonal projection onto L, and we have used that for any $\xi \in L$ it holds that $\langle \xi, x\rangle = \langle \xi, \Pi_L(x)\rangle$. Letting now $\theta_1 = \xi/\sigma^2$, $\theta_2 = 1/\sigma^2$, $t(x) = (\Pi_L(x), -\|x\|^2/2)$, and introducing the inner product $\langle\cdot,\cdot\rangle_{\text{rep}}$ as

$$\langle \theta, t(x)\rangle_{\text{rep}} = \langle \theta_1, t_1(x)\rangle + \theta_2 t_2(x),$$

we can write the density as

$$f_\theta(x) = \exp\{\langle \theta, t(x)\rangle_{\text{rep}} - \psi(\theta)\}$$

with the cumulant function being given as

$$\psi(\theta) = -\log c - \frac{d}{2}\log\theta_2 + \frac{\|\theta_1\|^2}{2\theta_2}. \tag{3.4}$$

This is a direct generalization of what we found in Example 3.5. Thus we have represented the general linear model as an exponential family with base measure λ_V, canonical parameter space $\Theta = L \times \mathbb{R}_+ \subseteq L \times \mathbb{R}$, and canonical statistic $t(x) = (\Pi_L(x), -\|x\|^2/2)$. □

For the sake of completeness, we conclude this section by identifying some families that are *not* exponential families in the sense we have defined here.

Example 3.7. [Models that are not exponential] We first consider the *uniform model* in Example 1.8. This is not an exponential family since the support of the uniform density is $[0, \theta]$ and thus depends on the unknown parameter θ, contradicting Remark 3.2.

Further, also the *Cauchy model* is not an exponential model. Although this is not so easy to show formally, there is no way that the family of densities

$$f_{\alpha,\beta}(x) = \frac{\beta}{\pi((x-\alpha)^2 + \beta^2)}, \quad x \in \mathbb{R}$$

for $\alpha, \beta \in \mathbb{R} \times \mathbb{R}_+$ can be represented in the required form.

Finally, certain subfamilies of regular exponential families are not necessarily regular exponential families themselves. Consider, for example the subfamily of the normal family determined by the relation $\xi = \sigma$, i.e. that the standard deviation is equal to the mean. In terms of the canonical parameters (θ_1, θ_2), this is the submodel given by $\theta_2 = \theta_1^2$ and the problem is that the set

$$\Theta_0 = \{\theta \in \Theta \mid \theta_2 = \theta_1^2\}$$

is *not* an open subset of Θ. We shall later—in Section 3.6—look at more general families, known as *curved* exponential families and show that such families share many—but not all—properties with those that are regular. The model determined by Θ_0 is an example of a curved exponential family. □

3.3 Properties of exponential families

Exponential families have many nice properties, including the following important collection of results:

Theorem 3.8. *Let X be a random variable and assume that X follows a distribution from a k-dimensional regular and minimally represented exponential family with representation space $(\mathcal{X}, \mathbb{E})$, canonical parameter space $\Theta \subseteq \mathbb{R}^k$, canonical statistic $t(X)$, and base measure μ. Then $t(X)$ has moments of any order.*

Proof. For $\theta \in \Theta$ and $\theta \pm h \in \Theta$ we have for all $n \in \mathbb{N}$ that

$$\mathbf{E}_\theta\{|h^\top t(X)|^n\} < \infty$$

by Lemma B.1. Choosing $h = (h_1, \ldots, h_k)$ so that $h_i = \varepsilon$ and $h_j = 0$ for $j \neq i$ yields that for ε sufficiently small it holds for all $n \in \mathbb{N}$ that

$$\mathbf{E}_\theta\{\varepsilon^n |t_i(X)|^n\} < \infty$$

and hence all moments of $t(X)$ are finite. □

Theorem 3.9. *In the same situation as specified above, the normalizing constant $c(\theta)$ is a smooth function of θ, and we have*

$$\frac{\partial^{m_1+\cdots+m_k}}{\partial\theta_1^{m_1}\cdots\partial\theta_k^{m_k}}c(\theta) = \int_{\mathcal{X}} \prod_{i=1}^k t_i(x)^{m_i} e^{\theta^\top t(x)}\, d\mu(x) = c(\theta)\mathbf{E}_\theta\left\{\prod_{i=1}^k t_i(X)^{m_i}\right\}$$

for all $m_1, \ldots, m_k \in \mathbb{N}_0$.

Proof. The relation follows from Lemma B.2 since this allows to switch the order of differentiation and integration (Schilling, 2017, Theorem 11.5):

$$\begin{aligned}
\frac{\partial^{m_1+\cdots+m_k}}{\partial\theta_1^{m_1}\cdots\partial\theta_k^{m_k}}c(\theta) &= \frac{\partial^{m_1+\cdots+m_k}}{\partial\theta_1^{m_1}\cdots\partial\theta_k^{m_k}}\int_{\mathcal{X}} e^{\theta^\top t(x)}\, d\mu(x) \\
&= \int_{\mathcal{X}} \frac{\partial^{m_1+\cdots+m_k}}{\partial\theta_1^{m_1}\cdots\partial\theta_k^{m_k}} e^{\theta^\top t(x)}\, d\mu(x) \\
&= \int_{\mathcal{X}} \prod_{i=1}^k t_i(x)^{m_i} e^{\theta^\top t(x)}\, d\mu(x) \\
&= c(\theta)\mathbf{E}_\theta\left\{\prod_{i=1}^k t_i(X)^{m_i}\right\},
\end{aligned}$$

as desired. □

And, importantly, an exponential family satisfies the regularity conditions needed for score and information to be well-defined.

Theorem 3.10. *A regular and minimally represented exponential family is smooth and stable.*

Proof. Since an exponential function is smooth and the normalization constant $c(\theta)$ is smooth by Theorem 3.9, we just have to establish stability. For such an exponential family, we get

$$\frac{\partial}{\partial \eta_i} L_x(\eta) = \left(t_i(x) - \frac{c_i(\eta)}{c(\eta)} \right) \frac{e^{\eta^\top t(x)}}{c(\eta)} \tag{3.5}$$

where $c_i(\eta) = \partial c(\eta)/\partial \eta_i$, and further

$$\begin{aligned} \frac{\partial^2}{\partial \eta_i \partial \eta_j} L_x(\eta) &= \left(t_i(x) - \frac{c_i(\eta)}{c(\eta)} \right) \left(t_j(x) - \frac{c_j(\eta)}{c(\eta)} \right) \frac{e^{\eta^\top t(x)}}{c(\eta)} \\ &\quad + \left(\frac{c_i(\eta) c_j(\eta)}{c(\eta)^2} - \frac{c_{ij}(\eta)}{c(\eta)} \right) \frac{e^{\eta^\top t(x)}}{c(\eta)}, \end{aligned} \tag{3.6}$$

where $c_{ij}(\eta) = \partial^2 c(\eta)/\partial \eta_i \partial \eta_j$. For any $m = (m_1, \ldots, m_k)$ with $m_i \in \mathbb{N}_0$, Lemma B.2 yields that

$$\left| \prod_{j=1}^{k} t_i(X)^{m_i} e^{\eta^\top t(x)} \right| \leq h_\theta(x)$$

for all η in an open neighbourhood U_θ around θ where h_θ is μ- integrable. Also $c(\eta)$, $c(\eta)^{-1}$, $|c_i(\eta)|$, $|c_j(\eta)|$, and $|c_{ij}(\eta)|$ are all bounded above in such a neighbourhood. Thus, the derivatives in (3.5) and (3.6) are locally bounded by integrable functions, as desired. □

3.4 Constructing exponential families

There are numerous ways of constructing new exponential families from other exponential families. We mention a few of the simplest here.

3.4.1 *Product families*

Consider an exponential family, $\mathcal{P} = \{P_\theta \,|\, \theta \in \Theta\}$, having representation space $(\mathcal{X}, \mathbb{E})$, base measure μ, and canonical statistic $s : \mathcal{X} \mapsto \mathbb{R}^k$, and another exponential family, $\mathcal{Q} = \{Q_\xi \,|\, \xi \in \Xi\}$, with representation space $(\mathcal{Y}, \mathbb{F})$, base measure ν, and canonical statistic $t : \mathcal{Y} \mapsto \mathbb{R}^m$. We may then form the *outer product family*

$$\mathcal{P} \odot \mathcal{Q} = \{P_\theta \otimes Q_\xi \,|\, (\theta, \xi) \in \Theta \times \Xi\}.$$

This again becomes an exponential family, and if each of the original families were regular and minimal, so is the product family. Indeed, the representation space is

$$(\mathcal{X} \times \mathcal{Y}, \mathbb{E} \otimes \mathbb{F}) \subseteq (\mathbb{R}^{k+m}, \mathbb{B}(\mathbb{R}^{k+m})),$$

the base measure is $\mu \otimes \nu$, the canonical parameter space is $\Theta \times \Xi$, and the canonical statistic is

$$u : \mathcal{X} \times \mathcal{Y} \mapsto \mathbb{R}^{k+m}; \quad u(x, y) = (s(x), t(y)),$$

so the dimension of the model is $k + m$. The corresponding family of densities with respect to $\mu \otimes \nu$ may now be written as

$$f_{\theta,\xi}(x, y) = \frac{1}{c(\theta)d(\xi)} e^{\theta^\top s(x) + \xi^\top t(y)},$$

where $c(\theta)$ and $d(\xi)$ are the normalizing constants in the two families considered.

This construction corresponds to the situation where we simply have two completely independent and unrelated statistical models, but we wish to consider them together for some reason. We note that the log-likelihood function for the product model is simply the sum of the log-likelihood functions for the factors

$$\ell_{x,y}(\theta, \xi) = \ell_x(\theta) + \ell_y(\xi). \tag{3.7}$$

A different form of product occurs when the parameter spaces coincide, i.e. when $\Xi = \Theta$ and the canonical statistics s and t map into the same space $\mathbb{R}^k$. We may then form the *direct product family*

$$\mathcal{P} \otimes \mathcal{Q} = \{P_\theta \otimes Q_\theta \,|\, \theta \in \Theta\}$$

which now representation space $(\mathcal{X} \times \mathcal{Y}, \mathbb{E} \otimes \mathbb{F})$, base measure is $\mu \otimes \nu$, canonical parameter space Θ, and canonical statistic is $u(x, y) = s(x) + t(y)$ since the corresponding family of densities with respect to $\mu \otimes \nu$ may be written as

$$f_\theta(x, y) = \frac{1}{c(\theta)d(\theta)} e^{\theta^\top (s(x) + t(y))},$$

where $c(\theta)$ and $d(\theta)$ are the normalizing constants in the two families considered. Again, if the two families are regular and minimally represented, so is the direct product family and the dimension of the direct product is the same as for the constituents and equal to k; the log-likelihood functions are again obtained by simple addition

$$\ell_{x,y}(\theta) = \ell_x(\theta) + \ell_y(\theta). \tag{3.8}$$

This construction corresponds to combining information from two independent statistical experiments to obtain inference on the *same common parameter* θ. Section 8.3.1 provides examples of these product constructions and their use.

3.4.2 *Repeated observations*

An important instance of the direct product above corresponds to independent repetitions. More precisely, consider a sample $X_1, \ldots, X_n$ from an exponential family $\mathcal{P}$ on the representation space $(\mathcal{X}, \mathbb{E})$ with canonical parameter space Θ and canonical statistic t.

The model corresponding to repeated observations will be the n-fold direct product and have representation space $(\mathcal{X}^n, \mathbb{E}^{\otimes n})$ and associated family

$$\mathcal{P}^{\otimes n} = \{P_\theta^{\otimes n}, \theta \in \Theta\}.$$

Theorem 3.11. *Assume that $\mathcal{P}$ is a minimally represented and regular exponential family as above. Then the n-fold direct product $\mathcal{P}^{\otimes n}$ is a minimally represented regular exponential family with the same parameter space and base measure $\mu^{\otimes n}$. The canonical statistic t_n is determined as*

$$t_n(x_1, \dots, x_n) = t(x_1) + \cdots + t(x_n)$$

and the normalizing constant c_n and cumulant function ψ_n satisfy

$$c_n(\theta) = c(\theta)^n, \quad \psi_n(\theta) = n\psi(\theta).$$

Proof. The density of $P_\theta^{\otimes n}$ with respect to $\mu^{\otimes n}$ is for $x = (x_1, \dots, x_n)$

$$f_\theta^{\otimes n}(x) = \prod_{i=1}^n \frac{1}{c(\theta)} e^{\langle \theta, t(x_i) \rangle} \, d\mu(x_i) = \frac{1}{c(\theta)^n} e^{\langle \theta, t(x_1) + \cdots + t(x_n) \rangle} \, d\mu^{\otimes n}(x),$$

which yields the relations mentioned since

$$\psi_n(\theta) = \log c_n(\theta) = n \log c(\theta) = n\psi(\theta).$$

We need to establish is that the family is minimally represented. But assume for contradiction that $\langle \lambda, t_n(X) \rangle$ is constant almost surely. Then, by independence,

$$0 = \mathbf{V}_\theta\{\langle \lambda, t_n(X) \rangle\} = n\mathbf{V}_\theta\{\langle \lambda, t(X_1) \rangle\}$$

which yields $\langle \lambda, t(X_1) \rangle$ constant, contradicting that the family $\mathcal{P}$ was minimally represented. □

3.4.3 *Transformations*

In general, if an exponential model has been set up for a random variable X, and $Y = s(X)$ is a function of X, the family of distributions of Y will typically not be an exponential family. An exception is when X is transformed by the canonical statistic, so that $Y = t(X)$. This leads again to an exponential family with the same parameter space and the lifted base measure. We formulate that as a theorem.

Theorem 3.12. *Assume that $\mathcal{P} = \{P_\theta \,|\, \theta \in \Theta\}$ is a minimally represented and regular exponential family with canonical parameter space Θ, canonical statistic t, and base measure μ. Then the family $\tilde{\mathcal{P}}$ of distributions of $Y = t(X)$*

$$\tilde{\mathcal{P}} = \{Q_\theta \,|\, \theta \in \Theta\}$$

where $Q_\theta = t(P_\theta)$, is a regular and minimally represented exponential family with canonical parameter space Θ, base measure $\nu = t(\mu)$, and canonical statistic $\tilde{t} : \mathbb{R}^k \mapsto \mathbb{R}^k$ given as $\tilde{t}(y) = y$. The cumulant function is unchanged.

Proof. This follows directly from using the transformation formula ensuring that

$$Q_\theta(A) = P_\theta\{t^{-1}(A)\} = \int_{t^{-1}(A)} e^{\langle\theta,t(x)\rangle-\psi(\theta)}\,\mu(dx) = \int_A e^{\langle\theta,y\rangle-\psi(\theta)}\,t(\mu)(dy).$$

Preservation of the other properties is obvious. □

This fact is typically exploited in connection with repeated observations, where the observations $X_1,\ldots,X_n$ are replaced by the sum $Y = t(X_1) + \cdots + t(X_n)$. An exponential family with the identity $t(y) = y$ as canonical statistic is known as a *natural exponential family* (NEF).

Example 3.13. [Binomial model] Consider the Bernoulli model with $X_1,\ldots,X_n$ independent and identically distributed as

$$P_\mu(X_i = x) = \mu^x(1-\mu)^{1-x}, \quad x \in \{0,1\}$$

represented as an exponential family using the parameter $\theta = \log\{\mu/(1-\mu)\}$ with densities

$$f_\theta(x_1,\ldots,x_n) = \frac{e^{\theta\sum_i x_i}}{(1+e^\theta)^n}$$

with respect to counting measure on $\{0,1\}^n$. We may instead consider the distribution of the sum $Y = X_1 + \cdots + X_n$ directly yielding the binomial family

$$\tilde{g}_\theta(y) = \binom{n}{y}\frac{e^{\theta y}}{(1+e^\theta)^n}$$

having density

$$g_\theta(y) = \frac{e^{\theta y}}{(1+e^\theta)^n}$$

with respect to $\nu = \binom{n}{y} \cdot m$ where m is counting measure on $\mathcal{Y} = \{0,\ldots,n\}$. Thus, the binomial family is an example of a NEF. □

If we use the canonical statistic to transform the observations, the log-likelihood function of the transformed family is identical to the original log- likelihood function since when $y = t(x)$

$$\tilde{\ell}_y(\theta) = \theta^\top y - \psi(\theta) = \theta^\top t(x) - \psi(\theta) = \ell_x(\theta).$$

So any inference based on the log-likelihood function is unaffected by the transformation $y = t(x)$. This property is known as *sufficiency* of the transformation t; *no information about* θ *in* x *is lost* by using $y = t(x)$ instead of x. The notion of sufficiency is yet another fundamental concept introduced by R. A. Fisher, but the concept will not be discussed in detail in this book.

3.4.4 *Affine subfamilies*

We next consider exponential families given by assuming that the canonical parameter is contained in a subset $\Theta_0 \subseteq \Theta$ of the form

$$\Theta_0 = \Theta \cap (L + a),$$

where $L \subseteq V$ is a linear subspace of V of dimension $\dim L = d$ and $a \in V$ or, in other words, the parameter is restricted to an affine subspace $L + a$ of the original parameter space. To ensure this is interesting, we shall assume that Θ_0 is not empty. We now have

Theorem 3.14. *Assume that $\mathcal{P} = \{P_\theta \mid \theta \in \Theta\}$ is a minimally represented and regular exponential family with canonical parameter space Θ and Θ_0 is a non-empty affine subset of Θ as above. Then the subfamily*

$$\tilde{\mathcal{P}} = \{P_{\theta+a} \mid \theta \in \tilde{\Theta}\}$$

is an exponential family with canonical parameter space $\tilde{\Theta} = \Theta_0 - a \subseteq L$. The canonical statistic $\tilde{t} : \mathcal{X} \mapsto L$ may be represented as $\tilde{t}(x) = \Pi_L(t(x))$, where Π_L is the projection onto L. The base measure and cumulant function is

$$\tilde{\mu}(dx) = e^{\langle a,t(x)\rangle}\,\mu(dx), \quad \tilde{\psi}(\tilde{\theta}) = \psi(\tilde{\theta} + a)$$

and the family $\tilde{\mathcal{P}}$ is a regular and minimally represented exponential family of dimension $d = \dim L$.

Proof. We consider the exponent in the representation of the larger family and have for $\theta \in \Theta_0$ and thus $\tilde{\theta} = \theta - a \in L$:

$$\begin{aligned} \langle \theta, t(x)\rangle &= \langle \theta - a, t(x)\rangle + \langle a, t(x)\rangle \\ &= \langle \theta - a, \Pi_L(t(x))\rangle + \langle a, t(x)\rangle \\ &= \langle \tilde{\theta}, \tilde{t}(x)\rangle + \langle a, t(x)\rangle, \end{aligned}$$

where $\tilde{t}(x) = \Pi_L(t(x))$. So if we let $\tilde{\mu}(dx) = e^{\langle a,t(x)\rangle}\mu(dx)$ we may write the density $g_{\tilde{\theta}}$ of $Q_{\tilde{\theta}} = P_{\tilde{\theta}+a}$ with respect to $\tilde{\mu}$ as

$$g_{\tilde{\theta}}(x) = e^{\langle \tilde{\theta}, \tilde{t}(x)\rangle - \psi(\tilde{\theta}+a)}.$$

Now, since Θ was an open and convex subset of V, $\tilde{\Theta}$ is an open and convex subset of L so the family is regular. It is also minimally represented since for any $\lambda \in L$ we have

$$\langle \lambda, \tilde{t}(X)\rangle = \langle \lambda, \Pi_L(t(X))\rangle = \langle \lambda, t(X)\rangle$$

and thus, if $\langle \lambda, \tilde{t}(x)\rangle$ is a.e. constant with respect to $\tilde{\mu}$, then also $\langle \lambda, t(x)\rangle$ is a.e. constant with respect to μ, contradicting that $\mathcal{P}$ was minimally represented. □

Note that we could have represented the family with the canonical statistic $t(x)$ instead of $\tilde{t}(x) = \Pi_L(t(x))$ since

$$\langle \tilde{\theta}, \tilde{t}(x)\rangle = \langle \tilde{\theta}, \Pi_L(t(x))\rangle = \langle \tilde{\theta}, t(x)\rangle$$

but then the representation would not be regular since $\tilde{\Theta}$ is not an open subset of V.

Example 3.15. Consider two independent Poisson random variables X and Y with expectations λ and μ, respectively, and corresponding canonical parameters $\theta = \log \lambda$ and $\xi = \log \mu$, as in Example 3.4. We first consider the outer product family which is the family of densities

$$f_{\lambda,\mu}(x,y) = \frac{\lambda^x}{x!}e^{-\lambda}\frac{\mu^y}{y!}e^{-\mu} = e^{\theta x + \xi y - (e^\theta + e^\xi)}\frac{1}{x!y!}$$

with canonical parameter $(\theta, \xi)^\top$, canonical statistic $t(x,y) = (x,y)^\top$, cumulant function $\psi(\theta,\xi) = e^\theta + e^\xi$, and base measure equal to $\frac{1}{x!y!} \cdot m$ where m is counting measure on $\mathbb{N}_0^2$.

Next consider the submodel determined by $\mu = \lambda^2$. This in an affine (in fact linear) submodel of the outer product above, where

$$L = \{(\theta,\xi) \mid \xi = 2\theta\}.$$

The space L is spanned by the vector $v = (1,2)^\top$, so the projection onto L of $t = (t_1,t_2)^\top$ is given as

$$\Pi_L(t) = \frac{\langle t, v\rangle}{\|v\|^2}v = \frac{t_1 + 2t_2}{5}\begin{pmatrix}1\\2\end{pmatrix}$$

corresponding to the relation

$$\theta x + 2\theta y = \theta(x + 2y) = \langle v, \Pi_L(t(x,y))\rangle.$$

Alternatively we may simply represent the family of densities as

$$g_\theta(x,y) = f_{\lambda,\lambda^2}(x,y) = \frac{\lambda^{x+2y}}{x!y!}e^{-\lambda-\lambda^2} = e^{\theta(x+2y)-(e^\theta+e^{2\theta})}\frac{1}{x!y!}$$

showing that the new cumulant function becomes $\tilde{\psi}(\theta) = \psi(\theta, 2\theta) = e^\theta + e^{2\theta}$. The corresponding model is an example of a *log-linear* Poisson model. □

In this example, the affine submodel was determined as the pre-image of a linear map, *in casu* as

$$\Theta_0 = h^{-1}(0), \quad h(\theta,\xi) = \xi - 2\theta.$$

We shall also be interested in the case, where Θ_0 is given as the *image* of an affine map, i.e. where Θ_0 has the form

$$\Theta_0 = \{a + A\beta \mid \beta \in B\} \tag{3.9}$$

where $B \subseteq \mathbb{R}^d$ is open and convex. We have the following variant of Theorem 3.14:

Theorem 3.16. *Assume that $\mathcal{P} = \{P_\theta \mid \theta \in \Theta\}$ is a minimally represented and regular exponential family with canonical parameter space Θ and Θ_0 is a non-empty affine subset of Θ given as (3.9) where A is injective and $B \subseteq \mathbb{R}^d$ is open and convex. Then the subfamily*

$$\tilde{\mathcal{P}} = \{P_{A\beta+a} \mid \beta \in B\}$$

is a regular exponential family with canonical parameter space B. The canonical statistic $\tilde{t}: \mathcal{X} \mapsto \mathbb{R}^d$ may be represented as $\tilde{t}(x) = A^* t(x)$, where A^* is the adjoint of A. The base measure and cumulant function is

$$\tilde{\mu}(dx) = e^{\langle a, t(x) \rangle} \mu(dx), \quad \tilde{\psi}(\beta) = \psi(A\beta + a)$$

and the family $\tilde{\mathcal{P}}$ is a regular and minimally represented exponential family of dimension $d = \dim L$.

Proof. We consider the exponent in the representation of the larger family as

$$\begin{aligned} \langle A\beta + a, t(x) \rangle &= \langle A\beta, t(x) \rangle + \langle a, t(x) \rangle \\ &= \langle \beta, A^* t(x) \rangle + \langle a, t(x) \rangle \\ &= \langle \beta, \tilde{t}(x) \rangle\rangle + \langle a, t(x) \rangle. \end{aligned}$$

So if we let $\tilde{\mu}(dx) = e^{\langle a, t(x) \rangle} \mu(dx)$, we may write the density g_β of $P_{A\beta + a}$ with respect to $\tilde{\mu}$ as

$$g_\beta(x) = e^{\langle \beta, \tilde{t}(x) \rangle - \psi(A\beta + a)}.$$

Now, since B was an open subset of $\mathbb{R}^d$, the family is regular. It is also minimally represented since

$$\langle \lambda, \tilde{t}(X) \rangle = \langle A\lambda, t(X) \rangle$$

and thus if $\langle \lambda, \tilde{t}(X) \rangle$ is a.s. constant with respect to $\tilde{\mu}$, then also $\langle A\lambda, t(X) \rangle$ is a.s. constant with respect to μ, implying that $A\lambda = 0$. But since A was assumed injective, this implies $\lambda = 0$. □

3.5 Moments, score, and information

One important feature of an exponential family is that the mean and variance of the canonical statistic may be calculated by differentiation of the cumulant function, and the score function and information has a particularly simple form, as we shall show in this section.

Theorem 3.17. *In a minimally represented and regular exponential family with canonical parameter space Θ and canonical statistic t, the cumulant function $\psi(\theta)$ is smooth and it holds that*

$$\mathbf{E}_\theta\{t(X)\} = \nabla\psi(\theta) = \tau(\theta), \quad \mathbf{V}_\theta\{t(X)\} = D^2\psi(\theta) = D\tau(\theta) = \kappa(\theta).$$

Further, the map $\theta \mapsto \tau(\theta)$ is smooth and injective, the map $\theta \mapsto \kappa(\theta)$ is smooth, and $\kappa(\theta)$ is positive definite for all $\theta \in \Theta$.

Proof. Theorem 3.9 yields that $c(\theta)$ is smooth and hence $\psi(\theta) = \log c(\theta)$ is smooth. Letting $\tau(\theta) = \mathbf{E}_\theta\{t(X)\}$, we get from Theorem 3.9 that

$$\nabla\psi(\theta)_i = \frac{1}{c(\theta)} \frac{\partial c(\theta)}{\partial \theta_i} = \mathbf{E}_\theta\{t(X)_i\} = \tau(\theta)_i$$

whence τ is smooth as a derivative of a smooth function.

Letting $\kappa(\theta) = D^2\psi(\theta)$ yields that κ is smooth; differentiating once more and using Theorem 3.9 yields

$$\begin{aligned}\kappa(\theta)_{ij} &= \frac{\partial^2\psi(\theta)}{\partial\theta_i\partial\theta j} = \frac{1}{c(\theta)}\frac{\partial^2 c(\theta)}{\partial\theta_i\partial\theta_j} - \frac{1}{c(\theta)^2}\frac{\partial c(\theta)}{\partial\theta_i}\frac{\partial c(\theta)}{\partial\theta_i}\\ &= \mathbf{E}_\theta\{t_i(X)t_j(X)\} - \mathbf{E}_\theta\{t_i(X)\}\mathbf{E}_\theta\{t_j(X)\} = \mathbf{V}_\theta\{t_i(X), t_j(X)\}.\end{aligned}$$

We first argue that $\kappa(\theta)$ is positive definite. Since $\Sigma_\theta = \mathbf{V}_\theta\{t(X)\}$ is a covariance matrix, it is positive semidefinite. Suppose there is a vector $\lambda \in \mathbb{R}^k$ such that $\lambda^\top\Sigma_\theta\lambda = 0$. Then we must have

$$\mathbf{V}_\theta\{\lambda^\top t(X)\} = \lambda^\top\Sigma_\theta\lambda = 0$$

implying that $\lambda^\top t(X)$ is almost surely constant. But this contradicts that the family was assumed minimal. Hence Σ_θ must be positive definite.

Next we argue that τ is injective. Consider $\theta_1, \theta_2 \in \Theta$ with $\theta_1 \neq \theta_2$ and let for $\alpha \in [0, 1]$

$$\theta_\alpha = \alpha\theta_1 + (1 - \alpha)\theta_2.$$

Then $\theta_\alpha \in \Theta$ since Θ is convex. Next, define $g : [0, 1] \mapsto \mathbb{R}$ as

$$g(\alpha) = (\theta_1 - \theta_2)^\top\tau(\theta_\alpha).$$

Composite differentiation with respect to $\alpha \in (0, 1)$ yields

$$g'(\alpha) = (\theta_1 - \theta_2)^\top D\tau(\theta_\alpha)(\theta_1 - \theta_2) = (\theta_1 - \theta_2)^\top\kappa(\theta_\alpha)(\theta_1 - \theta_2) > 0$$

since $\kappa(\theta_\alpha)$ is positive definite. Thus $g(\alpha)$ is strictly increasing on $(0, 1)$ and continuous on $[0, 1]$ so we must have $g(1) > g(0)$, i.e.

$$g(1) = (\theta_1 - \theta_2)^\top\tau(\theta_1) > (\theta_1 - \theta_2)^\top\tau(\theta_2) = g(0)$$

which can be rearranged to give

$$(\theta_1 - \theta_2)^\top(\tau(\theta_1) - \tau(\theta_2)) > 0$$

whence we conclude that $\tau(\theta_1) \neq \tau(\theta_2)$ so τ is injective. □

We shall illustrate this in a few examples.

Example 3.18. [Poisson moments] Consider the simple Poisson model, represented as a regular exponential family in Example 3.4 with canonical parameter $\theta = \log\lambda$ as

$$f_\theta(x) = \frac{1}{x!}e^{\theta x - e^\theta}$$

with cumulant function $\psi(\theta) = e^\theta$. We get by direct differentiation that

$$\mathbf{E}_\theta(X) = \psi'(\theta) = e^\theta = \lambda, \quad \mathbf{V}_\theta(X) = \psi''(\theta) = e^\theta = \lambda$$

identifying $\lambda = e^\theta$ as the mean of the Poisson distribution. □

Example 3.19. [Bernoulli moments] Consider the Bernoulli model as an exponential family in Example 3.3 with canonical parameter $\theta = \log(\mu/(1-\mu))$, i.e. with density

$$f_\theta(x) = \mu^x(1-\mu)^{(1-x)} = \frac{e^{\theta x}}{1+e^\theta}, \quad x \in \{0,1\}.$$

Here the cumulant function is $\psi(\theta) = \log(1+e^\theta)$ so we get for the moments that

$$\mathbf{E}_\theta(X) = \psi'(\theta) = \frac{e^\theta}{1+e^\theta} = \mu, \quad \mathbf{V}_\theta(X) = \psi''(\theta) = \frac{e^\theta}{(1+e^\theta)^2} = \mu(1-\mu)$$

as we know well. □

Example 3.20. As another small example, let us consider the model in Example 3.15 with two related Poisson distributions. Here we derived the cumulant function to be $\tilde{\psi}(\theta) = e^\theta + e^{2\theta}$ and we thus get by differentiation that

$$\mathbf{E}_\theta(X+2Y) = \tilde{\psi}'(\theta) = e^\theta + 2e^{2\theta} = \lambda + 2\lambda^2,$$
$$\mathbf{V}_\theta(X+2Y) = \tilde{\psi}''(\theta) = e^\theta + 4e^{2\theta} = \lambda + 4\lambda^2$$

which we of course could also derive directly from the fact that X and Y are independent Poisson variables with means λ and λ^2. □

For a more complicated example, we consider the linear normal model:

Example 3.21. [Moments in linear normal model] In the linear normal model we may also derive the moments from the cumulant function (3.4). We get

$$\mathbf{E}\{\Pi_L(X)\}_i = \frac{\partial\psi(\theta)}{\partial\theta_{1i}} = \frac{\theta_{1i}}{\theta_2} = \frac{\xi_i/\sigma^2}{1/\sigma^2} = \xi_i,$$

where we have identified θ_1 and ξ with their coordinates in an orthonormal basis for L, and further

$$\mathbf{E}\{-\|X\|^2/2\} = \frac{\partial\psi(\theta)}{\partial\theta_2} = -\frac{d}{2\theta_2} - \frac{\|\theta_1\|^2}{\theta_2^2} = -\frac{d\sigma^2}{2} - \frac{\|\xi\|^2}{2},$$

or, equivalently

$$\mathbf{E}\{\|X\|^2\} = d\sigma^2 + \|\xi\|^2.$$

Differentiating a second time yields

$$\mathbf{V}\{\Pi_L(X)\}_{ij} = \frac{\partial^2\psi(\theta)}{\partial\theta_{1i}\partial\theta_{1j}} = \frac{\delta_{ij}}{\theta_2} = \sigma^2\delta_{ij}$$

where δ_{ij} is Kronecker's delta:

$$\delta_{ij} = \begin{cases} 1 & \text{if } i = j \\ 0 & \text{otherwise,} \end{cases}$$

and

$$\mathbf{V}\{\Pi_L(X)_i, -\|X\|^2/2\}_{ij} = \frac{\partial^2\psi(\theta)}{\partial\theta_{1i}\partial\theta_2} = -\frac{\theta_{i1}}{\theta_2^2} = \frac{-\xi_i/\sigma^2}{1/\sigma^4} = -\sigma^2\xi_i$$

as well as

$$\mathbf{V}\{-\|X\|^2/2\} = \frac{\partial^2\psi(\theta)}{\partial\theta_2^2} = \frac{d}{2\theta_2^2} + \frac{\|\theta_1\|^2}{\theta_2^3} = d\sigma^4/2 + \sigma^2\|\xi\|^2$$

so that

$$\mathbf{V}\{\|X\|^2\} = 2\sigma^4 d + 4\sigma^2\|\xi\|^2,$$

is the variance of $\|X\|^2$. □

As a corollary, we obtain the following relations for score and information in exponential families.

Corollary 3.22. *In a minimally represented and regular exponential family with canonical parameter space Θ, the score, information function, Fisher information, and quadratic score satisfy*

$$S(x,\theta)^\top = t(x) - \tau(\theta), \quad I(x,\theta) = \kappa(\theta) = i(\theta),$$
$$Q(x,\theta) = (t(x) - \tau(\theta))^\top \kappa(\theta)^{-1}(t(x) - \tau(\theta)).$$

Proof. This is obtained by direct differentiation. □

Remark 3.23. *It is important to realize that the relations for the score and information in Corollary 3.22 are specific to the canonical parametrization with Θ as parameter space. For an arbitrary smooth reparametrization, Theorem 1.30 or Corollary 1.33 must be used. The quadratic score is invariant under reparametrizations, but the others are not.*

For example, since the map τ is smooth and injective, we can define a new parametrization by letting $\eta = \tau(\theta)$ with parameter space $C = \tau(\Theta)$. This parametrization is known as the *mean value parametrization* of the exponential family. Thus in this mean value parametrization we have from Theorem 1.30 that

$$S(x,\eta)^\top = \kappa(\theta)^{-1}(t(x) - \tau(\theta)) = \kappa(\theta)^{-1}(t(x) - \eta)),$$
$$i(\eta) = \mathbf{V}_\eta\{S(X,\eta)^\top\} = \kappa(\theta)^{-1}\kappa(\theta)\kappa(\theta)^{-1} = \kappa(\theta)^{-1}.$$

In other words, *the information about the mean value parameter η is the* inverse *of the information about the canonical parameter θ.*

3.6 Curved exponential families

Sometimes it is too restrictive to assume that the parameter set Θ is open and convex to cover a range of cases of interest to us. Instead we consider the following.

Definition 3.24. A *curved exponential family* of *dimension* m and *order* k is a family of the form

$$\mathcal{P} = \{P_{\phi(\beta)}, \beta \in B\} \subseteq \mathcal{Q} = \{P_\theta, \theta \in \Theta\}$$

where

a) $\mathcal{Q}$ is a k-dimensional regular and minimally represented exponential family with canonical parameter space Θ.

b) $B \subseteq \mathbb{R}^m$ is open and $m \leq k$.

c) $\phi : B \to \Theta$ is smooth.

d) The Jacobian matrix

$$J(\beta) = D\phi(\beta) = \left\{\frac{\partial \phi_i}{\partial \beta_j}(\beta)\right\}$$

has full rank m for all $\beta \in B$.

e) ϕ is a homeomorphism onto its image.

The larger exponential family $\mathcal{Q}$ shall be termed the *ambient* exponential family.

We recall that ϕ is a *homeomorphism* onto its image if and only if ϕ is injective, continuous, and has a continuous inverse, i.e. if it holds that

$$\lim_{n\to\infty} \phi(\beta_n) = \phi(\beta) \iff \lim_{n\to\infty} \beta_n = \beta.$$

Figure 3.1 yields an example where ϕ is not a homeomorphism, and one that is. It is worth mentioning that *all results following remain true if ϕ is just twice continuously differentiable* rather than smooth. The proofs then just need to be adapted by using versions of the implicit and inverse function theorems that only assume twice continuous differentiability and have similar weaker conclusions.

Remark 3.25. *Note that if the ambient exponential family is reparametrized with new parameter space Λ and $\lambda = \rho(\theta)$ with ρ smooth and injective having a regular Jacobian, then $\phi : B \to \Theta$ satisfies the regularity conditions if and only if $\tilde{\phi} = \rho \circ \phi$ does. This means that we can verify whether a family is a curved exponential family in any valid parametrization.*

Figure 3.1 – The curve to the right does not satisfy the requirement of a homeomorphism, whereas the curve to the left does.

We illustrate the use of this remark in a simple example.

Example 3.26. Consider the family of distributions on $\mathbb{R}^2_+$ determined by X and Y being independent and exponentially distributed with expectations $\mathbf{E}(X) = \lambda_1$ and $\mathbf{E}(Y) = \lambda_2$, where $\lambda \in \Lambda = \mathbb{R}^2_+$, i.e. the family of distributions with densities

$$f_\lambda(x, y) = \frac{1}{\lambda_1\lambda_2} e^{-x/\lambda_1 - y/\lambda_2}$$

with respect to Lebesgue measure on $\mathbb{R}^2_+$. As an outer product (see Section 3.4.1) of regular exponential families, this is a regular exponential family with canonical parameter $\theta = (\theta_1, \theta_2)^\top = (1/\lambda_1, 1/\lambda_2)^\top \in \Theta = \mathbb{R}^2_+$.

Now consider the subfamily determined by the relation $\lambda_2 = \lambda_1^2$. This is given as the image of the map $\tilde{\phi} : \mathbb{R}_+ \mapsto \Lambda$ determined as $\lambda = \tilde{\phi}(\beta) = (\beta, \beta^2)^\top$. The map $\tilde{\phi}$ is clearly an injective homeomorphism and it has Jacobian

$$\tilde{J}(\beta) = D\tilde{\phi}(\beta) = \begin{pmatrix} 1 \\ 2\beta \end{pmatrix}$$

which has full rank for all β. The point here is that we may verify the regularity conditions without having to reparametrize using the canonical parameter θ and the map $\theta = \phi(\beta) = (1/\beta, 1/\beta^2)^\top$ into the canonical parameter space Θ. □

Clearly, a *minimal and regular exponential family is itself a curved exponential family* with $m = k$, $B = \Theta$, and $\phi(\beta) = \beta$.

Example 3.27. [Affine subfamilies] Another simple example of a curved exponential family is an affine subfamily as discussed in Section 3.4.4. There we considered a submodel determined by intersecting the canonical parameter space with an affine subspace, i.e. we considered

$$\Theta_0 = \Theta \cap (L + a)$$

where L is a subspace of $\mathbb{R}^k$ of dimension m and $a \in \mathbb{R}^k$. We may parametrize L by a linear map $A : \mathbb{R}^m \mapsto \mathbb{R}^k$ and this yields a curved exponential model with $B = \phi^{-1}(\Theta_0)$ and

$$\phi(\beta) = A\beta + a.$$

Then ϕ satisfies the conditions above if and only if $A = D\phi$ has full rank m, i.e. if A is injective.

Another instance of a curved family appears if the affine restriction is used on a parametrization which is not canonical. For example, if we consider the mean value parameter $\eta = \tau(\theta) \in M = \tau(\Theta)$ and now restrict η to an affine subspace of M

$$M_0 = M \cap (L + a)$$

as above. In most cases, this will not itself be a regular exponential family, but it will satisfy the conditions for a curved exponential family. □

Example 3.28. [Bivariate normal with mean on semi-circle] A simple example of a proper curved subfamily of an exponential family is obtained by assuming

$$X \sim \mathcal{N}_2(\theta, I_2).$$

If $\Theta = \mathbb{R}^2$, this is a regular exponential family of dimension 2 with $t(X) = X$ as canonical statistic, since we can write

$$\begin{aligned} f(x;\theta) &= \frac{1}{2\pi} e^{-((x_1-\theta_1)^2+(x_2-\theta_2)^2)/2} \\ &= \frac{1}{2\pi} e^{\theta_1 x_1+\theta_2 x_2 - x_1^2/2 - x_2^2/2 - \theta_1^2/2 - \theta_2^2/2} \\ &= \frac{e^{\theta_1 x_1 + \theta_2 x_2}}{e^{(\theta_1^2+\theta_2^2)/2}} \frac{1}{2\pi} e^{-(x_1^2+x_2^2)/2}. \end{aligned}$$

We have then represented this family as a regular exponential family with base measure $\mathcal{N}_2(0, I_2)$ and cumulant function $\psi(\theta) = (\theta_1^2 + \theta_2^2)/2$. The log-likelihood function, ignoring additive constants is

$$\ell_x(\theta) = \theta^\top x - \frac{1}{2}\|\theta\|^2.$$

Assume now further that the mean θ lies on a semi-circle in the right half-plane $\{\theta : \theta_1 > 0\}$:

$$\theta = \phi(\beta) = \begin{pmatrix} \cos\beta \\ \sin\beta \end{pmatrix},$$

where $\beta \in B = (-\pi/2, \pi/2)$. Here ϕ is a smooth and injective homeomorphism with Jacobian

$$J(\beta) = \begin{pmatrix} -\sin\beta \\ \cos\beta \end{pmatrix}$$

which is never zero, hence has full rank. This is a curved exponential family of dimension 1 and order 2. The parameter space is displayed in Figure 3.2, and the log-likelihood function now reduces to

$$\ell_x(\theta) = \phi(\beta)^\top x. \tag{3.10}$$

since $\|\phi(\beta)\|^2 = 1$ for all β so the cumulant function becomes an additive constant and may be ignored. □

Another example is obtained by assuming the coefficient of variation to be fixed and known:

Example 3.29. [Fixed coefficient of variation] Consider the model determined by $X \sim \mathcal{N}(\beta, \beta^2)$ where $\beta \in B = (0, \infty)$ is unknown. In other words, the normal distribution is considered to have a fixed and known *coefficient of variation*

$$\mathbf{C}_\beta\{X\} = \frac{\sqrt{\mathbf{V}_\beta\{X\}}}{\mathbf{E}_\theta\{X\}} = 1.$$

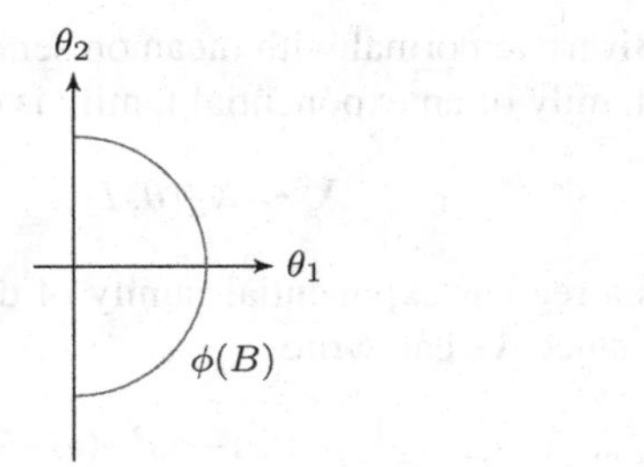

Figure 3.2 – The curved subfamily in Example 3.28 determined by the mean being on a semi-circle.

This is a curved subfamily of the normal exponential family

$$\mathcal{P} = \{\mathcal{N}(\xi, \sigma^2), \xi \in \mathbb{R}, \sigma^2 > 0\}$$

with canonical parameters

$$\theta = \begin{pmatrix} \theta_1 \\ \theta_2 \end{pmatrix} = \begin{pmatrix} \xi\sigma^{-2} \\ \sigma^{-2} \end{pmatrix}$$

and hence the family considered is given by the map

$$\phi(\beta) = \begin{pmatrix} \beta\beta^{-2} \\ \beta^{-2} \end{pmatrix} = \begin{pmatrix} \beta^{-1} \\ \beta^{-2} \end{pmatrix}.$$

The map ϕ is clearly a smooth and injective homomorphism with Jacobian

$$J(\beta) = \begin{pmatrix} \beta^{-2} \\ -2\beta^{-3} \end{pmatrix}$$

having full rank $m = 1$ for all β. The curved parameter space is displayed in Fig. 3.3. □

We now have:

Theorem 3.30. *A curved exponential family is smooth and stable.*

Proof. The proof is essentially identical to that of Lemma 1.29 where a bijective reparametrization was considered. Theorem 3.10 ensures the larger family $\mathcal{Q}$ to have smooth and suitably bounded derivatives. Smoothness as a function of β now follows by composition of functions. We let $L_x(\theta)$ and $\tilde{L}_x(\beta)$ denote the likelihood functions in the regular and curved family so that

$$\tilde{L}_x(\beta) = L_x(\phi(\beta)).$$

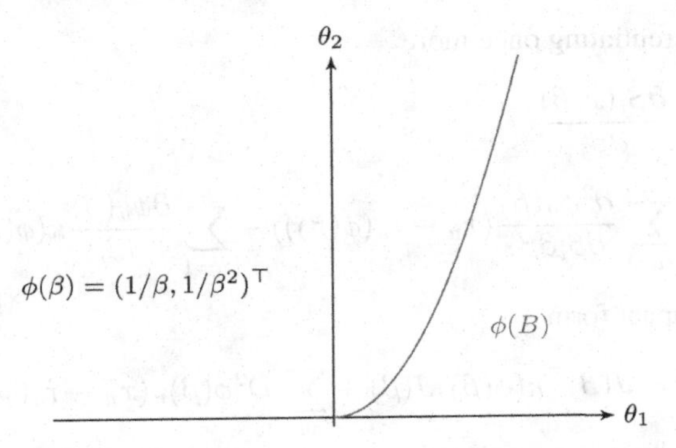

Figure 3.3 – The curved parameter space of the model in Example 3.29 determined by normal distributions with a fixed coefficient of variation.

Using composite differentiation (Rudin, 1976, Theorem 9.15) yields

$$\frac{\partial \tilde{L}_x(\beta)}{\partial \beta_i} = \sum_u \frac{\partial L_x(\theta)}{\partial \theta_u} \frac{\partial \phi_u(\beta)}{\partial \beta_i}$$

or, in matrix form

$$D\tilde{L}_x(\beta) = DL_x(\phi(\beta))D\phi(\beta),$$

where $D\phi(\theta)$ is the Jacobian of ϕ. So the left-hand side is locally bounded by integrable functions if $DL_x(\phi(\beta))$ is.

Differentiating a second time yields

$$\frac{\partial^2 \tilde{L}_x(\beta)}{\partial \beta_i \partial \beta_j} = \sum_{uv} \frac{\partial^2 L_x(\theta)}{\partial \theta_u \partial \theta_v} \frac{\partial \phi_u(\beta)}{\partial \beta_i} \frac{\partial \phi_v(\beta)}{\partial \beta_i} + \sum_u \frac{\partial L_x(\theta)}{\partial \theta_u} \frac{\partial^2 \phi_u(\beta)}{\partial \beta_i \partial \beta_j}$$

which again is locally bounded by integrable functions if the derivatives on the right hand side are. □

Proposition 3.31. *The score, Fisher information, and quadratic score in a curved exponential family are given as*

$$S(x, \beta) = (t(x) - \tau(\phi(\beta)))^\top J(\beta), \quad i(\beta) = J(\beta)^\top \kappa(\phi(\beta))J(\beta),$$
$$Q(x, \beta) = (t(x) - \tau(\phi(\beta)))^\top J(\beta)i(\beta)^{-1}J(\beta)^\top (t(x) - \tau(\phi(\beta))).$$

Proof. This is obtained by composite differentiation of the log-likelihood function. We get

$$S(x, \beta) = \frac{\partial}{\partial \beta} \left(\phi(\beta)^\top t(x) - \psi(\phi(\beta))\right) = (t(x) - \tau(\phi(\beta)))^\top J(\beta)$$

and further, differentiating once more

$$\begin{aligned} I_{ij}(x,\beta) &= \frac{\partial S_i(x,\beta)}{\partial \beta_j} \\ &= \sum_{u=1}^{k} \frac{\partial^2 \phi_u(\beta)}{\partial\beta_i \partial\beta_j}(x_u - \tau_u(\phi(\beta))) + \sum_{u=1}^{k} \frac{\partial \phi_u(\beta)}{\partial \beta_i} \kappa(\phi(\beta))_{uv} \frac{\partial \phi_v(\beta)}{\partial \beta_j}, \end{aligned}$$

or in a more compact form

$$I(x,\beta) = J(\beta)^\top \kappa(\phi(\beta)) J(\beta) + \sum_u D^2\phi(\beta)_u (x_u - \tau_u(\phi(\beta))).$$

We get the Fisher information by taking expectations and noticing that the last term has expectation zero whence

$$i(\beta) = J(\beta)^\top \kappa(\phi(\beta)) J(\beta). \tag{3.11}$$

The expression for the quadratic score now follows. □

Remark 3.32. *Note the analogy between the expression (3.11) and the corresponding expression for the linear normal model in (2.10).*

3.7 Exercises

Exercise 3.1. Let X and Y be independent and exponentially distributed random variables with $\mathbf{E}(X) = \mu$ and $\mathbf{E}(Y) = 2\mu$, where $\mu > 0$. Represent the family of joint distributions of (X, Y) as a one-dimensional regular exponential family, identify the base measure, the canonical parameter, canonical statistic, and cumulant function.

Exercise 3.2. Consider the family of negative binomial distributions, i.e. distributions of a random variable X with densities

$$P_\mu(X = x) = \binom{x + r - 1}{r - 1} (1 - \mu)^r \mu^x$$

with respect to counting measure on $\mathbb{N}_0 = 0, 1, \ldots$. Here $r \in \mathbb{N}$ is considered fixed and known, whereas $\mu \in (0, 1)$ is unknown.

a) Represent this family as a regular exponential family of dimension 1 and identify base measure, canonical parameter, canonical statistic, and cumulant function.

b) Find the mean and the variance in the distribution $\mathbf{E}_\mu(X)$ and $\mathbf{V}_\mu(X)$.

Exercise 3.3. Consider the family of Pareto distributions with densities

$$f_\alpha(x) = \alpha x^{-\alpha-1} \mathbf{1}_{(1,\infty)}(x)$$

with respect to standard Lebesgue measure on $\mathbb{R}$, where $\alpha > 0$ is unknown. Represent this family as a regular exponential family of dimension 1 and identify base measure, canonical parameter, canonical statistic, and cumulant function.

Exercise 3.4. The *inverse normal distribution* has density

$$f_{\mu,\lambda}(x) = \sqrt{\frac{\lambda}{2\pi x^3}} \exp\left\{\frac{-\lambda(x-\mu)^2}{2\mu^2 x}\right\}$$

with respect to standard Lebesgue measure on $\mathbb{R}_+$. You may without proof assume that $\int_0^\infty f_{\mu,\lambda}(x)\,dx = 1$ for all $(\mu,\lambda) \in \mathbb{R}_+^2$. Let now

$$\mathcal{P} = \{P_{\mu,\lambda}, \mu > 0, \lambda > 0\}$$

where $P_{\mu,\lambda}$ has density function $f_{\mu,\lambda}$ as above.

a) Represent the family $\mathcal{P}$ as an exponential family of dimension 2 and identify the base measure, canonical parameter, canonical statistic, and cumulant function.

b) Show that the mean and variance in the family is given as

$$\mathbf{E}_{\mu,\lambda}(X) = \mu, \quad \mathbf{V}_{\mu,\lambda}(X) = \frac{\mu^3}{\lambda}.$$

Now consider the subfamily of inverse normal distributions where the mean is equal to the variance, i.e. where $\lambda = \mu^2$. This is known as the *standard* inverse normal distributions.

c) Argue that this family is an affine subfamily of the full family and thus forms a regular exponential family of dimension 1;

d) Identify the base measure, canonical parameter, canonical statistic, and cumulant function in this subfamily.

e) Determine the mean and variance of $Y = X^{-1}$.

Exercise 3.5. Let (X, Y) be random variables taking values in $\mathbb{N}_0 \times \mathbb{R}_+$ with a distribution determined as follows: X is drawn from a Poisson distribution with mean λ yielding the value x and subsequently Y is drawn from a gamma distribution with shape parameter $x + 1$ and scale parameter β. In other words, the joint distribution of (X, Y) has density

$$f_{(\lambda,\beta)}(x,y) = \frac{\lambda^x}{x!}e^{-\lambda}\frac{y^x}{\beta^{x+1}x!}e^{-y/\beta}, \quad y > 0,\ x = 0, 1, \ldots$$

with respect to $m \times \nu$, where m is counting measure on $\mathbb{N}_0$ and ν is the standard Lebesgue measure on $\mathbb{R}_+$.

a) Argue that the family of distributions with unknown $(\lambda, \beta) \in \mathbb{R}_+^2$, may be represented as a minimal and regular two-dimensional exponential family, determine the canonical parameters, the canonical parameter space, associated canonical statistics, and cumulant function.

b) Find the mean and covariance matrix for $(X, Y)^\top$.

c) Show that the subfamily given by the restriction $\lambda = \beta$ is a minimal and regular one-dimensional family and determine the canonical parameters, associated canonical statistics, and cumulant function.

Exercise 3.6. Let (X, Y) be random variables taking values in $\mathbb{R}_+ \times \mathbb{N}_0$ with a distribution determined as follows: X is drawn from an exponential distribution with mean λ yielding the value x, and subsequently Y is drawn from a Poisson distribution with mean βx. In other words, the joint distribution of (X, Y) has density

$$f_{(\lambda,\beta)}(x, y) = \frac{1}{\lambda} e^{-x/\lambda} \frac{(\beta x)^y}{y!} e^{-\beta x} \quad x > 0,\ y = 0, 1, \ldots$$

with respect to $\nu \times m$, where ν is the standard Lebesgue measure on $\mathbb{R}_+$ and m is counting measure on $\mathbb{N}_0$.

a) Argue that the family of distributions with unknown $(\lambda, \beta) \in \mathbb{R}_+^2$, may be represented as a minimal and regular two-dimensional exponential family, determine the canonical parameters, the canonical parameter space, and associated canonical statistics, and cumulant function.

b) Find the mean and covariance matrix for $(X, Y)^\top$.

c) Show that the subfamily given by the restriction $\lambda = \beta$ is a curved exponential family of dimension one and order two.

d) Find the log-likelihood function, score function, Fisher information, and quadratic score for β in this subfamily.

Exercise 3.7. Consider the family of gamma distributions with identical parameters for shape and scale, i.e. with densities

$$f_\beta(x) = \frac{x^{\beta-1} e^{-x/\beta}}{\Gamma(\beta)\beta^\beta}, \quad \beta \in \mathbb{R}_+$$

with respect to standard Lebesgue measure on $\mathbb{R}_+$.

a) Represent this family as a curved exponential family of dimension one and order two.

b) Find the log-likelihood function, score function, Fisher information, and quadratic score for the family.

Exercise 3.8. Let X and Y be independent random variables with X Poisson distributed with mean λ and Y exponentially distributed with rate λ, where $\lambda > 0$.

a) Represent the family of joint distributions of (X, Y) as a curved exponential family of dimension one and order two.

b) Find the log-likelihood function, score function, Fisher information, and quadratic score for the family.

Exercise 3.9. Let X and Y be independent and exponentially distributed random variables with $\mathbf{E}(X) = \beta$ and $\mathbf{E}(Y) = 1/\beta$ where $\beta > 0$.

a) Represent the family of joint distributions of (X, Y) as a curved exponential family of dimension one and order two.

b) Find the log-likelihood function, score function, Fisher information, and quadratic score for the family.

Exercise 3.10. Let $(X_1, Y_1, \ldots, Y_n, X_n)$ be independent and exponentially distributed random variables where X_j and Y_j have densities with respect to standard Lebesgue measure:

$$f_j(x; \theta) = j\theta e^{-j\theta x}, \; x > 0, \quad g_j(y; \lambda) = \lambda e^{-\lambda y}, \; y > 0, \quad j = 1, \ldots, n,$$

and $(\theta, \lambda) \in \mathbb{R}_+^2$ both unknown. Note that $Y_1, \ldots, Y_n$ are identically distributed whereas this is not the case for $X_1, \ldots, X_n$.

a) Argue that this specifies a minimal and regular exponential family of dimension two, identify the base measure, canonical statistic, and cumulant function.

b) Consider the subfamily determined by the restriction $(\theta, \lambda) = (\beta, 1/\beta)$ and show that this is a curved exponential family of dimension one and order two.

c) Find the log-likelihood function, score function, Fisher information, and quadratic score in this subfamily.

Chapter 4

Estimation

4.1 General concepts and exact properties

An *estimator* is a measurable function $t : \mathcal{X} \to \Theta$ from the sample space to the parameter space; the value $t(x)$ of the estimator is an *estimate* based on data x. The estimator represents a guess on the value θ when we assume that x represents an observed outcome $X = x$ of the random variable X having distribution $P_\theta \in \mathcal{P}$. We often write

$$\hat{\theta} = \hat{\theta}(X) = t(X)$$

emphasizing the fact that also the estimate $\hat{\theta}$ is a random variable, assuming that Θ has been equipped with a measurable structure so that $(\Theta, \mathbb{T})$ is a measurable space. In most of this book we consider the case where the parameter space is a subset $\Theta \subseteq \mathbb{R}^k$ for some k, so Θ inherits the Borel-σ-algebra from $\mathbb{R}^k$.

Also, we may in general not be interested in guessing θ, but only a specific *parameter function of interest* $\phi : \Theta \mapsto \Lambda \subseteq \mathbb{R}^d$; so our estimator will be of the form

$$\hat{\lambda} = \widehat{\phi(\theta)} = t(X),$$

where $t : \mathcal{X} \to \Lambda$. For a number of reasons, it is convenient to allow the estimator to take values outside Λ, so we define formally:

Definition 4.1. Consider a parametrized statistical model on the representation space $(\mathcal{X}, \mathbb{E})$ with associated family $\mathcal{P} = \{P_\theta \mid \theta \in \Theta\}$ and a parametric function $\phi : \Theta \mapsto \Lambda \subseteq \mathbb{R}^d$. An *estimator* of λ is a measurable map $t : \mathcal{X} \to \mathbb{R}^d$. We say that the estimator is *well-defined* on $B = \{x : t(x) \in \Lambda\} \subseteq \mathcal{X}$ and the set $A = \mathcal{X} \setminus B = t^{-1}(\mathbb{R}^d \setminus \Lambda)$ is the *exceptional set* of the estimator.

Clearly we would wish A to have small probability under P_θ as our guess on λ would otherwise mostly be very bad. But we would also be more ambitious and wish to ensure that $\hat{\lambda}$ is a 'good guess' and therefore not too far from the value $\lambda = \phi(\theta)$ corresponding to P_θ. In other words, we would wish that the distribution $t(P_\theta)$ is concentrated around $\phi(\theta)$.

One way of measuring this concentration is the *mean square error* (MSE) which is defined as

$$\mathbf{MSE}_\theta(\hat{\lambda}) = \mathbf{E}_\theta(\|\hat{\lambda} - \lambda\|^2) = \mathbf{E}_\theta(\|t(X) - \lambda\|^2)$$

DOI: 10.1201/9781003272359-4

where $\hat{\lambda} = t(X)$. We would also in general worry if our estimator was 'systematically wrong' in some way and introduce the notion of *bias* of an estimator as the difference between its mean and the value of the parameter it is supposed to estimate:

$$\mathbf{B}_\theta(\hat{\lambda}) = \mathbf{E}_\theta(\hat{\lambda} - \lambda) = \mathbf{E}_\theta(t(X) - \lambda)$$

and note that we then have

$$\begin{aligned} \mathbf{MSE}_\theta(\hat{\lambda}) &= \mathbf{E}_\theta(\|\hat{\lambda} - \lambda\|^2) \\ &= \mathbf{E}_\theta(\|\hat{\lambda} - \mathbf{E}_\theta(\hat{\lambda})\|^2) + \mathbf{E}_\theta(\|\mathbf{E}_\theta(\hat{\lambda} - \lambda)\|^2) \\ &= \operatorname{tr}(\mathbf{V}_\theta(\hat{\lambda})) + \|\mathbf{B}_\theta\|^2, \end{aligned}$$

i.e. *the MSE of an estimator is the sum of the trace of its variance and the squared norm of its bias.* Here the *trace* of a symmetric matrix is $\operatorname{tr}(A) = \sum_i a_{ii}$, i.e. the sum of its diagonal elements, see Section A.5 for further details.

If $\lambda \in \mathbb{R}$ is one-dimensional, it is customary to use the term *standard error* for the square root of the variance

$$\mathbf{SE}_\theta(\hat{\lambda}) = \sqrt{\mathbf{V}_\theta(\hat{\lambda})}$$

so we then have

$$\mathbf{MSE}_\theta(\hat{\lambda}) = \mathbf{SE}_\theta(\hat{\lambda})^2 + \mathbf{B}_\theta(\hat{\lambda})^2.$$

Beware the difference between *standard deviation* and standard error; the first is used to measure the spread of the distribution of X, whereas the second is used for the variability of the estimate of a parameter. The standard deviation of X when $X \sim \mathcal{N}(\xi, \sigma^2)$ is σ, whereas the standard error of the mean $\bar{X}_n = (X_1 + \cdots + X_n)/n$ is then

$$\mathbf{SE}_{\xi,\sigma^2}(\bar{X}_n) = \frac{\sigma}{\sqrt{n}}.$$

Ideally, when constructing an estimator, we would wish to reduce both bias and variance, but this is not always possible; if bias is minimized, variance will often increase and *vice versa* so there could be a trade-off. An estimator is said to be *unbiased* if $\mathbf{B}_\theta(\hat{\lambda}) = 0$.

Example 4.2. [Uniform distribution] Consider the problem of estimating the unknown parameter in the uniform distribution on $(0, \theta)$ where $\theta \in \Theta = \mathbb{R}_+$, corresponding to the model in Example 1.8. We could consider the estimators

$$\tilde{\theta}_n = \frac{2}{n}(X_1 + \cdots + X_n), \quad \hat{\theta}_n = \max(X_1, \ldots, X_n)$$

where the rationale for the first estimator $\tilde{\theta}_n$ is that the expectation of a single observation is $\mathbf{E}_\theta(X_i) = \theta/2$, and the rationale for the second estimator $\hat{\theta}_n$ is that the largest value will be close to the true value. We first note that the estimator $\tilde{\theta}_n$ is unbiased, i.e.

$$\mathbf{E}_\theta(\tilde{\theta}_n) = \frac{2}{n}\, n\, \frac{\theta}{2} = \theta$$

so the MSE is equal to its variance

$$\mathbf{MSE}_\theta(\tilde{\theta}_n) = \mathbf{V}_\theta(\tilde{\theta}_n) = \frac{4}{n^2}\, n\, \frac{\theta^2}{12} = \frac{\theta^2}{3n}, \tag{4.1}$$

where we have used that $\mathbf{V}_\theta(X_i) = \mathbf{V}(\theta U) = \theta^2 \mathbf{V}(U)$ where U is uniform on $(0, 1)$ and

$$\mathbf{V}(U) = \mathbf{E}(U^2) - \mathbf{E}(U)^2 = \int_0^1 u^2\, du - \left(\int_0^1 u\, du\right)^2 = \frac{1}{3} - \frac{1}{4} = \frac{1}{12}.$$

Considering the second estimator, we first find its distribution function, exploiting that the maximum is smaller than x if and only if all of the variables are; so that for $x \in [0, \theta]$ we have

$$G_\theta^n(x) = P_\theta\{\hat{\theta}_n \leq x\} = \prod_{i=1}^{n} P_\theta\{X_i \leq x\} = \frac{x^n}{\theta^n}$$

yielding the density by differentiation

$$g_\theta(x) = \frac{nx^{n-1}}{\theta^n} \mathbf{1}_{(0,\theta)}(x).$$

We may now find the first two moments of the distribution:

$$\mathbf{E}_\theta(\hat{\theta}_n) = \int_0^\theta x \frac{nx^{n-1}}{\theta^n}\, dx = \frac{n}{n+1}\theta, \quad \mathbf{E}_\theta(\hat{\theta}_n^2) = \int_0^\theta x^2 \frac{nx^{n-1}}{\theta^n}\, dx = \frac{n}{n+2}\theta^2,$$

yielding the variance

$$\mathbf{V}_\theta(\hat{\theta}_n) = \frac{n}{n+2}\theta^2 - \left(\frac{n}{n+1}\theta\right)^2 = \frac{n\theta^2}{(n+2)(n+1)^2}.$$

In contrast to the unbiased estimator $\tilde{\theta}_n$, the estimator $\hat{\theta}_n$ has a negative bias

$$\mathbf{B}(\hat{\theta}_n) = \frac{n\theta}{n+1} - \theta = -\frac{\theta}{n+1}$$

reflecting that the estimator is systematically too small. However, the MSE of $\hat{\theta}_n$ is

$$\begin{aligned} \mathbf{MSE}_\theta(\hat{\theta}_n) &= \mathbf{V}_\theta(\hat{\theta}_n) + \mathbf{B}(\hat{\theta}_n)^2 \\ &= \frac{n\theta^2}{(n+2)(n+1)^2} + \frac{\theta^2}{n^2} = \frac{\theta^2(2n^3 + 4n^2 + 5n + 2)}{n^2(n+1)^2(n+2)} \end{aligned} \tag{4.2}$$

which approaches zero at the speed of n^{-2} which is much faster than the MSE of $\tilde{\theta}_n$, approaching zero at the speed of n^{-1}. The estimator $\hat{\theta}_n$ is biased, but it can be *bias corrected* by letting

$$\breve{\theta}_n = \frac{n+1}{n}\hat{\theta}_n = \frac{n+1}{n} \max(X_1, \ldots, X_n)$$

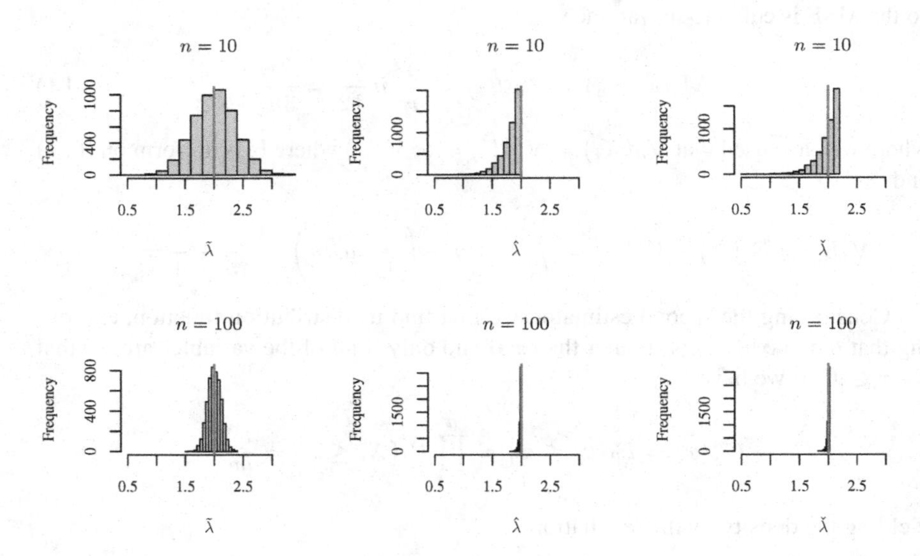

Figure 4.1 – Estimation of the range of a uniform distribution. The diagrams compare the three estimators in Example 4.2 for 5000 simulations of $n = 10$ observations (top row) and $n = 100$ observations (bottom row).

which is now unbiased and therefore has MSE equal to its variance

$$\mathbf{MSE}_\theta(\check{\theta}_n) = \mathbf{V}_\theta(\check{\theta}_n) = \frac{(n+1)^2}{n^2}\mathbf{V}_\theta(\hat{\theta}_n) = \frac{\theta^2}{n(n+2)}$$

which may be considerably smaller than both of $\mathbf{MSE}_\theta(\tilde{\theta}_n)$ as given in (4.1) and $\mathbf{MSE}_\theta(\hat{\theta}_n)$ as given in (4.2). Thus, the estimator $\check{\theta}_n$ appears preferable. These facts are illustrated in Fig. 4.1 where the estimators have been applied to simulated data with $n = 10$ and $n = 100$ observations. □

We should also note that the notion of unbiasedness is *not* equivariant, as the following example illustrates.

Example 4.3. [Estimating the variance in a normal distribution] Consider a sample $X = (X_1, \ldots, X_n)$ from a normal distribution $\mathcal{N}(\xi, \sigma^2)$, where $\xi \in \mathbb{R}$ and $\sigma^2 \in \mathbb{R}_+$ are both unknown. If our parameter of interest is $\phi(\xi, \sigma^2) = \sigma^2$, it is customary to estimate σ^2 as

$$\hat{\sigma}^2 = S^2 = \frac{\|X - \bar{X}\|^2}{n-1} = \frac{1}{n-1}\sum_{i=1}^{n}(X_i - \bar{X})^2$$

and since $Y = \|X - \bar{X}\|^2$ has a $\sigma^2\chi^2$-distribution with $n-1$ degrees of freedom (see Example 2.26), we have

$$\mathbf{E}_{\xi,\sigma^2}(\hat{\sigma}^2) = \frac{(n-1)\sigma^2}{n-1} = \sigma^2$$

so S^2 is indeed an unbiased estimator of σ^2. However, maybe we really would rather estimate the standard deviation σ and would then typically do so by taking the square root of the above estimate. Unfortunately, the estimate for σ is not unbiased as

$$\mathbf{E}_{\xi,\sigma^2}(\hat{\sigma}) = \frac{1}{\sqrt{n-1}}\mathbf{E}_{\xi,\sigma^2}(\sqrt{Y}) < \frac{1}{\sqrt{n-1}}\sqrt{\mathbf{E}_{\xi,\sigma^2}(Y)} = \sigma,$$

where we have used that for any positive random variable Z we have $(\mathbf{E}(\sqrt{Z}))^2 < \mathbf{E}(\sqrt{Z^2}) = \mathbf{E}(Z)$ with equality if and only if Z is almost surely constant. See also Exercise 4.2 for further aspects of this issue. □

An *unbiased estimator that has minimum variance* among all unbiased estimators is an *MVUE* (Minimum Variance Unbiased Estimator). Such estimators are usually considered attractive; however, in many examples there is only a single unbiased estimator and the property may then be less impressive.

Sometimes it may be of interest to limit the type of functions q when constructing estimators, e.g. to functions that are linear. If an estimator has minimal variance among all linear unbiased estimators it is a *best linear unbiased estimator* (BLUE).

The Fisher information is playing an important role in the sense that it gives a lower bound on the variance term for an estimator. This is known as the Cramér–Rao inequality. We first give this in the simplest case of a one-dimensional parameter.

Theorem 4.4 (Cramér–Rao inequality). *Consider a statistical model with a smooth and stable family of distributions with parameter space $\Theta \subseteq \mathbb{R}$ and let $\phi : \Theta \to \mathbb{R}$ be a smooth parameter function. Further, let $\hat{\lambda} = t(X)$ be an estimator of $\lambda = \phi(\theta)$ with expectation $\mathbf{E}_\theta(\hat{\lambda}) = g(\theta)$ where g is smooth. Suppose that every θ has a neighbourhood U_θ so that for all $\eta \in U_\theta$ we have*

$$|f_\eta(x) - f_\theta(x)| \leq f_\theta(x)|\eta - \theta|H_\theta(x) \tag{4.3}$$

where $\mathbf{E}_\theta(H_\theta(X)^2) < \infty$. Then

$$\mathbf{V}_\theta(\hat{\lambda}) \geq \frac{g'(\theta)^2}{i(\theta)}. \tag{4.4}$$

Proof. Consider an arbitrary $\eta \in \Theta$ and the random variable

$$Q_{\eta,\theta} = f_\eta(X)/f_\theta(X) = L_X(\eta)/L_X(\theta),$$

i.e. the ratio of the likelihood functions at η and θ. We then have

$$\mathbf{E}_\theta(Q_{\eta,\theta}) = \int_{\mathcal{X}} \frac{f_\eta(x)}{f_\theta(x)} f_\theta(x)\,d\mu(x) = \int_{\mathcal{X}} f_\eta(x)\,d\mu(x) = 1$$

and thus

$$\mathbf{V}_\theta(\hat{\lambda}, Q_{\eta,\theta}) = \int_{\mathcal{X}} t(x)\left(\frac{f_\eta(x)}{f_\theta(x)} - 1\right) f_\theta(x)\,d\mu(x) = g(\eta) - g(\theta).$$

By Cauchy–Schwarz' inequality, we now get

$$\begin{aligned}(g(\eta) - g(\theta))^2 &\leq \mathbf{V}_\theta(t(X)) \cdot \mathbf{V}_\theta(Q_{\eta,\theta}) \\ &= \mathbf{V}_\theta(t(X)) \int_{\mathcal{X}} \left(\frac{f_\eta(x) - f_\theta(x)}{f_\theta(x)}\right)^2 f_\theta(x)\,d\mu(x)\end{aligned}$$

Now divide both sides of the equation with $(\eta - \theta)^2$ and let $\eta \to \theta$. Dominated convergence assuming the regularity condition (4.3) yields the limit

$$g'(\theta)^2 \leq \mathbf{V}_\theta(\hat{\lambda})\mathbf{V}_\theta(S(X,\theta)) = \mathbf{V}_\theta(\hat{\lambda})i(\theta).$$

Dividing both sides of the inequality by $i(\theta)$ yields (4.4) as desired. □

The uniform distribution in Example 4.2 does *not* satisfy the conditions for the Cramér–Rao inequality to hold. However, we have

Proposition 4.5. *In a curved or regular exponential family, the conditions in Theorem 4.4 are satisfied.*

Proof. See Appendix B.1.2. □

Remark 4.6. *In the special case where $\lambda = \theta$ and $\hat{\theta}$ is an unbiased estimator so that $\gamma(\theta) = \mathbf{E}_\theta(\hat{\theta}) = \theta$ and thus $\gamma'(\theta) = 1$, we get the* simple Cramér–Rao inequality:

$$\mathbf{V}_\theta(\hat{\theta}) \geq i(\theta)^{-1}, \tag{4.5}$$

i.e. the inverse Fisher information is a lower bound for the variance of any unbiased estimator.

This implies the following:

Corollary 4.7. *If an unbiased estimator satisfies $\mathbf{V}_\theta(\hat{\theta}) = i(\theta)^{-1}$, it must be an MVUE.*

An MVUE with a variance that is optimal in the sense that it attains the Cramér–Rao bound $\mathbf{V}_\theta(\hat{\theta}) = i(\theta)^{-1}$ is said to be *efficient.*

Example 4.8. [Estimation in the exponential distribution] Consider estimation of the mean θ in an exponential distribution

$$f_\theta(x) = \frac{1}{\theta}e^{-x/\theta}, \quad x > 0$$

based on independent and identically distributed observations $X_1, \ldots, X_n$, and let us for simplicity assume that $n = 2m + 1$ is odd. We may consider the following three estimators

$$\hat{\theta}_n = \frac{X_1 + \cdots + X_n}{n}, \quad \tilde{\theta}_n = \frac{\mathrm{med}(X_1, \ldots, X_n)}{k(m)}, \quad \breve{\theta}_n = n\min(X_1, \ldots, X_n)$$

where

$$k(m) = \sum_{j=1}^{m+1} \frac{1}{2m - j + 2}.$$

We shall first argue that estimators are all unbiased.

This is obvious for the first estimator, $\hat{\theta}_n$ since $\mathbf{E}_\theta(X_i) = \theta$. To show that the second estimator is unbiased, we use a classical result by Rényi (1953), saying that the p-th order statistic $X_{(p)}$ from n independent and identically exponentially distributed random variables has the same distribution as a weighted sum of independent exponential random variables $Z_1, \ldots, Z_p$ with expectation $\mathbf{E}(Z_i) = 1$ as

$$X_{(p)} \overset{\mathcal{D}}{=} \theta \sum_{j=1}^{p} \frac{Z_j}{n - j + 1}.$$

It follows that the mean and variance of the sample median of an odd number $n = 2m + 1$ of exponentially distributed random variable is

$$\mathbf{E}_\theta(\text{med}(X_1, \ldots, X_n)) = \theta\, k(m), \quad \mathbf{V}_\theta(\text{med}(X_1, \ldots, X_n)) = \theta^2\, v(m)$$

where

$$v(m) = \sum_{j=1}^{m+1} \frac{1}{(2m - j + 2)^2}$$

and hence $\tilde{\theta}_n$ is unbiased. For the last estimator, we use that

$$P_\theta\{\min(X_1, \ldots, X_n) > x\} = \prod_{i=1}^{n} P_\theta\{X_i > x\} = e^{-nx/\theta}$$

showing that $Y_n = \min(X_1, \ldots, X_n)$ is again exponentially distributed but with mean θ/n, and hence

$$\mathbf{E}_\theta(\check{\theta}_n) = \mathbf{E}_\theta(nY_n) = n\,\frac{\theta}{n} = \theta$$

so $\check{\theta}_n$ is also an unbiased estimator.

Since all three estimators are unbiased, it makes sense to compare their variances. We have

$$\mathbf{V}_\theta(\hat{\theta}_n) = \frac{\theta^2}{n}, \quad \mathbf{V}_\theta(\tilde{\theta}_n) = \frac{\theta^2 v(m)}{k(m)^2}, \quad \mathbf{V}_\theta(\check{\theta}_n) = n^2 \left(\frac{\theta}{n}\right)^2 = \theta^2.$$

Clearly, the last estimator is really bad, as the variance does not even decrease with n. It may not be immediately obvious that $\tilde{\theta}_n$ has a larger variance than $\hat{\theta}_n$, i.e. that

$$\frac{v(m)}{k(m)^2} > \frac{1}{2m + 1}$$

but this is actually the case. Indeed we calculated the Fisher information for θ in Example 1.32 for a single observation to be $i(\theta) = \theta^{-2}$, and hence for n observations, the Fisher information is $i_n(\theta) = ni(\theta) = n\theta^{-2}$; thus the lower bound in the Cramér–Rao inequality (4.5) is θ^2/n, which is attained by $\hat{\theta}_n$.

In other words, $\hat{\theta}_n$ is an *efficient* estimator and an MVUE by Corollary 4.7 and therefore its variance and mean square error is at least as small as that of any other unbiased estimator.

To identify how much is actually gained by using $\hat{\theta}_n$ instead of $\tilde{\theta}_n$ for n large, we may find bounds for $k(m)$ and $v(m)$ as

$$\int_{m+1}^{2m+2} \frac{1}{x}\,dx = \log 2 < k(m) < \int_{m}^{2m+1} \frac{1}{x}\,dx = \log\frac{2m+1}{m}$$

$$\int_{m+1}^{2m+2} \frac{1}{x^2}\,dx = \frac{1}{2m+2} < v(m) < \int_{m}^{2m+1} \frac{1}{x^2}\,dx = \frac{m+1}{m(2m+1)},$$

where we have compared the sums defining $k(m)$ and $v(m)$ to Riemann sums for the corresponding integrals. We have thus

$$\operatorname{reff}(\tilde{\theta}_n, \hat{\theta}_n) = \lim_{n\to\infty} \frac{\mathbf{V}_\theta(\tilde{\theta}_n)}{\mathbf{V}_\theta(\hat{\theta}_n)} = \lim_{m\to\infty} \frac{(2m+1)v(m)}{k(m)^2} = \frac{1}{(\log 2)^2} = 2.08$$

implying that we need about twice as many observations using $\tilde{\theta}_n$ instead of $\hat{\theta}_n$ to get the same precision for our estimate. The quantity $\operatorname{reff}(\tilde{\theta}_n, \hat{\theta}_n)$ calculated above is known as the *asymptotic relative efficiency* of $\tilde{\theta}_n$ to $\hat{\theta}_n$.

The findings above are illustrated in Fig. 4.2 where the estimators have been applied to simulated data with $n = 11$ and $n = 51$ observations. □

For the sake of completeness, we also give Cramér's version of the theorem with a slightly different regularity condition, which in contrast to the above involves the estimator t itself. The proof is simpler, but the conclusion of the theorem is also weaker.

Theorem 4.9 (Cramér's inequality). *Consider a statistical model with a smooth and stable family of distributions with parameter space $\Theta \subseteq \mathbb{R}$, let $\phi : \Theta \to \mathbb{R}$ be a smooth parameter function and $\hat{\lambda} = t(x)$ an estimator of $\lambda = \phi(\theta)$ with expectation $\mathbf{E}_\theta(\hat{\lambda}) = \gamma(\theta)$ and assume*

$$\int |t(x)| g_\theta(x)\,\mu(dx) < \infty \tag{4.6}$$

where g_θ is μ-integrable and dominates $\partial f_\eta(x)/\partial\eta$ for $\eta \in U_\theta$, where U_θ is a neighbourhood of θ. Then γ is differentiable and

$$\mathbf{V}_\theta(\hat{\lambda}) \geq \frac{\gamma'(\theta)^2}{i(\theta)}.$$

Proof. The condition (4.6) ensures γ is differentiable with derivative found by differentiation under the integral sign so that

$$\gamma'(\theta) = \int t(x)\frac{\partial f_\theta(x)}{\partial\theta}\,\mu(dx) = \mathbf{E}_\theta(t(X)S(X,\theta)) = \mathbf{E}_\theta((t(X) - \gamma(\theta))S(X,\theta))$$

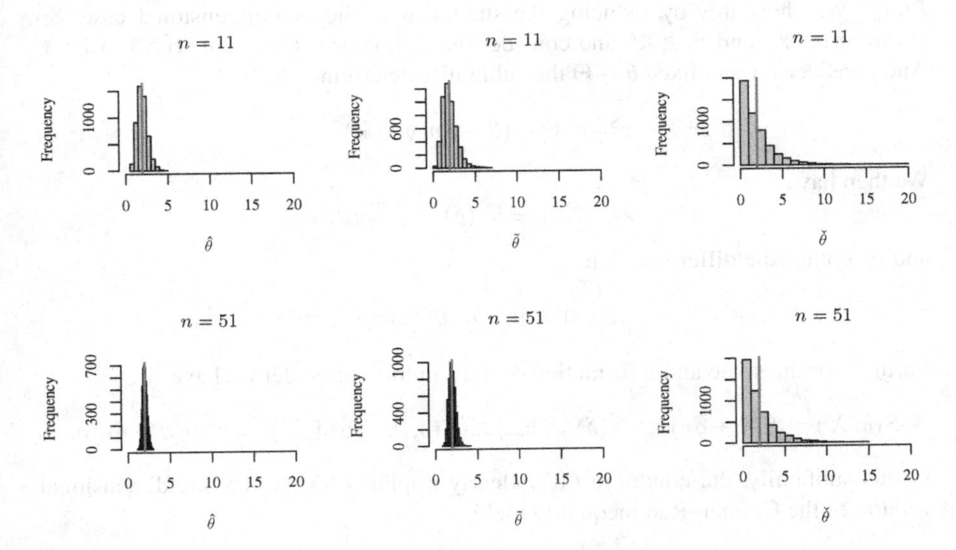

Figure 4.2 – Estimation of the mean of an exponential distribution. The diagrams compare the three estimators in Example 4.8 for 5000 simulations of $n = 11$ observations (top row) and $n = 51$ observations (bottom row).

where we have exploited that $\mathbf{E}_\theta(S(X,\theta)) = 0$. Now the Cauchy–Schwarz inequality yields

$$\gamma'(\theta)^2 \leq \mathbf{V}_\theta(t(X))\mathbf{V}_\theta(S(X,\theta)) = \mathbf{V}_\theta(\hat{\lambda})i(\theta)$$

and the inequality follows. □

The Cramér–Rao inequality has also a multivariate version.

Corollary 4.10 (Multivariate Cramér–Rao). *Consider a statistical model with a smooth and stable family of distributions with parameter space $\Theta \subseteq \mathbb{R}^k$, let $\phi : \Theta \to \mathbb{R}^d$ be a smooth parameter function, and $\hat{\lambda} = t(X)$ an estimator of λ with expectation $\mathbf{E}_\theta(\hat{\lambda}) = g(\theta)$, where g is smooth. Suppose further that every θ has a neighbourhood U_θ so that for all $\eta \in U_\theta$ we have*

$$|f_\eta(x) - f_\theta(x)| \leq f_\theta(x)H_\theta(x)||\eta - \theta|| \tag{4.7}$$

where $\mathbf{E}_\theta(H_\theta(X)^2) < \infty$. Then its covariance matrix $\mathbf{V}_\theta(\hat{\lambda})$ satisfies

$$\mathbf{V}_\theta(\hat{\lambda}) - Dg(\theta)i(\theta)^{-1}Dg(\theta)^\top \textit{ is positive semidefinite,} \tag{4.8}$$

where $Dg(\theta)$ is the Jacobian matrix with entries $\partial g_i(\theta)/\partial\theta_j$. In other words, it holds for all $u \in \mathbb{R}^d$ that

$$u^\top \mathbf{V}_\theta(\hat{\lambda})u \geq u^\top Dg(\theta)i(\theta)^{-1}Dg(\theta)^\top u. \tag{4.9}$$

Proof. We show this by reducing the statement to the one-dimensional case. So choose $u \in \mathbb{R}^d$ and $v \in \mathbb{R}^k$ and consider the estimator $t_u(X) = u^\top t(X)$ of $u^\top \lambda$. And consider for any fixed $\theta \in \Theta$ the subfamily determined by

$$\Theta_v = \Theta \cap (\theta + \delta v, \delta \in \mathbb{R}).$$

We then have

$$\mathbf{E}_{\theta+\delta v}(t_u(X)) = h_u(\delta) = u^\top g(\theta + \delta v)$$

and by composite differentiation

$$h'_u(\delta) = u^\top Dg(\theta + \delta v)v.$$

Further, for the score and information $\tilde{S}$ and $\tilde{i}$ in this submodel we have

$$\tilde{S}(\delta, X) = S(\theta + \delta v)v, \quad \tilde{i}(\delta) = \mathbf{E}_{\theta+\delta v}(\tilde{S}(\delta, X)^\top \tilde{S}(\delta, X)) = v^\top i(\theta + \delta v)v.$$

In this subfamily, the condition (4.7) clearly implies (4.3) so the one-dimensional version of the Cramér–Rao inequality yields

$$\mathbf{V}_{\theta+\delta v}(t_u(X)) \geq \frac{(u^\top Dg(\theta + \delta v)v)^2}{v^\top i(\theta + \delta v)v}.$$

Letting $\delta = 0$ and using that $\mathbf{V}_{\theta+\delta v}(t_u(X)) = u^\top \mathbf{V}_\theta(t(X))u$ we get

$$u^\top \mathbf{V}_\theta(t(X))u \geq \frac{(u^\top Dg(\theta)v)^2}{v^\top i(\theta)v}. \tag{4.10}$$

This inequality holds for any choice of $v \in \mathbb{R}^k$; in particular we may choose

$$v = i(\theta)^{-1} Dg(\theta)^\top u$$

and get

$$u^\top Dg(\theta)v = u^\top Dg(\theta)i(\theta)^{-1}Dg(\theta)^\top u$$

and also

$$v^\top i(\theta)v = u^\top Dg(\theta)i(\theta)^{-1}i(\theta)i(\theta)^{-1}Dg(\theta)^\top u = u^\top Dg(\theta)^\top i(\theta)^{-1}Dg(\theta)^\top u.$$

Inserting the last two relations into (4.10) yields (4.9), as required. □

4.2 Various estimation methods

In this section we briefly describe a few important and classical methods for constructing good estimators. This book shall primarily be concerned with the *method of maximum likelihood* as this is almost universally applicable and based on a simple, yet ingenious principle. However, some of the methods described below may be used as a supplement or substitute when the method of maximum likelihood for some reason fails or leads to difficult computational problems.

4.2.1 *The method of least absolute deviations*

The method of least absolute deviations is one of the earliest estimation methods and goes back to at least *Galileo Galilei* (1564–1642) and was also used extensively by *Pierre Simon Laplace* (1749–1827); see for example Hald (1990, p. 49 ff) and Hald (1998, p. 112 ff) for further details. For an observation $x \in \mathbb{R}^d$, an unknown parameter θ is sought estimated as

$$\hat{\theta}_{\mathrm{LD}} = \mathrm{argmin}_{\theta\in\Theta}\|x - \mathbf{E}_\theta(X)\|_1 = \mathrm{argmin}_{\theta\in\Theta}\sum_{i=1}^{d}|x_i - \mathbf{E}_\theta(X_i)|.$$

A main reason for this method to have come out of fashion in the early 19th century was associated with computational difficulties. There was typically no simple and explicit solution of the minimization problems, even if the expectation $\mathbf{E}_\theta(X)$ was a linear function of θ, and it was quickly taken over by the method of least squares, where minimization could be performed by solving a system of linear equations; see the next subsection.

However, it has had a renaissance in modern times, not least because of progress in the theory and practice of *convex optimization*; if $\mathbf{E}_\theta(X) = A\theta$ is a linear function, the function to be minimized

$$g(\theta, x) = \|x - A\theta\|_1$$

is a *convex* function of θ and good methods now exist for minimizing such functions; see e.g. Boyd and Vandenberghe (2004).

Example 4.11. Let us assume that $x = (x_1, \ldots, x_n)$ is a sample from a distribution on $\mathbb{R}$ with $\mathbf{E}_\theta(X_i) = \theta$. Now estimating θ by the method of least absolute deviations leads to determining the estimate as

$$\hat{\theta}_{\mathrm{LD}} = \mathrm{argmin}_{\theta\in\Theta}\sum_{i=1}^{n}|x_i - \theta|.$$

Let $x_{(i)}$ denote the ith *order statistic*, i.e. the sample is ordered so that

$$x_{(1)} \leq x_{(2)} \leq \cdots \leq x_{(n)}$$

and note that then the objective function may be written as

$$g_n(\theta) = \sum_{i=1}^{n}|x_i - \theta| = \sum_{i=1}^{n}|x_{(i)} - \theta|.$$

If $x_{(r)} \leq \theta \leq x_{(r+1)}$ we further get

$$\begin{aligned} g_n(\theta) &= \sum_{i=1}^{r}(\theta - x_{(i)}) + \sum_{i=r+1}^{n}(x_{(i)} - \theta) \\ &= \sum_{i=r+1}^{n} x_{(i)} - \sum_{i=1}^{r} x_{(i)} + (2r - n)\theta, \end{aligned}$$

so in this interval we have $g'_n(\theta) = 2r - n$. Thus the objective function is decreasing in θ for $r < n/2$ and increasing in θ for $r > n/2$. It follows that if $n = 2p + 1$ is odd, the function has a unique minimum at $\hat{\theta}_{LD} = x_{(p+1)}$ or if $n = 2p$ is even, any value in the interval $[x_{(p)}, x_{(p+1)}]$ is a minimizer of $g_n(\theta)$. In other words

$$\hat{\theta}_{LD} = \text{med}(x_1, \ldots, x_n),$$

i.e. *the LD-estimator is the sample median*, defined as the interval $[x_{(p)}, x_{(p+1)}]$ if $n = 2p$ is even.

An example of the LD function for a simulation of 51 exponentially distributed observations with mean $\theta = 2$ is displayed in Fig. 4.3. □

Generally, an advantage of the method of least absolute deviations is that it tends to be relatively robust to data errors. For example, if one or two of the data points are perturbed due to, say, recording errors, the median will only be vaguely affected and may not be affected at all. However, it may not always be a good estimator unless the distribution of X is roughly symmetric around θ as illustrated in Example 4.12 below.

Example 4.12. [LD estimator in exponential distribution] Consider estimation of the mean θ of an exponential distribution based on $n = 2m + 1$ observations, as in Example 4.8. In this example we calculated that the sample median (i.e. the LD estimator of θ) had mean and variance

$$\mathbf{E}_\theta(\hat{\theta}_{LD}) = \theta k(m) \approx \theta \log 2 = 0.693 \times \theta, \quad \mathbf{V}_\theta(\hat{\theta}_{LD}) = \theta^2 v(m) \approx \frac{\theta^2}{n}$$

yielding the MSE

$$\mathbf{MSE}_\theta(\hat{\theta}_{LD}) \approx \theta^2 \left(\frac{1}{n} + (\log 2 - 1)^2 \right)$$

so the LD estimator is systematically far too small and the bias is dominating the error as n increases.

To make this estimator work reasonably, it is necessary to apply a bias correction as in Example 4.8 to obtain $\tilde{\theta}_n = \hat{\theta}_{LD}/k(m)$ as we have seen. The phenomenon is illustrated for 21 simulated observations in Fig. 4.3. □

4.2.2 The method of least squares

The method of least squares—or *ordinary least squares* (OLS)—estimates the unknown parameter by minimizing the sum of the squared deviations of the observations from their expectation:

$$\hat{\theta}_{\text{OLS}} = \text{argmin}_{\theta \in \Theta} \|x - \mathbf{E}_\theta(X)\|_2^2 = \text{argmin}_{\theta \in \Theta} \sum_{i=1}^{n} (x_i - \mathbf{E}_\theta(X_i))^2 .$$

The method is usually attributed to *Carl Friedrich Gauss* (1777–1855) although it was first published (in 1805) by *Adrien Marie Legendre* (1752–1833), see Hald

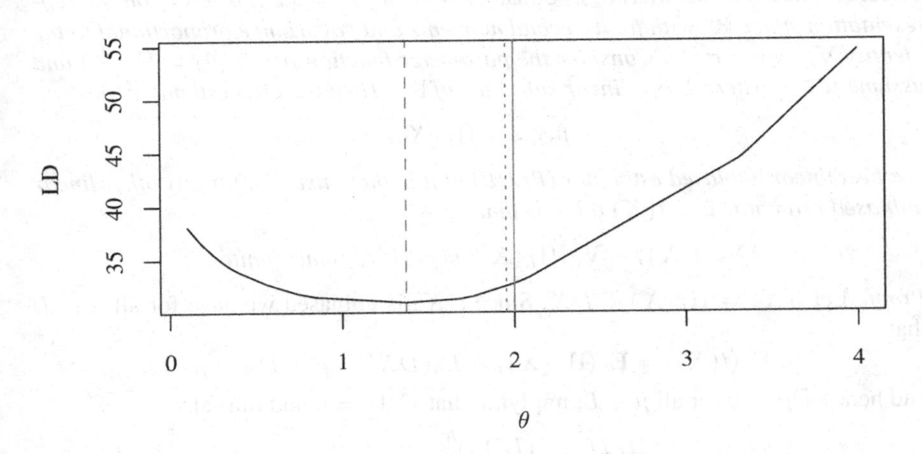

Figure 4.3 – The LD function $g_n(\theta) = \sum_i^n |x_i - \theta|$ for a sample of $n = 21$ simulated independent and exponentially distributed random variables with mean $\theta = 2$. The true value is indicated by a solid line, and the $\hat{\theta}_{LD}$ by a dashed line. The bias corrected estimator $\tilde{\theta}_n$ is indicated by a dotted line.

(1998, Ch. 19 and 21). However, there is no question that Gauss made a comprehensive study of the method and its properties—including Theorem 4.13 below—and he was a wizard in using it in every thinkable way.

One main advantage of OLS is that it has a simple computational solution in the case when $\mu = \mathbf{E}_\theta(X) = A\theta$ is a linear function of the parameter θ, $\theta \in \Theta = \mathbb{R}^k$, and A is a $d \times k$ matrix. Then an alternative formulation of the minimization problem is to find

$$\hat{\mu}_{\mathrm{OLS}} = \operatorname{argmin}_{\mu \in L} \|x - \mu\|_2^2$$

where $L = (y \in \mathbb{R}^d \mid \exists \theta \in \mathbb{R}^k : y = A\theta)$ is the linear subspace of $\mathbb{R}^d$ determined as the image of the map $\theta \mapsto A\theta$. Thus the minimizer is simply given as the orthogonal projection onto L of the observation x:

$$\hat{\mu}_{\mathrm{OLS}} = A\hat{\theta}_{\mathrm{OLS}} = \Pi_L(x)$$

where Π_L is the projection onto L. Further, this projection can be calculated by solving the linear equation system

$$A^\top A\theta = A^\top x$$

known as *the normal equations* and thus the computational issues are benign and simple, and extensively studied.

In addition, the OLS has the following optimality property, known as the *Gauss–Markov theorem:*

Theorem 4.13 (Gauss–Markov). *Consider a model $\mathcal{P} = \{P_\theta \mid \theta \in \Theta\}$ on the representation space $\mathbb{R}^d$ with finite second moments and covariance proportional to the identity $\mathbf{V}_\theta(X) = \sigma^2 I_d$. Consider the parameter function $\mu = \mu(\theta) = \mathbf{E}_\theta(X)$ and assume $\mu \in L$ where L is a linear subspace of $\mathbb{R}^d$. Then the OLS estimator*

$$\hat{\mu}_{\mathrm{OLS}} = \Pi_L(X),$$

is a best linear unbiased estimator (BLUE) of μ in the sense that for any other linear unbiased estimator $\tilde{\mu} = t(X)$ it holds that

$$\mathbf{V}_\theta(t(X)) - \mathbf{V}_\theta(\Pi_L(X)) \textit{ is positive semidefinite.}$$

Proof. Let $t(X) = \Pi_L(X) + DX$. Since $t(X)$ is unbiased we have for all $\mu \in L$ that

$$\mathbf{E}_\theta(t(X)) = \mathbf{E}_\theta(\Pi_L(X)) + E_\theta(DX) = \mu + D\mu = \mu$$

and hence $D\mu = 0$ for all $\mu \in L$, implying that $D\Pi_L = 0$ and thus also

$$\Pi_L D^\top = (D\Pi_L)^\top = 0^\top = 0.$$

But then, using that Π_L is idempotent and symmetric, we have

$$\begin{aligned} \mathbf{V}_\theta(t(X)) &= \sigma^2(\Pi_L + D)(\Pi_L + D)^\top \\ &= \sigma^2(\Pi_L + \Pi_L D^\top + D\Pi_L + DD^\top) \\ &= \sigma^2\Pi_L + \sigma^2 DD^\top = \mathbf{V}_\theta(\hat{\mu}_{\mathrm{OLS}}) + \sigma^2 DD^\top. \end{aligned}$$

Since $\sigma^2 DD^\top$ is positive semidefinite, the proof is complete. □

In the case where the covariance $\mathbf{V}_\theta(X) = \Sigma$ is not proportional to the identity we would use *weighted least squares* (WLS) and determine the estimate as

$$\hat{\mu}_{\mathrm{WLS}} = \mathrm{argmin}_{\mu \in L_A} \|x - \mu\|_W^2 = \mathrm{argmin}_{\mu \in L_A} (x - \mu)^\top W (x - \mu),$$

where $W = \Sigma^{-1}$ is the *weight matrix*. The normal equations for WLS are still linear:

$$A^\top W A\theta = A^\top W x$$

with the solution $\hat{\theta} = (A^\top W A)^{-1} A^\top W x$ and hence

$$A\hat{\theta} = A(A^\top W A)^{-1} A^\top W x = \Pi_L(x)$$

since this expression is indeed identified as the matrix for the projection onto L with respect to the inner product $\langle x, y\rangle_W = x^\top W y$; see also Proposition 2.21.

The conclusion in the Gauss–Markov theorem also holds in this generality—as also shown by Gauss—and the proof is essentially identical to the proof above. Note that *in the case where $X \sim \mathcal{N}_d(\mu, \Sigma)$ with $\mu = A\theta$, the WLS estimate is not only BLUE, but in fact MVUE* as the variance attains the Cramér–Rao lower bound. Indeed, we have

$$\mathbf{V}_\theta(\hat{\theta}) = (A^\top W A)^{-1} A^\top W \Sigma W A (A^\top W A)^{-1} = (A^\top W A)^{-1} = i(\theta)^{-1},$$

where the last equality is from (2.10) .

4.2.3 M-estimators

A large class of estimators that include the LD and OLS estimators are constructed in the following way. For $\mathcal{X} = \mathbb{R}^k$ and $\Theta \subseteq \mathbb{R}^k$ we consider a function $g : \mathcal{X} \times \Theta \mapsto \mathbb{R}$. Then for a sample $(x_1, \ldots, x_n)$, we consider the estimator

$$\hat{\theta}_g = \operatorname{argmin}_{\theta \in \Theta} \sum_{i=1}^{n} g(\theta, x_i).$$

Often it is assumed that the function g is *convex* in θ and for every x has a unique minimum for $\theta = x$ or when $\mathbf{E}_\theta(X) = x$. Examples of such functions that we have seen already are $g(\theta, x) = |x - \theta|$, and $g(\theta, x) = (x - \theta)^2$. Indeed the LD and OLS estimators are examples of estimators constructed in this way.

An estimator of this kind is known as an *M-estimator*. We shall not discuss these in any detail in these lecture notes but point out that many estimators used in modern machine learning are of this kind and under suitable smoothness assumptions they tend to have benign properties for large n.

4.2.4 The method of moments

In this section we shall discuss estimators that are constructed by matching the empirical averages of specific functions to their expectations, or moments. We consider a sample $X_1 \ldots, X_n$ from a parametrized family $\mathcal{P} = \{P_\theta \mid \theta \in \Theta\}$ on the representation space $(\mathcal{X}, \mathbb{E})$. In addition we consider a statistic $t : \mathcal{X} \mapsto \mathbb{R}^k$ with finite expectation for all $\theta \in \Theta$.

Definition 4.14. The *moment function* $m : \Theta \mapsto \mathbb{R}^k$ of the statistic t is the function

$$m(\theta) = \mathbf{E}_\theta(t(X)).$$

Typical choices for $\mathcal{X} = \mathbb{R}$ would be $t(x)^\top = (x, x^2)$ corresponding to the first two moments of the distribution P_θ, but many other choices of t are possible. If the moment function is injective, we define the moment estimator as follows.

Definition 4.15. Let $X_1, \ldots X_n$ be a sample from a parametrized family $\mathcal{P} = \{P_\theta \mid \theta \in \Theta\}$ and $t : \mathcal{X} \mapsto \mathbb{R}^k$ a statistic with an injective moment function $m : \Theta \to \mathbb{R}^k$ and let

$$\bar{T}_n = \frac{1}{n} \sum_{i=1}^{n} t(X_i).$$

The *moment estimator* $\tilde{\theta}_{\text{mom}}$ of θ based on m is well-defined if $\bar{T}_n \in m(\Theta)$ and is then obtained by equating the empirical and theoretical moments:

$$m(\tilde{\theta}_{\text{mom}}) = \frac{1}{n} \sum_{i=1}^{n} t(X_i) = \bar{T}_n.$$

In other words, we have $\tilde{\theta}_{\text{mom}} = m^{-1}(\bar{T}_n)$.

The method of moments was advocated by *Karl Pearson* (1857–1936), who used the method with moment function $t(x) = (x, x^2, x^3, x^4)^\top$, i.e. based on the first four moments of the distribution; in addition he introduced a 'system of curves', i.e. a four parameter family of distributions that could then be fitted with the method of moments; see Hald (1998, p. 721). Later it has been established that this method is not always good, as the empirical moments of high order tend to be very unstable and have a huge variability.

Example 4.16. [Moment estimators in the normal model] Consider the simple normal model determined by $X_1, \dots, X_n$ being independent and normally distributed as $\mathcal{N}(\xi, 1)$, where $\xi \in \mathbb{R}$ is unknown. There are a variety of possible moment estimators of ξ, depending on what we choose as moment function. We could, for example, consider

$$t_1(x) = x, \quad t_2(x) = \mathbf{1}_{(0,\infty)}(x), \quad t_3(x) = x^3$$

with corresponding moment functions

$$m_1(\xi) = \xi, \quad m_2(\xi) = P_\xi(X > 0) = \Phi(\xi), \quad m_3(\xi) = 3\xi + \xi^3$$

since $X \stackrel{\mathcal{D}}{=} Z + \xi$ where Z has a standard normal distribution and thus

$$\mathbf{E}_\xi(X^3) = \mathbf{E}((Z+\xi)^3) = \mathbf{E}(Z^3) + 3\xi\mathbf{E}(Z^2) + 3\xi^2\mathbf{E}(Z) + \xi^3 = 3\xi + \xi^3.$$

These moment functions are all injective; the image of m_1 and m_3 is all of $\mathbb{R}$, so the moment estimators $\hat{\xi}_1$ and $\hat{\xi}_3$ are always well-defined; whereas m_2 maps $\mathbb{R}$ injectively to $(0,1)$ so $\hat{\xi}_2$ is well defined if and only if the observed sample $x_1, \dots, x_n$ contains both positive and negative values, which of course will happen with high probability if n is large. If we let

$$A_n = \frac{1}{n}\sum_{i=1}^n X_i, \quad B_n = \frac{1}{n}\sum_{i=1}^n \mathbf{1}_{(0,\infty)}(X_i), \quad C_n = \frac{1}{n}\sum_{i=1}^n X_i^3,$$

the corresponding moment estimators are

$$\hat{\xi}_{1n} = A_n, \quad \hat{\xi}_{2n} = \Phi^{-1}(B_n), \quad \hat{\xi}_{3n} = g(C_n).$$

Here Φ is the standard normal distribution function and

$$g(y) = 2\sinh\left(\frac{1}{3}\sinh^{-1}\left(\frac{y}{2}\right)\right)$$

is the unique real root of the third-degree equation

$$x^2 + 3x + y = 0$$

expressed in terms of the hyperbolic sine function sinh

$$\sinh(x) = \frac{e^x - e^{-x}}{2}.$$

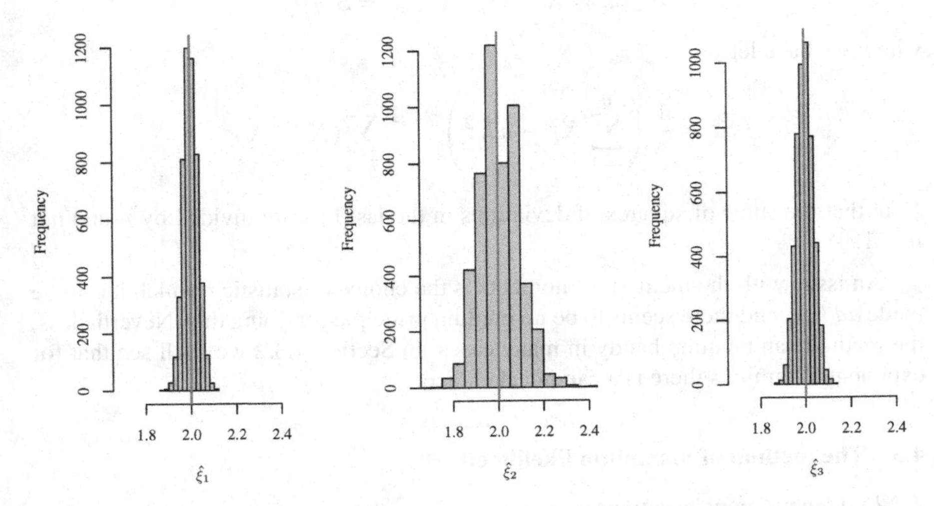

Figure 4.4 – Simulation of 5000 estimates of ξ with the three moment estimators in Example 4.16. The standard average to the left is clearly the preferable estimator and the estimator in the middle is the worst.

We shall later, in Section 5.2, return to this example and discuss the properties of these estimators for large n. At this point we restrict ourselves to showing the results of a simulation where we have estimated ξ 5000 times using these three methods for a sample size of $n = 100$. The outcome of the simulation is displayed in Fig.4.4; the plots indicate that $\hat{\xi}_1$ is the best, whereas $\hat{\xi}_2$ the worst of these estimators. □

Moment estimators tend to be sensible, are often easy to construct and compute, and as such they can be quite useful; however, they may also be very inefficient. The following is an example of a moment estimator that is quite often used in practice.

Example 4.17. Consider the family of gamma distributions, with densities with respect to Lebesgue measure

$$f(x; \alpha, \beta) = \frac{x^{\alpha-1}}{\beta^\alpha \Gamma(\alpha)} e^{-x/\beta}, \quad x > 0,$$

where $\alpha > 0$ and $\beta > 0$. For this family we have the first and second moments

$$\mathbf{E}_{\alpha,\beta}(X) = \alpha\beta, \quad \mathbf{E}_{\alpha,\beta}(X^2) = \alpha(\alpha+1)\beta^2$$

and thus $m(\alpha, \beta)^\top = (\alpha\beta, \alpha(\alpha+1)\beta^2)$. Thus the moment estimation equation is

$$\bar{X} = \alpha\beta, \quad \frac{1}{n}\sum_i X_i^2 = \alpha(\alpha+1)\beta^2$$

with the solution

$$\hat{\alpha}_{\text{mom}} = \bar{X}^2/\tilde{S}^2, \quad \hat{\beta}_{\text{mom}} = \tilde{S}^2/\bar{X}$$

where we have let

$$\tilde{S}^2 = \frac{1}{n}\left(\sum_{i=1}^{n} X_i^2 - n\bar{X}^2\right) = \frac{1}{n}\sum_{i=1}^{n}(X_i - \bar{X})^2.$$

Note that the sums of squares of deviations in the last line are divided by n and not $n-1$. □

An issue with the method of moments is the choice of statistic t which has to be made *ad hoc*, and there seems to be no guiding principle for doing this. Nevertheless, the method can be quite handy in many cases. In Section 4.3.2 we shall see that for exponential families there is a canonical choice.

4.3 The method of maximum likelihood

4.3.1 General considerations

The estimation methods mentioned above have all been *ad hoc* and have not been exploiting properties of the specific model although Example 4.12 clearly shows that estimation methods may be very bad if applied to models that do not go well with the methods.

In contrast, the *method of maximum likelihood* as invented by R. A. Fisher yields a universal estimation method for any dominated statistical model which is specifically designed to that model.

More precisely, for any parametrized and dominated statistical model $\mathcal{P}$ on a representation space $(\mathcal{X}, \mathbb{E})$ and associated family of densities $\mathcal{F} = \{f_\theta \mid \theta \in \Theta\}$, we define the *maximum-likelihood estimator* (MLE) as

$$\hat{\theta}_{ML} = \text{argmax}_{\theta\in\Theta}\, \ell_x(\theta)$$

provided this is well-defined. So if we have a sample $(X_1, \ldots, X_n)$ from $\mathcal{P}$, the MLE is defined as

$$\hat{\theta}_{ML} = \text{argmax}_{\theta\in\Theta} \sum_{i=1}^{n} \ell_{X_i}(\theta)$$

so *the MLE is an M-estimator* based on the negative of the log-likelihood function $g(\theta, x) = -\ell_x(\theta)$. Indeed, the LD, OLS, or WLS estimators are all maximum-likelihood estimators corresponding to specific models as we shall see.

Example 4.18. [Laplace model] Consider a sample $x = (x_1, \ldots, x_n)$ from the Laplace distribution with density

$$f_\theta(x) = \frac{1}{2}e^{-|x-\theta|}$$

with respect to standard Lebesgue measure on $\mathbb{R}$. Here the log-likelihood function becomes

$$\ell_{x_1,\ldots,x_n}(\theta) = -\sum_{i=1}^{n} |x_i - \theta|$$

and this is clearly maximized in θ if and only if the sum $\sum_i |x_i - \theta|$ is minimized. In other words, from Example 4.11 we conclude that the MLE is the *median*

$$\hat{\theta}_{ML} = \hat{\theta}_{LD} = \text{med}(x_1, \ldots, x_n).$$

Thus *within this model*, the method of least absolute deviations coincides with the method of maximum likelihood. □

If the family is smooth, the maximizer of ℓ must be a stationary point of the log-likelihood function, i.e. it satisfies the equation

$$D\ell_x(\hat{\theta}_{ML}) = 0$$

or equivalently

$$S(x, \hat{\theta}_{ML}) = 0 \tag{4.11}$$

and this equation is known as the *likelihood equation* or *score equation*. For a sample $(x_1, \ldots, x_n)$, the score equation takes the form

$$\sum_{i=1}^{n} S(x_i, \hat{\theta}_{ML}) = 0.$$

A solution of the score equation is not necessarily the MLE, as a stationary point of the log-likelihood function may not be a local nor a global maximum. The *observed information*

$$I(x, \hat{\theta}_{ML}) = -DS(x, \hat{\theta}_{ML}) = -D^2\ell_x(\hat{\theta}_{ML})$$

is positive definite at the solution if and only if the solution corresponds to a local maximum and if this holds and the solution is unique, it must be a global maximum. But generally an additional argument is needed to establish that a solution of the score equation is also the MLE.

In the following *we shall omit the subscript of the estimator so that $\hat{\theta}$ always refers to the MLE unless specified otherwise.* We hasten to provide some examples of maximum-likelihood estimation.

Example 4.19. [MLE in Poisson model] Consider the simple Poisson model with unknown mean $\lambda \in \Lambda = (0, \infty)$ and density

$$f_\lambda(x) = \frac{\lambda^x}{x!} e^{-\lambda}, \quad x \in \mathbb{N}_0.$$

The associated log-likelihood function and score function is

$$\ell_x(\lambda) = x \log \lambda - \lambda, \quad S(x, \lambda) = \frac{x}{\lambda} - 1$$

so the score equation has a unique solution $\hat{\lambda} = x$ if and only if $x > 0$. For $x = 0$ there is no solution within the parameter space. The information function is

$$I(x, \lambda) = \frac{x}{\lambda^2}$$

which is strictly positive, ensuring that the solution corresponds to a local maximum of the likelihood function and since there is only one solution, it is a global maximum. Thus the MLE is $\hat{\lambda} = x$. □

Example 4.20. [MLE in exponential distribution] Consider a sample $(x_1, \ldots, x_n)$ from an exponential distribution with mean $\theta \in \Theta = (0, \infty)$, i.e. with density

$$f_\theta(x) = \prod_{i=1}^{n} \frac{1}{\theta} e^{-x_i/\theta} = \frac{1}{\theta^n} e^{-\sum_i x_i/\theta}, \quad x \in \mathbb{R}_+^n.$$

The log-likelihood and score functions become

$$\ell_x(\theta) = -n \log \theta - \frac{\sum_i x_i}{\theta}, \quad S_x(\theta) = \frac{-n}{\theta} + \frac{\sum_i x_i}{\theta^2}$$

and the score equation has a unique solution for

$$\hat{\theta}_n = \frac{x_1 + \cdots + x_n}{n}.$$

The observed information is

$$I(x, \hat{\theta}_n) = -\frac{n}{\hat{\theta}_n^2} + \frac{2\sum_i x_i}{\hat{\theta}_n^3} = \frac{n^3}{(\sum_i x_i)^2} > 0$$

so the unique solution is indeed the MLE. Note that this estimator is also the MVUE as considered in Example 4.8. □

Example 4.21. [MLE in uniform model] We calculated the likelihood function in the uniform model in Example 1.18 and from the display in Fig. 1.2 we conclude that the MLE of θ based on a sample $(X_1, \ldots, X_n)$ is

$$\hat{\theta}_n = \max(X_1, \ldots, X_n).$$

Since this model is not smooth, we cannot obtain the estimate from the score equation. Note that even though the MLE here is quite reasonable, we may prefer the bias corrected version

$$\check{\theta}_n = \frac{n+1}{n} \hat{\theta}_n = \frac{n+1}{n} \max(X_1, \ldots, X_n)$$

as discussed further in Example 4.2. □

We conclude this section with the well-known estimation problem in the simple normal model.

Example 4.22. [MLE in the simple normal model] We consider a sample from the simple normal model where $X_1, \dots X_n$ are independent and identically normally distributed as $\mathcal{N}(\xi, \sigma^2)$, where $\theta = (\xi, \sigma^2) = \Theta = \mathbb{R} \times \mathbb{R}_+$ is unknown. The log-likelihood function becomes

$$\ell_{x_1,\dots,x_n}(\xi, \sigma^2) = -\frac{n}{2}\log\sigma^2 - \frac{1}{2}\sum_{i=1}^{n}\frac{(x_i - \xi)^2}{\sigma^2}.$$

For fixed σ^2, this is maximized in ξ for $\hat{\xi}_n = \bar{x}_n$, and thus

$$\ell_{x_1,\dots,x_n}(\bar{x}_n, \sigma^2) = -\frac{n}{2}\log\sigma^2 - \frac{1}{2}\sum_{i=1}^{n}\frac{(x_i - \bar{x}_n)^2}{\sigma^2} = -\frac{n}{2}\left(\log\sigma^2 - \frac{(n-1)s_n^2}{n\sigma^2}\right).$$

Differentiation w.r.t. σ^2 shows that this function is maximized by

$$\hat{\sigma}_n^2 = \frac{1}{n}\sum_{i=1}^{n}(x_i - \bar{x}_n)^2 = \frac{n-1}{n}s_n^2.$$

We note that the MLE divides the sum of squares of deviation from the mean with n rather than $n-1$. It is customary to use the bias corrected MLE $\check{\sigma}_n^2 = s_n^2$. □

This example is a special instance of that of a linear normal model, where the details and likelihood analysis is quite similar.

Example 4.23. [Linear normal model] Consider the linear normal model as in Example 3.6, determined by the family of densities

$$f_{(\xi,\sigma^2)}(x) = \frac{1}{(2\pi\sigma^2)^{d/2}}e^{\frac{-\|x-\xi\|^2}{2\sigma^2}}$$

with respect to standard Lebesgue measure λ_V on a d-dimensional Euclidean space V, where $(\xi, \sigma^2) \in L \times \mathbb{R}_+$ for L being an m-dimensional linear subspace of V. Ignoring irrelevant additive constants, the log-likelihood becomes

$$\ell_x(\xi, \sigma^2) = -\frac{d}{2}\log\sigma^2 - \frac{-\|x-\xi\|^2}{2\sigma^2}.$$

for fixed σ^2, this is maximized in ξ by minimizing $\|x-\xi\|^2$ over L. The miminimizer is the projection $\Pi_L(x)$ onto L and hence $\hat{\xi} = \Pi_L(x)$. Inserting this into ℓ yields the *profile likelihood* for σ^2

$$\ell_x(\hat{\xi}, \sigma^2) = -\frac{d}{2}\log\sigma^2 - \frac{-\|x-\Pi_L(x)\|^2}{2\sigma^2},$$

which—as in the previous example—is maximized by

$$\hat{\sigma}^2 = \frac{1}{d}\|x - \Pi_L(x)\|^2.$$

From Theorem 2.24(d) we have that $\|X - \Pi_L(X)\|^2/\sigma^2 \sim \chi^2(d-m)$, where $d = \dim L$. This implies that $\hat{\sigma}^2$ may be heavily biased if m is relatively large compared to d since

$$\mathbf{E}_{(\xi,\sigma^2)}(\hat{\sigma}^2) = \frac{d-m}{d}\sigma^2$$

and it is therefore common to use the bias-corrected version

$$\tilde{\sigma}^2 = \frac{1}{d-m}\|x - \Pi_L(x)\|^2$$

for estimating σ^2. But beware that $\tilde{\sigma}$ is not unbiased for σ nor is $\tilde{\sigma}^{-2}$ unbiased for σ^{-2}; see Exercise 4.2 . □

One important property of the method of maximum likelihood is that it is equivariant under reparametrizations of the model, in contrast to most other estimation methods. We formulate this important result as a theorem.

Theorem 4.24 (Equivariance of the MLE). *Consider a parametrized statistical model $\mathcal{P} = \{P_\theta \mid \theta \in \Theta\}$ with associated family of densities $\mathcal{F} = \{f_\theta \mid \theta \in \Theta\}$ and a bijective reparametrization $\phi : \Theta \mapsto \Lambda$ so an alternative representation of the family is $\mathcal{F} = \{g_\lambda \mid \lambda \in \Lambda\}$. Then the MLE $\hat{\theta}$ of θ based on an observation $X = x$ is well-defined if and only the MLE $\hat{\lambda}$ of λ is well-defined and then $\hat{\lambda} = \phi(\hat{\theta})$.*

Proof. This is a simple consequence of the equivariance of the likelihood function itself since if $\lambda = \phi(\theta)$ we have

$$\tilde{\ell}_x(\theta) = f_\theta(x) = g_{\phi(\theta)}(x) = g_\lambda(x) = \ell_x(\lambda)$$

and thus

$$\tilde{\ell}_x(\hat{\theta}) \geq \tilde{\ell}_x(\theta), \theta \in \Theta \iff \ell_x(\hat{\lambda}) \geq \ell_x(\lambda), \lambda \in \Lambda.$$

This completes the proof. □

4.3.2 Maximum likelihood in regular exponential families

We have seen that the MLE is always an M-estimator and in specific cases corresponds to methods of least deviations or least squares. Here we shall establish that for regular exponential families, the MLE is also a moment estimator.

Theorem 4.25. *Let $(X_1, \ldots, X_n)$ be a sample from a minimal and regular exponential family with canonical parameter $\theta \in \Theta$ and canonical statistic t. Then the MLE $\hat{\theta}_n$ based on this sample is a moment estimator with respect to the canonical statistic t and moment function*

$$m(\theta) = \tau(\theta) = \nabla\psi(\theta) = \mathbf{E}_\theta(t(X)).$$

If the moment equation has a solution, this solution is unique.

Proof. Let $x = (x_1, \ldots, x_n)$ and $\bar{t}_n = (t(x_1) + \cdots + t(x_n))/n$. The log-likelihood function is determined as

$$\bar{\ell}_n(\theta) = \frac{1}{n} \log L(\theta; x) = \theta^\top \bar{t}_n - \psi(\theta).$$

Differentiation yields the score equation

$$\tau(\theta) = \bar{t}_n$$

where $\tau(\theta) = \nabla\psi(\theta) = \mathbf{E}_\theta(t(X))$, so the score equation is exactly the moment equation for the moment estimator based on the canonical statistic t.

Since τ is injective by Theorem 3.17, there is at most one point satisfying the score equation. Differentiating a second time yields

$$D^2\bar{\ell}_n(\theta) = -\kappa(\theta) = -\mathbf{V}_\theta(t(X)) = -\Sigma_\theta.$$

Since the family is minimal, Σ_θ is positive definite; so if the score equation has a solution, it is a maximizer and hence the moment estimator is also the MLE. □

Example 4.26. [The gamma family] Consider again the family of gamma distributions in Example 4.17.

This can be represented as a minimal and regular exponential family of dimension 2, with canonical parameters

$$\theta = (\theta_1, \theta_2)^\top = (\alpha, 1/\beta)^\top,$$

canonical parameter space $\Theta = (0, \infty)^2$, canonical statistic $t(x) = (\log x, -x)^\top$, cumulant function

$$\psi(\theta) = \log c(\theta) = \log \Gamma(\theta_1) - \theta_1 \log(\theta_2),$$

and base measure $\mu(dx) = x^{-1}\, dx$ on $(0, \infty)$. The exponential representation of the density becomes

$$f(x; \theta) = \frac{e^{\theta_1 \log x + \theta_2(-x)}}{(\theta_2)^{-\theta_1}\Gamma(\theta_1)}.$$

For the mean we get by differentiation of the cumulant function

$$\begin{aligned} \tau(\theta)^\top &= \mathbf{E}_\theta((\log X, -X)) \\ &= (\Psi(\theta_1) - \log(\theta_2), -\theta_1/\theta_2)) = (\Psi(\alpha) + \log(\beta), -\alpha\beta)) \end{aligned}$$

where $\Psi(\alpha) = D \log \Gamma(\alpha)$ is known as the *digamma function*.

The likelihood equations or moment equations for these statistics become

$$\Psi(\alpha) + \log \beta = \overline{\log x}, \quad \alpha\beta = \bar{x},$$

or, equivalently,

$$\log \alpha - \Psi(\alpha) = \log \bar{x} - \overline{\log x}; \quad \beta = \bar{x}/\alpha. \tag{4.12}$$

The function $\alpha \to \log \alpha - \Psi(\alpha)$ is strictly decreasing in α since it has derivative

$$\frac{1}{\alpha} - \Psi_1(\alpha) = -\sum_{i=1}^{\infty} \frac{1}{(\alpha+i)^2} < 0.$$

It can also be shown that for all $\alpha > 0$ it holds that

$$\frac{1}{2\alpha} < \log \alpha - \Psi(\alpha) < \frac{1}{\alpha}$$

and thus the first of the equations in (4.12) has a unique solution $\hat{\alpha}$ with $\alpha \in (0, \infty)$ if and only if $\overline{\log x} < \log \bar{x}$ which holds if and only if at least two of $x_1, \dots, x_n$ are different. Then $\hat{\beta}$ is determined from $\hat{\alpha}$ and $\bar{x}$ via the second equation in (4.12).

We note that although the MLE is a moment estimator it uses *a different moment function* than in Example 4.17. □

4.4 Exercises

Exercise 4.1. Let X and Y be independent and exponentially distributed with $\mathbf{E}(X) = \lambda$ og $\mathbf{E}(Y) = 3\lambda$. Consider the following two estimators of λ:

$$\hat{\lambda} = (3X + Y)/6, \quad \tilde{\lambda} = \sqrt{XY/3}.$$

In the following you may without proof use that $\Gamma(1.5) = \Gamma(0.5)/2 = \sqrt{\pi}/2$, where

$$\Gamma(y) = \int_0^\infty u^{y-1} e^{-u}\, du$$

is the gamma function.

a) Which of these estimators are unbiased estimators of λ?

b) Determine the variance for both of these estimators;

c) Compare the variances to the Cramér–Rao lower bound and comment on the result;

d) Which of the estimators has the smallest mean square error?

e) Confirm the findings above via a simulation experiment.

Exercise 4.2. The standard estimator

$$S_n^2 = \frac{1}{n-1} \sum_{i=1}^{n} (X_i - \bar{X})^2$$

in the normal distribution is distributed as $S_n^2 \sim \sigma^2 \chi^2(n-1)/(n-1)$ and thus unbiased for the variance σ^2, but not for the precision $\rho^2 = 1/\sigma^2$, nor for the standard deviation σ, as discussed in Example 4.3.

a) Find and compare the mean square errors of S_n and the MLE $\hat{\sigma}_n = S_n\sqrt{(n-1)/n}$ as estimates of the standard deviation σ.

b) Determine a constant a_n such that

$$\check{\rho}_n^2 = \frac{a_n}{S_n^2}$$

is an unbiased estimator of the precision ρ^2 for $n > 3$.

Exercise 4.3. Let $X_1, \ldots, X_n$ be independent with means $\mathbf{E}(X_i) = \mu + \beta_i$ and variances $\mathbf{V}(X_i) = \sigma_i^2$. Such a situation could, for example occur when X_i are estimators of μ obtained from independent sources and β_i is the *bias* of the estimator X_i.

We now consider pooling the estimators of μ into a common estimator by using a linear combination:

$$\tilde{\mu} = w_1 X_1 + w_2 X_2 + \cdots + w_n X_n.$$

a) If the estimators are unbiased, i.e. if $\beta_i = 0$ for all i, show that a linear combination $\tilde{\mu}$ as above is unbiased if and only if $\sum w_i = 1$;
b) In the case when $\beta_i = 0$ for all i, show that an unbiased linear combination has minimum variance when the weights w_i are inversely proportional to the variances σ_i^2;
c) Show that the variance of $\tilde{\mu}$ for optimal weights w_i is $\mathbf{V}(\tilde{\mu}) = 1/\sum_i \sigma_i^{-2}$;
d) Next, consider the case where the estimators may be biased so we could have $\beta_i \neq 0$. Find the mean square error of the optimal linear combination obtained above, and compare its behaviour as $n \to \infty$ in the biased and unbiased case, when $\sigma_i^2 = \sigma^2, i = 1, \ldots, n$.

Exercise 4.4. Let $X = (X_1, \ldots, X_n)$ be a sample of size n from the uniform distribution on the interval $(\mu - \delta, \mu + \delta)$ with density

$$f_{\mu,\delta}(x) = \frac{1}{2\delta}\mathbf{1}_{(\mu-\delta,\mu+\delta)}$$

with respect to standard Lebesgue measure on $\mathbb{R}$, where $\theta = (\delta, \mu) \in \Theta = \mathbb{R}_+ \times \mathbb{R}$ with δ and μ both unknown.

a) Determine the moment estimator of $\theta = (\delta, \mu)$ based on $t(x) = (x, x^2)$ and denote the estimator by $\tilde{\theta} = (\tilde{\mu}, \tilde{\delta})$;
b) Consider also the estimator $\hat{\theta}$ given by

$$\hat{\mu} = (X_{(1)} + X_{(n)})/2, \quad \hat{\delta} = (X_{(n)} - X_{(1)})/2;$$

and show that $\hat{\mu}$ is an unbiased estimator of μ;
c) Consider also the estimator

$$\check{\mu} = \text{med}(X_1, \ldots, X_n)$$

and show that also this is an unbiased estimator of μ. *Hint:* Use that $X_1, \ldots, X_n$ has the same distribution as $\mu + \delta U_1, \ldots \mu + \delta U_n$, where U_i is independent and identically uniformly distributed on $(-1, 1)$;

d) Compare the variances of $\tilde{\mu}$, $\hat{\mu}$, and $\check{\mu}$ by simulation;

e) Compare the mean square errors of $\tilde{\delta}$ and $\hat{\delta}$ by simulation.

Exercise 4.5. Let $X = (X_1, \ldots, X_n)$ be independent and Poisson distributed with

$$\mathbf{E}(X_j) = \lambda N_j$$

where $\lambda > 0$ is unknown and N_j are known constants. Models of this type arise, for example, in risk studies where N_j is the number of individuals at risk in group j, X_j the number of events, e.g. accidents or casualties, and λ is the risk rate.

a) Show that the above model defines a regular exponential family with canonical parameter $\theta = \log \lambda$;

b) Identify the canonical statistic and find its mean and variance;

c) Find the maximum likelihood estimate $\hat{\lambda}$ of λ;

d) Find the Fisher information for λ and the variance of $\hat{\lambda}$.

An alternative estimator is the *average rate*

$$\tilde{\lambda} = \frac{1}{n} \sum_{j=1}^{n} \frac{X_j}{N_j}.$$

e) Show that $\tilde{\lambda}$ is an unbiased estimator of λ and determine its variance;

f) Compare the variances of the estimators $\hat{\lambda}$ and $\tilde{\lambda}$ to the Cramér–Rao lower bound;

g) How are the findings above related to those in Exercise 4.3?

Exercise 4.6. Consider a sample $Z_1, \ldots, Z_n$ of independent and identically distributed random variables, where $Z_i = X_i - Y_i$ with X_i and Y_i independent and exponentially distributed with mean θ.

a) Determine a moment estimator $\tilde{\theta}$ of θ based on the statistic $t(z) = z^2$.

b) Show that if we also had observed both of X_i and Y_i, the MLE would be

$$\hat{\theta} = \frac{1}{2n} \sum_{i=1}^{n} (X_i + Y_i);$$

c) Compare the estimators $\tilde{\theta}$ and $\hat{\theta}$ via a suitable simulation study.

Exercise 4.7. Show that the estimator $\hat{\theta}$ under b) in Exercise 4.4 is a MLE.

Exercise 4.8. Consider the family of gamma distributions with identical parameters for shape and scale, i.e. with densities

$$f_\beta(x) = \frac{x^{\beta-1} e^{-x/\beta}}{\Gamma(\beta)\beta^\beta}, \quad \beta \in \mathbb{R}_+$$

with respect to standard Lebesgue measure on $\mathbb{R}_+$ as in Exercise 3.7. Show that the MLE of β is unique and well-defined as a solution to the score equation. *Hint:* Use that the *trigamma function* $\Psi_1(\beta) = D^2 \log \Gamma(\beta)$ is strictly positive on $\mathbb{R}_+$.

Exercise 4.9. Let X and Y be independent random variables with X Poisson distributed with mean λ and Y exponentially distributed with rate λ, where $\lambda > 0$ as in Exercise 3.8. Determine the MLE of λ and identify when it is well-defined.

Exercise 4.10. Let X and Y be independent and exponentially distributed random variables with $\mathbf{E}(X) = \beta$ and $\mathbf{E}(Y) = 1/\beta$ where $\beta > 0$ as in Exercise 3.9. Determine the MLE of β and identify when it is well-defined.

Chapter 5

Asymptotic Theory

We consider a sample $(X_1, \ldots, X_n)$ of n observations from a parametrized family of distributions $\mathcal{P} = \{P_\theta \mid \theta \in \Theta\}$ and an associated estimator $\hat{\theta}_n$:

$$\hat{\theta}_n = h_n(X_1, \ldots, X_n)$$

where $h_n : \mathcal{X}^n \to \Theta$ is measurable.

The method of maximum likelihood is an almost universally applicable method that yields marvellously good estimators in a large number of cases. In most of these cases, however, the maximum-likelihood estimator (MLE) can only be computed numerically and its exact distributional properties are difficult to obtain. Therefore, it is important to have good and practical ways to assess the distributional properties approximately.

It is common folklore that *under suitable and not very strong regularity conditions, it holds that for samples of sufficiently large size n, the MLE is approximately normally distributed with the inverse Fisher information as approximate covariance:*

$$\hat{\theta}_n \overset{\text{as}}{\sim} \mathcal{N}_k(\theta, i_n(\theta)^{-1}). \tag{5.1}$$

Here

$$i_n(\theta) = \mathbf{E}_\theta(-D^2\ell_n(\theta; X_1, \ldots, X_n)) = \mathbf{V}_\theta(S_n(X_1, \ldots, X_n; \theta)^\top)$$

where $S_n = D\ell_n$ is the score statistic and D represents differentiation with respect to the parameters $\theta \in \Theta \subseteq \mathbb{R}^k$.

If $X_1, \ldots, X_n$ are independent and identically distributed, it holds that $i_n(\theta) = ni(\theta)$, where $i(\theta)$ is the Fisher information in a single observation.

In other words, the MLE is approximately an unbiased estimator and since its approximate variance achieves the lower bound in the Cramér–Rao inequality, it is generally difficult, if not impossible, to find an estimator of higher asymptotic precision. This is usually expressed by saying that *MLE is asymptotically efficient*. We refrain from giving a precise definition of this concept and a corresponding proof.

Similarly, it is also common folklore that *if $\Theta_0 \subseteq \Theta$ represents a sufficiently nice hypothesis about the parameters of the distribution, the maximized-likelihood ratio statistic Λ_n satisfies*

$$\Lambda_n = -2\log\frac{\sup_{\theta\in\Theta_0} L(\theta; X_1, \ldots, X_n)}{\sup_{\theta\in\Theta} L(\theta; X_1, \ldots, X_n)} \overset{\text{as}}{\sim} \chi^2(m - d), \tag{5.2}$$

DOI: 10.1201/9781003272359-5

where $\chi^2(m-d)$ is the χ^2-distribution with degrees of freedom $m-d$, where m is the number of free parameters needed to describe Θ, and d is the number of free parameters needed to describe Θ_0. If $\Theta_0 = \{\theta\}$ consists of a single point, we let $d = 0$. In this chapter we shall rigorously establish the folklore as in (5.1) and in (5.2) for repeated samples in a number of cases, including curved exponential families. We begin by establishing some preliminary results and facts that we need for further developments.

5.1 Asymptotic consistency and normality

As we are interested in the behaviour of estimators when the sample size is large, we shall consider a *sequence* of estimators $(\hat{\theta}_n)_{n\in\mathbb{N}}$ based on samples $(X_1,\ldots,X_n), n \in \mathbb{N}$. As mentioned earlier, it is convenient to allow the estimator to hit points outside of Θ, as in the following simple example.

Example 5.1. Consider a sample $X = (X_1,\ldots,X_n)$ from the simple Poisson model

$$P_\lambda(X_i = x_i) = \frac{\lambda^{x_i}}{x_i!}e^{-\lambda}, \quad \lambda \in \Lambda = (0,\infty).$$

We have seen that a sensible estimator is

$$\hat{\lambda}_n = \bar{X}_n = (X_1 + \ldots + X_n)/n$$

but it may happen that $\hat{\lambda}_n = 0 \notin \Lambda$. However, this happens with very low probability if n is large and hence we should be able to ignore it for large n. □

This motivates the following definition:

Definition 5.2. A sequence $\hat{\theta}_n = h_n(X_1,\ldots,X_n), n \in \mathbb{N}$ of estimators is *asymptotically well-defined* if there is a sequence $A_n \subseteq \mathcal{X}^n$ such that $h_n(A_n) \subseteq \Theta$ and

$$\lim_{n\to\infty} P_\theta\{(X_1,\ldots,X_n) \in A_n\} = 1$$

for all $\theta \in \Theta$.

Clearly, we would not only like that our sequence of estimators is well-defined for large n, but also that it is close to the 'true' value of θ that has generated our sample. This concept is called asymptotic consistency. More precisely, we define:

Definition 5.3. A sequence $\hat{\theta}_n = h_n(X_1,\ldots,X_n), n \in \mathbb{N}$ of estimators is said to be *asymptotically consistent* if it is asymptotically well-defined and it holds that

$$\text{plim}_{n\to\infty}\, \hat{\theta}_n = \theta$$

with respect to P_θ for all $\theta \in \Theta$.

We may often omit the prefix 'asymptotically' in front of consistent when it is obvious from the context. In other words $\hat{\theta}_n, n \in \mathbb{N}$ is consistent if and only if for all $\theta \in \Theta$ and $\epsilon > 0$ it holds that

$$P_\theta\{\|\hat{\theta}_n - \theta\| > \epsilon\} \to 0 \text{ for } n \to \infty.$$

Results about consistency of estimators are therefore based on variants of the Law of Large Numbers.

Finally, we shall also be interested in quantifying how close the estimate is to the value of θ generating the data. To do this, we need the following.

Definition 5.4. A sequence $\hat{\theta}_n = h_n(X_1, \ldots, X_n), n \in \mathbb{N}$ of estimators is said to be *asymptotically normal* with *asymptotic mean* θ *and asymptotic variance* $\Sigma(\theta)/n$, if it is asymptotically well-defined and it holds that

$$\sqrt{n}(\hat{\theta}_n - \theta) \overset{\mathcal{D}}{\to} \mathcal{N}(0, \Sigma(\theta))$$

for all $\theta \in \Theta$.

Here $\overset{\mathcal{D}}{\to}$ denotes convergence in distribution, see Section A.2.2 for further details on that concept. Note in particular that an asymptotically normal estimator with asymptotic mean and variance as above is also *asymptotically consistent*; see Corollary A.17. As results on consistency were based on variants of the Law of Large Numbers, results on asymptotic normality are based on variants of the Central Limit Theorem.

5.2 Asymptotics of moment estimators

In this section we show that under suitable regularity conditions, moment estimators are consistent and asymptotically normal. More precisely, we have

Theorem 5.5. *Let $X_1, \ldots, X_n$ be independent and identically distributed observations from a parametrized family $\mathcal{P} = \{P_\theta \mid \theta \in \Theta\}$ with $\Theta \subseteq \mathbb{R}^k$ open. Let further $t : \mathcal{X} \mapsto \mathbb{R}^k$ be a statistic with moment function $m(\theta) = \mathbf{E}_\theta\{t(X)\}$, where m is smooth and injective with* $\det(Dm(\theta)) \neq 0$ *for all $\theta \in \Theta$. Then the moment estimator $\tilde{\theta}_n$ is consistent.*

Proof. We must first establish that the estimator is asymptotically well-defined. So let $X = (X_1, \ldots, X_n)$ and $\bar{T}_n = (t(X_1) + \cdots + t(X_n))/n$. The moment estimator is the solution to the equation

$$m(\theta) = \bar{T}_n.$$

Since m is injective, there is at most one point satisfying the moment equation. A solution exists if and only if $\bar{T}_n$ is in the image of m. The law of large numbers (Theorem A.9) ensures that

$$\bar{T}_n \overset{P}{\to} m(\theta) \text{ for } n \to \infty$$

and thus for any neighbourhood U of $m(\theta)$ it holds that

$$P_\theta\{\bar{T}_n \in U\} \to 1 \text{ for } n \to \infty.$$

Now the inverse function theorem (Theorem A.20) ensures that $m(\Theta)$ is open, and its inverse m^{-1} is well-defined and smooth in U so we conclude that $\tilde{\theta}_n$ is asymptotically well-defined. Since we have

$$\tilde{\theta}_n = m^{-1}(\bar{T}_n) \tag{5.3}$$

and m^{-1} is continuous in U, we find that

$$\text{plim}_{n\to\infty}\,\tilde{\theta}_n = m^{-1}(m(\theta)) = \theta,$$

as desired. □

We can now relatively easy establish conditions for the moment estimator to be asymptotically normal. More precisely, we have

Theorem 5.6. *Let $(X_1,\ldots,X_n)$ be independent and identically distributed observations from the family $\mathcal{P} = \{P_\theta \mid \theta \in \Theta\}$ with $\Theta \subseteq \mathbb{R}^k$ open. Let further $t : \mathcal{X} \mapsto \mathbb{R}^k$ have moment function $m(\theta) = \mathbf{E}_\theta\{t(X)\}$, where m is smooth and injective with* $\det(Dm(\theta)) \neq 0$. *Assume further that the moment statistic satisfies*

$$\mathbf{E}_\theta(\|t(X)\|_2^2) < \infty.$$

Then the moment estimator $\tilde{\theta}_n$ is consistent and asymptotically normal

$$\tilde{\theta}_n \overset{\text{as}}{\sim} \mathcal{N}_k(\theta, \Sigma(\theta)/n).$$

with asymptotic covariance

$$\Sigma(\theta)/n = Dm(\theta)^{-1}\mathbf{V}_\theta(t(X))Dm(\theta)^{-\top}/n.$$

Proof. From the Central Limit Theorem (Theorem A.16), we get directly that

$$\bar{T}_n = (t(X_1) + \cdots + t(X_n))/n \overset{\text{as}}{\sim} \mathcal{N}_k(m(\theta), \mathbf{V}_\theta(t(X))/n).$$

Now we have $\tilde{\theta}_n = m^{-1}(\bar{T}_n)$, and the inverse function theorem (Theorem A.20) ensures that m^{-1} is smooth with derivative satisfying

$$Dm^{-1}(m(\theta)) = Dm(\theta)^{-1}.$$

Thus the delta method (Theorem A.19) yields

$$\begin{aligned}\tilde{\theta}_n &\overset{\text{as}}{\sim} \mathcal{N}_k\left(m^{-1}(m(\theta)), Dm^{-1}(m(\theta))\mathbf{V}_\theta(t(X))Dm^{-1}(m(\theta))/n\right)\\ &= \mathcal{N}_k\left(\theta, Dm(\theta)^{-1}\mathbf{V}_\theta(t(X))Dm(\theta)^{-\top}/n\right).\end{aligned}$$

This concludes the proof. □

Example 5.7. [Moment estimator asymptotics in simple normal model] In Example 4.16 we found three moment estimators $\hat{\xi}_{in}, i = 1,2,3$ for an unknown mean based on a sample $X_1,\ldots,X_n$ from a normal distribution $\mathcal{N}(\xi, 1)$ with known variance equal to one. These were based on the statistics

$$t_1(x) = x, \quad t_2(x) = \mathbf{1}_{(0,\infty)}(x), \quad t_3(x) = x^3$$

with corresponding moment functions

$$m_1(\xi) = \xi, \quad m_2(\xi) = P_\xi(X > 0) = \Phi(\xi), \quad m_3(\xi) = 3\xi + \xi^3.$$

Since $\mathbf{E}_\xi(t_i(X)^2) < \infty$ for all these statistics and the moment functions are smooth with non-zero derivatives

$$m_1'(\xi) = 1, \quad m_2'(\xi) = \frac{1}{\sqrt{2\pi}} e^{-\xi^2/2}, \quad m_3'(\xi) = 3(1+\xi^2),$$

we conclude from Theorem 5.6 that the corresponding moment estimators are all asymptotically normally distributed with the correct asymptotic mean ξ.

To calculate their asymptotic variance, we find the variance of the first two statistics

$$\mathbf{V}_\xi(t_1(X)) = 1, \quad \mathbf{V}_\xi(t_2(X)) = \Phi(\xi)(1-\Phi(\xi)),$$

where we have used that $\mathbf{1}_{(0,\infty)}(X)$ follows a Bernoulli distribution with success probability $\Phi(\xi)$. Further, for the third statistic, we get

$$\begin{aligned} \mathbf{V}_\xi(t_3(X)) &= \mathbf{E}((Z+\xi)^6) - (\mathbf{E}_\xi(X^3))^2 \\ &= 15 + 45\xi^2 + 15\xi^4 + \xi^6 - (3\xi + \xi^3)^2 = 15 + 36\xi^2 + 9\xi^4. \end{aligned}$$

It follows that the asymptotic variances are $\sigma_i^2(\xi)/n$, where

$$\sigma_1^2(\xi) = 1, \quad \sigma_2^2(\xi) = 2\pi\Phi(\xi)(1-\Phi(\xi))e^{\xi^2}, \quad \sigma_3^2(\xi) = \frac{5 + 12\xi^2 + 3\xi^4}{3(1+\xi^2)^2}.$$

We note in particular that the last two are always larger than $\sigma_1^2(\xi) = 1$ and that $\sigma_2^2(\xi)$ may be huge for $|\xi|$ large. For $\xi = 2$ as used in the simulations behind Fig. 4.4 we get

$$\sigma_1^2(2) = 1, \quad \sigma_2^2(2) = 7.83, \quad \sigma_3^2(2) = 101/75 = 1.35$$

which conforms well with the plots, indicating that $\hat{\xi}_3$ in this case is only marginally worse than $\hat{\xi}_1$, whereas $\hat{\xi}_2$ is very bad. □

Example 5.8. [Gamma model: moment estimator asymptotics] For the moment estimator in Example 4.17, we get

$$Dm(\alpha, \beta) = \begin{pmatrix} \beta & \alpha \\ (2\alpha+1)\beta^2 & 2\alpha(\alpha+1)\beta \end{pmatrix}$$

and thus

$$Dm(\alpha, \beta)^{-1} = \frac{1}{\alpha\beta^2} \begin{pmatrix} 2\alpha(\alpha+1)\beta & -\alpha \\ -(2\alpha+1)\beta^2 & \beta \end{pmatrix}.$$

Further we have

$$\mathbf{V}_{\alpha,\beta}(t(X)) = \begin{pmatrix} \alpha\beta^2 & 2\alpha(\alpha+1)\beta^3 \\ 2\alpha(\alpha+1)\beta^3 & 2\alpha(\alpha+1)(2\alpha+3)\beta^4 \end{pmatrix}$$

yielding the asymptotic covariance of the moment estimator

$$\begin{aligned}\mathbf{V}_{\alpha,\beta}\left\{\begin{pmatrix}\hat{\alpha}_{\text{mom}}\\ \hat{\beta}_{\text{mom}}\end{pmatrix}\right\} &= \frac{1}{n}Dm(\alpha,\beta)^{-1}\mathbf{V}_{\alpha,\beta}(t(X))Dm(\alpha,\beta)^{-\top}\\ &= \frac{2(\alpha+1)}{n}\begin{pmatrix}\alpha & -\beta\\ -\beta & \frac{\alpha+3}{\alpha(\alpha+1)}\beta^2\end{pmatrix}.\end{aligned} \quad (5.4)$$

There are some calculations involved in getting to this... □

5.3 Asymptotics in regular exponential families

5.3.1 Asymptotic consistency of maximum likelihood

In regular exponential families we can now easily establish the asymptotic properties of the MLE using that in this case, MLE are moment estimators. Indeed we have:

Theorem 5.9. *Let $(X_1,\ldots,X_n)$ be a random sample from a minimal and regular exponential family with canonical parameter $\theta \in \Theta$. Then the MLE $\hat{\theta}_n$ is asymptotically consistent.*

Proof. This follows from Theorem 5.5 since the MLE is a moment estimator, and Theorem 3.17 states that $\tau(\theta)$ is injective and smooth and

$$D\tau(\theta) = D^2\psi(\theta) = \kappa(\theta) = \mathbf{V}_\theta(t(X))$$

is positive definite and hence $\det(D\tau(\theta)) > 0$ for all $\theta \in \Theta$. □

We should like to point out that the consistency of the MLE has only been shown for samples that are independent and identically distributed and it may fail, and badly so, if the situation is different, as the following example illustrates.

Example 5.10. [Double measurements] Consider the determination of the precision of a measuring instrument as in Example 2.30 and recall that we had independent observations

$$X_i, Y_i \sim \mathcal{N}(\xi_i, \sigma^2),\ i = 1,\ldots,n$$

where $\xi_i \in \mathbb{R}$ is characteristic for the ith unit and σ^2 is the precision of the instrument. Since this is a linear normal model, we readily conclude that the MLE for ξ, σ^2 is determined as

$$\hat{\xi}_i = \frac{X_i + Y_i}{2}, \quad i = 1,\ldots n;$$

$$\hat{\sigma}_n^2 = \frac{\sum_i (X_i - \hat{\xi}_i)^2 + \sum_i (Y_i - \hat{\xi}_i)^2}{2n} = \frac{\sum_i (X_i - Y_i)^2}{4n}.$$

These estimates are not consistent: in fact, $\hat{\xi}_i$ will for all i and n have variance $\sigma^2/2$ and even the maximum likelihood estimate of σ^2 goes astray since we have that $\mathbf{V}(X_i - Y_i) = 2\sigma^2$ and we thus get

$$\operatorname{plim}_{n\to\infty} \hat{\sigma}_n^2 = \lim_{n\to\infty} \frac{2n\sigma^2}{4n} = \frac{\sigma^2}{2}$$

which is only half of what it should be. This is one among other motivations for dividing the sum of squared deviations with the degrees of freedom (here n) instead of the number of observations (here $2n$) so as to achieve a consistent estimator. □

5.3.2 *Asymptotic normality of maximum likelihood*

The MLE in a regular exponential family is also asymptotically normal and its asymptotic variance achieves the Cramér–Rao lower bound. More precisely, we have

Theorem 5.11. *Let $(X_1, \dots, X_n)$ be a sample from a minimal and regular exponential family with canonical parameter $\theta \in \Theta$. Then the MLE $\hat{\theta}_n$ is asymptotically normally distributed with the inverse Fisher information as its asymptotic covariance matrix:*

$$\hat{\theta}_n \overset{\text{as}}{\sim} \mathcal{N}_k(\theta, i_n(\theta)^{-1}) = \mathcal{N}_k(\theta, \kappa(\theta)^{-1}/n).$$

Proof. Since the MLE is a moment estimator and $t(X)$ has moments of any order by Theorem 3.8, we can use Theorem 5.6 to conclude that the MLE is asymptotically normal. Recalling from Theorem 3.17 that

$$D\tau(\theta) = D^2\psi(\theta) = \kappa(\theta),$$

we find the asymptotic variance to be

$$\Sigma(\theta)/n = D\tau(\theta)^{-1}\kappa(\theta)D\tau(\theta)^{-T}/n = \kappa(\theta)^{-1}\kappa(\theta)\kappa(\theta)^{-1}/n = i_n(\theta)^{-1},$$

where $i_n(\theta) = ni(\theta)$ is the Fisher information in a sample of size n. This concludes the proof. □

Note that if we wish to use Theorem 5.11, it is a problem that we do not know θ and therefore cannot use the asymptotic variance $i_n(\theta)^{-1}$ to assess the accuracy of $\hat{\theta}_n$. The standard way out is to plug-in $\hat{\theta}_n$ for the unknown value and write

$$\hat{\theta}_n \overset{\text{as}}{\sim} \mathcal{N}_k(\theta, i_n(\hat{\theta}_n)^{-1}).$$

Whereas this does not immediately make sense—as the right-hand side now is random—it has the following formal meaning:

Corollary 5.12. *Let $X_1, \dots, X_n$ be a random sample from a minimal and regular exponential family with canonical parameter $\theta \in \Theta$. It then holds that*

$$\sqrt{ni(\hat{\theta}_n)}(\hat{\theta}_n - \theta) \overset{\mathcal{D}}{\to} \mathcal{N}_k(0, I_k) \text{ for } n \to \infty,$$

where $A = \sqrt{ni(\hat{\theta}_n)}$ is the unique positive definite matrix A satisfying $A^2 = ni(\hat{\theta}_n)$.

Proof. Since $\hat{\theta}_n$ is consistent and $A \to \sqrt{A}$ is continuous (also as a matrix function) it holds that $\sqrt{i(\hat{\theta}_n)} \overset{P}{\to} \sqrt{i(\theta)}$. Corollary A.12 in combination with Theorem 5.11 yields the result. □

Note that for large n the approximate variance of the MLE tends to zero at the rate of n^{-1} and scaling the deviation from the true value by $\sqrt{n}$ yields convergence in distribution. This is not necessarily the case for non-exponential models as the next example shows.

Example 5.13. [MLE in uniform model] We found in Example 1.18 that the MLE of the unknown parameter θ in the uniform distribution on the interval $(0, \theta)$ was

$$\hat{\theta}_n = \max(X_1, \dots, X_n),$$

and in Example 4.2 we found that its variance was tending to zero at the speed of n^{-2} i.e. much faster than in the regular case.

This is an example where the standard asymptotics fails and the MLE is *not* asymptotically normally distributed. Indeed, for large n the distribution of the scaled deviation from the true value $n(\theta - \hat{\theta}_n)$ is approximately exponential; more precisely, it holds that

$$n(\theta - \hat{\theta}_n) \xrightarrow{\mathcal{D}} \mathcal{E}_\theta \text{ for } n \to \infty$$

where $\mathcal{E}_\theta$ is the exponential distribution with mean θ. This may be seen in the following way. Take any $t > 0$; then we have

$$\begin{aligned}
\lim_{n\to\infty} P_\theta\{n(\theta - \hat{\theta}_n) > t\} &= \lim_{n\to\infty} P_\theta\left\{\hat{\theta}_n < \theta - \frac{t}{n}\right\} \\
&= \lim_{n\to\infty} \prod_{i=1}^{n} P_\theta\left\{X_i < \theta - \frac{t}{n}\right\} \\
&= \lim_{n\to\infty} \frac{(\theta - t/n)^n}{\theta^n} = \lim_{n\to\infty}\left(1 - \frac{t}{n\theta}\right)^n = e^{-t/\theta}
\end{aligned}$$

which is the upper tail of the exponential distribution function.

Note that here we must scale the deviation with n rather than $\sqrt{n}$ to obtain convergence in distribution. The approximate exponential shape of the distribution is also clearly visible in Fig. 4.1, even for $n = 10$. □

We note that an appropriate asymptotic result is valid also in any other smooth reparametrization of the exponential family.

Corollary 5.14. *Let $(X_1, \dots, X_n)$ be a random sample from a minimal and regular exponential family with canonical parameter $\theta \in \Theta$ and let $\lambda = \gamma(\theta)$ represent a reparametrization where $\gamma : \Theta \to \mathbb{R}^k$ is injective and smooth with a regular Jacobian. Then the MLE $\hat{\lambda}_n$ is asymptotically normally distributed with the inverse information as its asymptotic covariance:*

$$\hat{\lambda}_n \overset{\text{as}}{\sim} \mathcal{N}_k(\lambda, \tilde{i}_n(\lambda)^{-1}) = \mathcal{N}_k(\lambda, \tilde{i}(\lambda)^{-1}/n)$$

where $\tilde{i}(\lambda)$ is the Fisher information about λ in a single observation.

Proof. We first note that if $\lambda = \gamma(\theta)$ we have from Theorem 1.30

$$i(\theta) = D\gamma(\theta)^\top\, \tilde{i}(\lambda)\, D\gamma(\theta),$$

and thus

$$i(\theta)^{-1} = (D\gamma(\theta))^{-1}\,\tilde{i}(\lambda)^{-1}\,(D\gamma(\theta)^{\top})^{-1}.$$

whence

$$\tilde{i}(\lambda)^{-1} = D\gamma(\theta)\,i(\theta)^{-1}\,D\gamma(\theta)^{\top}.$$

Using the delta method on $\hat{\lambda}_n = \gamma(\hat{\theta}_n)$ in combination with Theorem 5.11, now yields

$$\hat{\lambda}_n \overset{\text{as}}{\sim} \mathcal{N}_k\left(\lambda, D\gamma(\theta)i(\theta)^{-1}D\gamma(\theta)^{\top}/n\right) = \mathcal{N}_k(\lambda, \tilde{i}(\lambda)^{-1}/n),$$

as required. □

Remark 5.15. *We shall later—in Theorem 5.29—establish this result for any curved exponential family.*

Example 5.16. [The gamma family] Consider again the family of gamma distributions as in Examples 4.17 and 4.26. Recall that this may be represented as a minimal and regular exponential family of dimension 2, with canonical parameters

$$\theta = (\theta_1, \theta_2)^{\top} = (\alpha, 1/\beta)^{\top},$$

canonical parameter space $\Theta = (0,\infty)^2$, canonical statistic $t(x) = (\log x, -x)^{\top}$, cumulant function

$$\psi(\theta) = \log c(\theta) = \log\Gamma(\theta_1) - \theta_1\log(\theta_2),$$

and base measure $\mu(dx) = x^{-1}\,dx$ on $(0,\infty)$. For the mean we differentiate the cumulant function

$$\tau(\theta) = \mathbf{E}_\theta\left\{\begin{pmatrix}\log X\\ -X\end{pmatrix}\right\} = \begin{pmatrix}\Psi(\theta_1) - \log(\theta_2)\\ -\theta_1/\theta_2\end{pmatrix} = \begin{pmatrix}\Psi(\alpha) + \log(\beta)\\ -\alpha\beta\end{pmatrix},$$

where $\Psi(\alpha) = D\log\Gamma(\alpha)$ is known as the *digamma function*. For the covariance matrix, we get by further differentiation

$$\mathbf{V}_\theta(t(X)) = \kappa(\theta) = \begin{pmatrix}\Psi_1(\theta_1) & -1/\theta_2\\ -1/\theta_2 & \theta_1/\theta_2^2\end{pmatrix} = \begin{pmatrix}\Psi_1(\alpha) & -\beta\\ -\beta & \alpha\beta^2\end{pmatrix}, \tag{5.5}$$

where now $\Psi_1(\alpha) = D^2\log\Gamma(\alpha) = \Psi'(\alpha)$ is the *trigamma function* satisfying

$$\Psi_1(z) = \sum_{i=0}^{\infty}\frac{1}{(z+i)^2} \quad \text{for } z \in \mathbb{C}\setminus\{0,-1,-2,\ldots\}.$$

The likelihood equations for (α,β) were obtained by equating the observed canonical statistics to their expectations as in (4.12). Expressing this in the canonical parameters, we get

$$\hat{\theta}_1 = \hat{\alpha}, \quad \hat{\theta}_2 = \hat{\alpha}/\bar{x}.$$

The asymptotic variance of this estimator is thus from (5.5)

$$i_n(\theta)^{-1} = \frac{1}{n} i(\theta)^{-1} = \frac{1}{n}\kappa(\theta)^{-1} = \frac{1}{n(\theta_1 \Psi_1(\theta_1) - 1)} \begin{pmatrix} \theta_1 & \theta_2 \\ \theta_2 & \Psi_1(\theta_1)\theta_2^2 \end{pmatrix},$$

which yields the full asymptotic distribution of $(\hat{\theta}_1, \hat{\theta}_2)$.

If we parametrize the gamma family in terms of (α, β), we have

$$\theta = \phi(\alpha, \beta) = (\alpha, 1/\beta)^\top$$

where $\phi : (0, \infty)^2 \to (0, \infty)^2$ is a smooth and injective homeomorphism with Jacobian

$$J(\alpha, \beta) = \begin{pmatrix} 1 & 0 \\ 0 & -\beta^{-2} \end{pmatrix}$$

which has full rank 2 for all $\beta > 0$

We can of course also determine the information matrix and the asymptotic distribution directly in terms of these. The information function is

$$I(x; \alpha, \beta) = -D^2\ell(\alpha, \beta; x) = \begin{pmatrix} \Psi_1(\alpha) & \frac{1}{\beta} \\ \frac{1}{\beta} & \frac{2x}{\beta^3} - \frac{\alpha}{\beta^2} \end{pmatrix}$$

and since $\mathbf{E}_{\alpha,\beta}(X) = \alpha\beta$ we get

$$i(\alpha, \beta) = \begin{pmatrix} \Psi_1(\alpha) & \frac{1}{\beta} \\ \frac{1}{\beta} & \frac{\alpha}{\beta^2} \end{pmatrix}$$

which has inverse

$$i(\alpha, \beta)^{-1} = \frac{1}{\alpha\Psi_1(\alpha) - 1} \begin{pmatrix} \alpha & -\beta \\ -\beta & \Psi_1(\alpha)\beta^2 \end{pmatrix}$$

yielding the asymptotic variance of $(\hat{\alpha}, \hat{\beta})$ when dividing by the number of observations n. We may compare this with the asymptotic variance (5.4) of the moment estimator:

$$2(\alpha + 1) \begin{pmatrix} \alpha & -\beta \\ -\beta & \frac{\alpha+3}{\alpha(\alpha+1)}\beta^2 \end{pmatrix}.$$

This is larger. Since it holds approximately that

$$\Psi_1(x) = \frac{1}{x} + \frac{1}{2x^2} + \cdots$$

we get

$$\frac{1}{\alpha\Psi_1(\alpha) - 1} \sim 2\alpha < 2(\alpha + 1)$$

and

$$\frac{2(\alpha + 3)(\alpha\Psi_1(\alpha) - 1)}{\alpha\Psi_1(\alpha)} \sim \frac{4\alpha(\alpha + 3)}{2\alpha + 1} < 1.$$

It demands a little more calculation to see that the difference between the covariance matrices is actually positive definite. □

5.3.3 *Likelihood ratios and quadratic forms*

As a direct consequence of Theorem 5.11 and Corollary 5.14 and basic properties of the multivariate normal distribution, we have a number of results concerning the asymptotic distribution of likelihood ratios and various quadratic forms.

Theorem 5.17. *Consider a random sample $Y_n = (X_1, \ldots, X_n)$ from a regular and minimally represented exponential family with $\Theta \subseteq \mathbb{R}^k$. Then the log-likelihood statistic $\Lambda_n = \Lambda_n(X_1, \ldots, X_n, \theta)$ satisfies*

$$\Lambda_n = 2\left(\ell_n(\hat{\theta}_n) - \ell_n(\theta)\right) \xrightarrow{\mathcal{D}} \chi^2(k)$$

with respect to P_θ. Here $\chi^2(k)$ is the χ^2-distribution with k degrees of freedom. Further, the log-likelihood ratio statistic is asymptotically equivalent to the statistics

$$\tilde{W}_n = n(\hat{\theta}_n - \theta)^\top i(\theta)(\hat{\theta}_n - \theta), \tag{5.6}$$

$$W_n = n(\hat{\theta}_n - \theta)^\top i(\hat{\theta}_n)(\hat{\theta}_n - \theta) \tag{5.7}$$

whereby these have the same asymptotic χ^2-distribution.

Note that here and in the following, we shall generally suppress the dependence of Λ_n on data and parameters to simplify notation.

Proof. First note that if $\lambda = \phi(\theta)$ is a smooth and injective reparametrization of the model, the log-likelihood ratio is invariant under reparametrization, since the likelihood itself is. We can therefore without loss of generality assume that we work in the canonical parametrization. We use Taylor's formula with remainder term in integral form (Theorem A.22) on the log-likelihood function $\ell_n(\theta)$ to obtain

$$\ell_n(\theta) = \ell_n(\hat{\theta}_n) + D\ell_n(\hat{\theta}_n)(\theta - \hat{\theta}_n) + \frac{n}{2}(\theta - \hat{\theta}_n)^\top K(\theta, \hat{\theta}_n)(\theta - \hat{\theta}_n),$$

with

$$\begin{aligned} K(\theta, \hat{\theta}_n) &= \frac{2}{n}\int_0^1 (1-t)D^2\ell_n(\theta + t(\hat{\theta}_n - \theta))\,dt \\ &= 2\int_0^1 (t-1)\kappa(\theta + t(\hat{\theta}_n - \theta))\,dt. \end{aligned}$$

We also note that $K(\cdot, \cdot)$ is continuous and $K(\theta, \theta) = \kappa(\theta) = i(\theta)$. Since $\hat{\theta}_n$ satisfies the score equation, we have $D\ell_n(\hat{\theta}_n) = 0$. We thus get

$$\Lambda_n = 2(\ell_n(\hat{\theta}_n) - \ell_n(\theta)) = n(\hat{\theta}_n - \theta)^\top K(\theta, \hat{\theta}_n)(\hat{\theta}_n - \theta).$$

Now $\hat{\theta}_n$ is consistent so $\hat{\theta}_n \xrightarrow{P} \theta$ and by continuity we also have $K(\theta, \hat{\theta}_n) \xrightarrow{P} i(\theta)$. Since $\hat{\theta}_n \overset{\text{as}}{\sim} \mathcal{N}_k(\theta, i(\theta)^{-1}/n)$ we conclude that

$$\tilde{W}_n = n(\hat{\theta}_n - \theta)^\top i(\theta)(\hat{\theta}_n - \theta) \xrightarrow{\mathcal{D}} \chi^2(k)$$

which is (5.6). The asymptotic equivalences now follow from Lemma A.18. □

Remark 5.18. *Most texts would use the Lagrange form of Taylor's theorem (Theorem A.23) in the above and similar proofs, leading to a remainder term with the quadratic form* $-D^2\ell(\theta^*)$ *instead of* $K(\theta, \hat{\theta}_n)$*, where* θ^* *is between* θ *and* $\hat{\theta}_n$*. Using the integral form of the remainder here and in subsequent proofs, we avoid discussing whether* θ^* *can be chosen as a measurable function of* $x = (x_1, \ldots, x_n)$*.*

We shall later—in Theorem 5.35—show a slightly more general version of Theorem 5.17, for curved exponential families. The proof is analogous, but we need to take more care, as the information function I_n is more difficult to control.

Remark 5.19. *The statistic* W_n *in (5.7) is known as the* Wald statistic. *We shall use the term Wald statistic for any statistic that is a quadratic form on the deviation of the estimator from the true value with the quadratic form being a consistent estimate of the asymptotic variance of the estimator, i.e. a statistic of the form*

$$\check{W}_n = W_n = n(\hat{\theta}_n - \theta)^\top \widehat{\Sigma_n(\theta)}^{-1} (\hat{\theta}_n - \theta) \tag{5.8}$$

where $\hat{\theta}_n \overset{\text{as}}{\sim} \mathcal{N}(\theta, \Sigma(\theta)/n)$ *and* $\operatorname{plim}_{n\to\infty} \widehat{\Sigma_n(\theta)} = \Sigma(\theta)$*. It follows from Lemma A.18 that any such statistic has the same asymptotic* χ^2*-distribution.*

Note also that whereas the Wald statistics are not equivariant under reparametrization, the quadratic score is. Indeed, if we are parametrizing the exponential family with the mean value parameter $\eta = \tau(\theta) = \mathbf{E}_\theta\{t(X)\}$, we have $\hat{\eta}_n = \bar{T}_n$ and hence the Wald statistic in this parametrization is identical to the quadratic score. The asymptotic distribution of the quadratic score was already established in Corollary 1.35. We formulate this as its own corollary:

Corollary 5.20. *Consider a sample* $Y_n = (X_1, \ldots, X_n)$ *from a regular and minimally represented exponential family of dimension* k *and let* $\eta = \tau(\theta) = \mathbf{E}_\theta\{t(X)\}$ *be the corresponding mean value parameter. Then the Wald statistic and quadratic score for* θ *are identical and*

$$W_n = Q_n = n(\hat{\eta}_n - \eta)^\top \mathbf{V}_\eta(t(X))^{-1}(\hat{\eta}_n - \eta) \xrightarrow{\mathcal{D}}_{n\to\infty} \chi^2(k). \tag{5.9}$$

Remark 5.21. *Note that since the log-likelihood ratio and quadratic score statistics are equivariant, it follows that these are also asymptotically equivalent in any smooth and injective reparametrization.*

The results may be further strengthened when submodels are considered. So we shall investigate the situation where $\tilde{\mathcal{P}} = \{P_\theta, \theta \in \Theta_0\}$ is an affine submodel of $\mathcal{P}$; we may without loss of generality assume that Θ_0 is represented as the image of an affine map as in (3.9), i.e. we have $\Theta_0 = AB + a$ where A is an injective and affine map and B is an open and convex subset of $\mathbb{R}^d$. We recall from Theorem 3.16 that $\tilde{\mathcal{P}}$ is again a regular exponential family with canonical statistic $\tilde{t}(x) = A^\top t(x)$ and the maximum-likelihood estimate under Θ_0 is then determined as the moment estimator, i.e. as the unique solution to the equation

$$\mathbf{E}_{A\hat{\beta}+a}(A^\top t(X)) = A^\top \tau(A\hat{\beta} + a) = A^\top t(X).$$

Letting now $\hat{\eta} = \tau(\hat{\theta})$ and $\hat{\hat{\eta}} = \tau(A\hat{\beta} + a)$ denote the estimates of the mean value parameter in the two cases, we have $\hat{\eta} = t(X)$ and thus

$$A^\top(\hat{\eta} - \hat{\hat{\eta}}) = 0. \tag{5.10}$$

To obtain a geometric interpretation of this equation, we note that by composite differentiation, we have

$$\frac{\partial \eta}{\partial \beta} = \frac{\partial \tau(A\beta + a)}{\partial \beta} = D\tau(A\beta + a)A = \kappa(A\beta + a)A = \Sigma_\beta A, \tag{5.11}$$

where we have let

$$\Sigma_\beta = \mathbf{V}_{A\beta+a}(t(X)).$$

Hence, we have for all $\beta \in B$ that

$$A = \Sigma_\beta^{-1}\frac{\partial \eta}{\partial \beta}.$$

Letting $\beta = \hat{\beta}$ we may rewrite equation (5.10) as

$$\left.\frac{\partial \eta}{\partial \beta}\right|_{\beta=\hat{\beta}}^\top \Sigma_{\hat{\beta}}^{-1}(\hat{\eta} - \hat{\hat{\eta}}) = \frac{\partial \eta}{\partial \hat{\beta}}^\top \Sigma_{\hat{\beta}}^{-1}(\hat{\eta} - \hat{\hat{\eta}}) = 0$$

i.e. *the residual $\hat{\eta} - \hat{\hat{\eta}}$ is orthogonal to the space spanned by the columns of $\partial\eta/\partial\beta$ at $\beta = \hat{\beta}$ with respect to the inner product determined by the inverse covariance $\Sigma_{\hat{\beta}}^{-1}$ of the canonical statistic.* We note that the space spanned by the partial derivatives is indeed the *space of tangents to the curved surface $\tau(AB + a)$.*

Letting now H_β denote the hat-matrix for this projection, we have from Proposition 2.21 that

$$H_\beta = \frac{\partial \eta}{\partial \beta}\left(\frac{\partial \eta}{\partial \beta}^\top \Sigma_\beta^{-1}\frac{\partial \eta}{\partial \beta}\right)^{-1}\frac{\partial \eta}{\partial \beta}^\top \Sigma_\beta^{-1} = \Sigma_\beta A\left(A^\top \Sigma_\beta A\right)^{-1}A^\top \tag{5.12}$$

where we have used (5.11). This fact now leads to the following decomposition theorem in analogy with Theorem 2.24 for the multivariate normal distribution:

Theorem 5.22. *Let $\tilde{\mathcal{P}} = \{P_\theta \mid \theta \in \Theta_0\}$ be a d-dimensional regular exponential subfamily of a regular exponential family $\mathcal{P} = \{P_\theta \mid \theta \in \Theta\}$ of dimension m and let $X_1, \ldots, X_n, \ldots$ be independent and identically distributed according to some $P_{A\beta+a} \in \tilde{\mathcal{P}}$.*

Further, let $\eta = \tau(A\beta + a)$, and let $\hat{\eta}_n = \bar{T}_n$ and $\hat{\hat{\eta}}_n$ denote the maximum likelihood estimates of η under $\mathcal{P}$ and $\tilde{\mathcal{P}}$ respectively. If H_β denotes the matrix for the projection onto the space spanned by the columns of $\partial\eta/\partial\beta$ with respect to the inner product determined by the inverse covariance Σ_β^{-1}, it holds that

$$\sqrt{n}\begin{pmatrix}\hat{\eta}_n - \hat{\hat{\eta}}_n \\ \hat{\hat{\eta}}_n - \eta\end{pmatrix} \overset{\text{as}}{=} \sqrt{n}\begin{pmatrix}(I - H_{\hat{\beta}_n})\bar{T}_n \\ H_{\hat{\beta}_n}\bar{T}_n - \eta\end{pmatrix} \overset{\mathcal{D}}{\to} \begin{pmatrix}(I - H_\beta)Z \\ H_\beta Z\end{pmatrix},$$

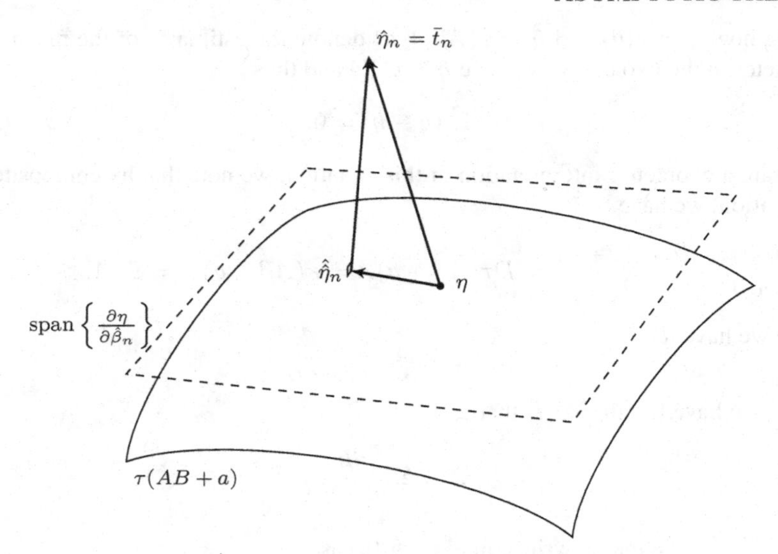

Figure 5.1 – Visualization of Theorem 5.22. The image under τ of the affine subspace $AB + a$ becomes a curved space, with tangent space (indicated by dashed lines) spanned by the columns of $\partial\eta/\partial\beta$. The residual $\hat{\eta}_n - \hat{\hat{\eta}}_n$ is orthogonal at $\hat{\hat{\eta}}_n$ to the tangent space with respect to the inner product determined by $\Sigma^{-1}_{\hat{\beta}_n}$.

where $Z \sim \mathcal{N}_k(0, \Sigma_\beta)$. It holds in particular that $\hat{\eta}_n - \hat{\hat{\eta}}_n$ and $\hat{\hat{\eta}}_n$ are asymptotically independent and normally distributed on appropriate subspaces of $\mathbb{R}^k$ with concentrations Σ^{-1}_β.

Proof. The statement about asymptotic equivalence follows directly from the delta method and the convergence in distribution from Corollary A.12. The last part of the conclusion follows from the normal decomposition theorem, Theorem 2.24. □

The situation is illustrated in Figure 5.1. We have a convenient corollary:

Corollary 5.23. *Under the assumptions in Theorem 5.22, the Wald statistics satisfy*

$$W_n = n(\hat{\eta}_n - \hat{\hat{\eta}}_n)^\top \hat{\hat{\Sigma}}_n^{-1}(\hat{\eta}_n - \hat{\hat{\eta}}_n) \xrightarrow{\mathcal{D}} \chi^2(m-d)$$
$$\tilde{W}_n = n(\hat{\eta}_n - \hat{\hat{\eta}}_n)^\top \hat{\Sigma}_n^{-1}(\hat{\eta}_n - \hat{\hat{\eta}}_n) \xrightarrow{\mathcal{D}} \chi^2(m-d),$$

where $\hat{\hat{\Sigma}}_n = \kappa(A\hat{\beta}_n + a)$, and $\hat{\Sigma}_n = \kappa(\hat{\theta}_n)$. Further, $W_n \overset{\text{as}}{=} \tilde{W}_n$.

Proof. This follows from Theorem 5.22 in combination with Corollary A.12 and Lemma A.18, since the estimates of the variances are consistent. □

Of these statistics, we would generally prefer W_n, since estimating the variance under the hypothesis yields a better estimate of the variance. However, we may occasionally wish to consider several submodels at the same time, for example when considering which regression coefficients are zero in a multiple regression with many explanatory variables. Then $\tilde{W}_n$ has the advantage that we do not have to calculate a new estimate of the covariance matrix for every submodel, but may use a single covariance estimate for all of them.

The likelihood ratio statistic at the submodel behaves in a way similar to the Wald statistics, as we shall now show.

Theorem 5.24. *Let $\tilde{\mathcal{P}} = \{P_\theta \mid \theta \in \Theta_0\}$ be a d-dimensional regular exponential subfamily of a regular exponential family $\mathcal{P} = \{P_\theta \mid \theta \in \Theta\}$ of dimension m and let $X_1, \ldots, X_n, \ldots$ be independent and identically distributed according to some $P_{A\beta+a} \in \tilde{\mathcal{P}}$. Then the log-likelihood ratio statistic Λ_n for the subfamily $\tilde{\mathcal{P}}$ satisfies*

$$\Lambda_n = 2\left(\ell_n(\hat{\theta}_n) - \ell_n(\hat{\hat{\theta}}_n)\right) \xrightarrow{\mathcal{D}} \chi^2(m-d)$$

with respect to any $P_0 \in \tilde{\mathcal{P}}$.

Proof. As before, the proof is based on a Taylor expansion. We let $\hat{\hat{\theta}}_n = A\hat{\beta}_n + a$ and have

$$\Lambda_n = 2\left(\ell_n(\hat{\theta}_n) - \ell_n(\hat{\hat{\theta}}_n)\right) = 2n\left(\psi(\hat{\hat{\theta}}_n) - \psi(\hat{\theta}_n)\right) + 2n(\hat{\theta}_n - \hat{\hat{\theta}}_n)^\top \bar{t}_n.$$

Next we apply Taylor's formula with the remainder term in integral form (Theorem A.22) on ψ to get

$$\begin{aligned}\psi(\hat{\hat{\theta}}_n) - \psi(\hat{\theta}_n) &= \tau(\hat{\theta}_n)^\top(\hat{\hat{\theta}}_n - \hat{\theta}_n) + \frac{1}{2}(\hat{\hat{\theta}}_n - \hat{\theta}_n)^\top K(\hat{\hat{\theta}}_n, \hat{\theta}_n)(\hat{\hat{\theta}}_n - \hat{\theta}_n)\\ &= (\hat{\hat{\theta}}_n - \hat{\theta}_n)^\top \bar{t}_n + \frac{1}{2}(\hat{\hat{\theta}}_n - \hat{\theta}_n)^\top K(\hat{\hat{\theta}}_n, \hat{\theta}_n)(\hat{\hat{\theta}}_n - \hat{\theta}_n)\end{aligned}$$

where we have used that $\tau(\hat{\theta}_n) = \bar{t}_n$ and

$$K(\hat{\hat{\theta}}_n, \hat{\theta}_n) = 2\int_0^1 (t-1)\kappa(\hat{\theta} + t(\hat{\hat{\theta}}_n - \hat{\theta}))\, dt.$$

Inserting this into the first expression for Λ_n yields

$$\Lambda_n = n(\hat{\hat{\theta}}_n - \hat{\theta}_n)^\top K(\hat{\hat{\theta}}_n, \hat{\theta}_n)(\hat{\hat{\theta}}_n - \hat{\theta}_n).$$

Now since $\hat{\theta}_n = \tau^{-1}(\hat{\eta}_n)$, $\hat{\hat{\theta}}_n = \tau^{-1}(\hat{\hat{\eta}}_n)$ are both asymptotically normally distributed and $D\tau^{-1} = \kappa^{-1}$, the delta method (Theorem A.19) gives

$$\sqrt{n}(\hat{\theta}_n - \hat{\hat{\theta}}_n) \overset{\text{as}}{=} \sqrt{n}\,\kappa(\theta)^{-1}(\hat{\eta}_n - \hat{\hat{\eta}}_n).$$

Hence,

$$\Lambda_n \overset{\text{as}}{=} n(\hat{\hat{\eta}}_n - \hat{\eta}_n)^\top \kappa(\theta)^{-1} K(\hat{\hat{\theta}}_n, \hat{\theta}_n)\kappa(\theta)^{-1}(\hat{\hat{\eta}}_n - \hat{\eta}_n).$$

Since both of $\hat{\theta}_n$ and $\hat{\hat{\theta}}_n$ are consistent and K is continuous with $K(\theta, \theta) = \kappa(\theta)$, we have

$$\kappa(\theta)^{-1} K(\hat{\hat{\theta}}_n, \hat{\theta}_n) \kappa(\theta)^{-1} \xrightarrow{P} \kappa(\theta)^{-1} = \mathbf{V}_\theta(t(X))^{-1}.$$

Now Corollary 5.23 and Lemma A.18 imply

$$\Lambda_n \overset{\text{as}}{=} n(\hat{\hat{\eta}}_n - \hat{\eta}_n)^\top \kappa(\theta)^{-1} (\hat{\hat{\eta}}_n - \hat{\eta}_n) \xrightarrow{\mathcal{D}} \chi^2(m-d),$$

as desired. □

5.4 Asymptotics in curved exponential families

5.4.1 Consistency of the maximum-likelihood estimator

In the following we shall demonstrate that the asymptotic properties of the MLE, likelihood ratios, etc. essentially are the same for curved and linear exponential families. The most difficult issue is to establish consistency of the MLE in the general curved case. This came essentially for free in the linear case, using the inverse function theorem and the fact that the MLE is a moment estimator; the latter may not hold in the general curved case.

Here we need to work harder to establish that the MLE is asymptotically well-defined and consistent. Once this has been established, the implicit function theorem in combination with the delta method yields relatively easily that the estimator is asymptotically normally distributed.

Note first that the log-likelihood function for such a family based on a sample $(X_1, \ldots, X_n)$ has the form

$$\bar{\ell}_n(\beta) = \phi(\beta)^\top \bar{T}_n - \psi(\phi(\beta)).$$

We let

$$\lambda(\eta, \beta) = \phi(\beta)^\top \eta - \psi(\phi(\beta))$$

and note that then $\bar{\ell}_n(\beta) = \lambda(\bar{T}_n, \beta)$. We now have the following crucial Lemma:

Lemma 5.25. *For a curved exponential family as defined above, there exists an open set O with $\tau(\phi(B)) \subseteq O \subseteq \tau(\Theta) \subseteq \mathbb{R}^k$ and a smooth function $g : O \to B$ such that*

$$\lambda(\eta, g(\eta)) > \lambda(\eta, \beta) \quad \textit{for all } \eta \in O, \beta \in B \setminus \{g(\eta)\}. \tag{5.13}$$

Further, g satisfies the equation:

$$g(\tau(\phi(\beta))) = \beta \textit{ for all } \beta \in B \tag{5.14}$$

and has derivative

$$Dg(\eta) = i(g(\eta))^{-1} J(g(\eta))^\top. \tag{5.15}$$

In particular, if $\bar{T}_n \in O$, the likelihood function attains its unique maximum over B at $\hat{\beta}_n = g(\bar{T}_n)$.

Proof. The proof is somewhat technical and deferred to Appendix B.2. □

In most literature, the first part of this lemma is stated as an assumption rather than a fact which essentially provides what you need, as the lemma gives no guidance in a specific case, where we typically need to establish (5.13) by other means. It just guarantees that our efforts in doing so will not be in vain for large enough samples.

The relation (5.14) is usually referred to as *Fisher consistency*. We can now show that the MLE is asymptotically consistent in a curved exponential family.

Theorem 5.26. *Let $X_1, \ldots, X_n$ be a sample from a curved exponential family with parameter space B. Then the maximum likelihood estimator $\hat{\beta}_n$ is asymptotically well-defined as a solution to the score equation*

$$S(X_1, \ldots, X_n; \beta)^\top = nS(\bar{T}_n, \beta)^\top = nJ(\beta)^\top \left(\bar{T}_n - \tau(\phi(\beta))\right) = 0. \quad (5.16)$$

Further, the MLE is asymptotically consistent and Fisher consistent so $\hat{\beta}_n = g(\bar{T}_n)$, where $g(\tau(\phi(\beta))) = \beta$.

Proof. Since O in Lemma 5.25 is open, there is a neighbourhood U with

$$\tau(\phi(\beta)) \in U \subseteq O.$$

The law of large numbers ensures that $\bar{T}_n \xrightarrow{P} \tau(\phi(\beta))$ and thus

$$\lim_{n\to\infty} P_{\phi(\beta)}(\bar{T}_n \in U) = 1.$$

But for $\bar{T}_n \in U$, Lemma 5.25 ensures that $\hat{\beta}_n$ is well-defined, maximizes the likelihood function, and solves the score equation (5.16). Thus the MLE is asymptotically well-defined and asymptotically consistent. The Fisher consistency is (5.14). □

The MLE is well-defined with a probability that tends to one when n becomes large. However, *in any specific case one must verify whether the likelihood function actually has a unique maximum for the actual observed value $\bar{T}_n = \bar{t}_n$*, i.e. check whether in fact it holds that $\bar{t}_n \in O$.

Example 5.27. [Continuation of Example 3.28] Consider the case when the mean of a bivariate normal distribution is assumed to be located on a semi-circle in the right half-plane. The score function is

$$\begin{aligned} S_n(\beta) &= nJ(\beta)^\top(\bar{t}_n - \tau(\phi(\beta))) \\ &= n(-\sin\beta, \cos\beta)\begin{pmatrix} \bar{x}_1 - \cos\beta \\ \bar{x}_2 - \sin\beta \end{pmatrix} \\ &= n(\bar{x}_2\cos\beta - \bar{x}_1\sin\beta). \end{aligned}$$

For $\bar{x}_1 = \bar{x}_2 = 0$ any β^* in the interval $B = (-\pi/2, \pi/2)$ satisfies the score equation, and for $\bar{x}_1 = 0 \neq \bar{x}_2$ the score equation does not have a solution in this interval. If $\bar{x}_1 \neq 0$, the score equation has a unique solution β^*

$$\beta^* = \tan^{-1}(\bar{x}_2/\bar{x}_1)$$

since $\tan(\beta) = \sin\beta/\cos\beta$ is a bijection from B to $\mathbb{R}$.

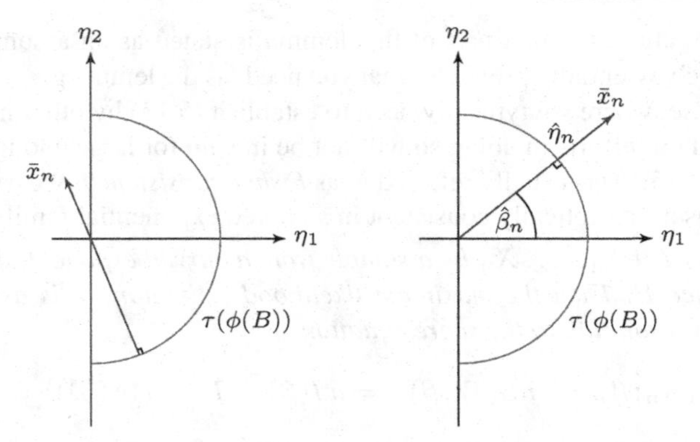

Figure 5.2 – The curved subfamily in Example 5.27 determined by the mean being on a semi-circle. The picture to the left shows a situation where the MLE does not exist and the solution to score equation maximizes the distance to the curve, hence minimizes the likelihood. In the right-most picture, the MLE exists and is determined as the point of intersection of the line to $\bar{x}_n$ and the semicircle. Note that the residual $\bar{x}_n - \hat{\eta}_n$ is orthogonal to the tangent of the circle at the MLE.

Note that β^* may be interpreted as the angle (measured with sign) between the abscissa and the vector from $(0, 0)$ to the observation $(\bar{x}_1, \bar{x}_2)$. This implies that we have

$$\cos \beta^* = |\bar{x}_1|/R, \quad \sin \beta^* = \text{sgn}(\bar{x}_1)\bar{x}_2/R, \tag{5.17}$$

where $R = ||\bar{x}|| = \sqrt{\bar{x}_1^2 + \bar{x}_2^2}$ is the length of the observation. Differentiating a second time and changing sign yields the information function

$$I_n(\beta) = -S_n'(\beta) = n(\bar{x}_1 \cos \beta + \bar{x}_2 \sin \beta)$$

so if β^* solves the score equation we get

$$\begin{aligned} I_n(\beta^*) &= n\left(\bar{x}_1 \frac{|\bar{x}_1|}{R} + \bar{x}_2 \frac{\text{sgn}(\bar{x}_1)\bar{x}_2}{R}\right) \\ &= n \,\text{sgn}(\bar{x}_1) \frac{\bar{x}_1^2 + \bar{x}_2^2}{R} = n \,\text{sgn}(\bar{x}_1) R \end{aligned}$$

and hence the solution only corresponds to a maximum if $\bar{x}_1 > 0$, i.e. if the observed canonical statistic $(\bar{x}_1, \bar{x}_2)$ is in the right half-plane. If $\bar{x}_1 < 0$, the solution is a minimum, so the MLE is not well-defined. These situations are illustrated in Figure 5.2.

Thus, in this example the set O in Lemma 5.25 is the right half plane $(0, \infty) \times \mathbb{R}$ and the MLE is only well-defined if the observed value of $\bar{x}$ is located there. Note that this will be happen with high probability if n is sufficiently large, although this probability depends drastically on β. For β close to either of $-\pi/2$ or $\pi/2$, n has to be massive to ensure this happens. □

Example 5.28. [Continuation of Example 3.29] Consider a sample of independent and identically distributed random variables $X_1, \dots, X_n$ with $X_i \sim \mathcal{N}(\beta, \beta^2)$, where $\beta \in B = (0, \infty)$ is unknown. The log-likelihood function becomes

$$\ell_n(\beta) = -\frac{SS_n}{2\beta^2} + \frac{S_n}{\beta} - n \log \beta$$

where $S_n = \sum_{i=1}^n X_i$ and $SS_n = \sum_{i=1}^n X_i^2$ are the sum and the sum of squares of the observations, corresponding to the canonical statistics in the larger family of all normal distributions. So here

$$\bar{T}_n = (\bar{T}_{1n}, \bar{T}_{2n})^\top = (S_n/n, SS_n/n)^\top$$

and the score statistic becomes

$$S_n(\bar{T}_n, \beta) = n\left(\frac{\bar{T}_{2n}}{\beta^3} - \frac{\bar{T}_{1n}}{\beta^2} - \frac{1}{\beta}\right).$$

Hence, the score equation is equivalent to the equation

$$\beta^2 + \bar{T}_{1n}\beta - \bar{T}_{2n} = 0$$

with roots

$$\alpha = \frac{-\bar{T}_{1n} \pm \sqrt{\bar{T}_{1n}^2 + 4\bar{T}_{2n}}}{2}$$

of which exactly one—corresponding to the plus sign—is positive unless it holds that $X_1 = X_2 = \cdots = X_n = 0$, which happens with probability 0. The information function is

$$I_n(\beta) = 3\frac{SS_n}{\beta^4} - 2\frac{S_n}{\beta^3} - \frac{n}{\beta^2}.$$

As there is only one stationary point, $\hat{\beta}_n$ is a global maximum if and only if the observed information $I_n(\hat{\beta}_n) > 0$. But if $\hat{\beta}_n$ is a solution to the score equation, we have $SS_n = S_n\hat{\beta}_n + n\hat{\beta}_n^2$ and therefore the observed information $I_n(\hat{\beta}_n)$ at the unique solution to the score equation satisfies

$$\begin{aligned}\hat{\beta}_n^4 I_n(\hat{\beta}_n) &= 3SS_n - 2S_n\hat{\beta}_n - n\hat{\beta}_n^2 \\ &= 3SS_n - 2SS_n + n\hat{\beta}_n^2 = SS_n + n\hat{\beta}_n^2 > 0.\end{aligned}$$

We thus conclude that the solution to the score equation is the MLE so

$$\hat{\beta}_n = \frac{-\bar{T}_{1n} + \sqrt{\bar{T}_{1n}^2 + 4\bar{T}_{2n}}}{2}. \tag{5.18}$$

In other words, we have in this example that $O = t(\mathbb{R}) = \{(x, y) \in \mathbb{R}^2 \mid y > 0\}$ and $P_\beta(O) = 1$, so the MLE is well-defined almost surely. □

5.4.2 *Asymptotic normality in curved exponential families*

To establish asymptotic normality of the MLE, we shall again use the delta method. For this, we exploit that we found the derivatives of $\hat{\beta}_n = g(\bar{t}_n)$ in (5.15).

Theorem 5.29. *Let $(X_1, \ldots, X_n)$ be a sample from a curved exponential family of dimension m with parameter $\beta \in B$. Then the MLE $\hat{\beta}_n$ is asymptotically normally distributed with the inverse information as its asymptotic covariance matrix:*

$$\hat{\beta}_n \overset{\text{as}}{\sim} \mathcal{N}_m(\beta, i_n(\beta)^{-1}) = \mathcal{N}_m\left(\beta, \frac{1}{n} i(\beta)^{-1}\right)$$

where $i(\beta)$ is the Fisher information about β for a single observation.

Proof. The central limit theorem implies that $\bar{T}_n \overset{\text{as}}{\sim} \mathcal{N}_k(\tau(\phi(\beta)), \kappa(\phi(\beta))/n)$. Since we have established in Theorem 5.26 that $\hat{\beta}_n = g(\bar{T}_n)$, where g is smooth, we can use the delta method (Theorem A.19) to conclude that

$$\hat{\beta}_n \overset{\text{as}}{\sim} \mathcal{N}_m\left(g(\tau(\phi(\beta))), \frac{1}{n} Dg(\tau(\phi(\beta)))\kappa(\phi(\beta))Dg(\tau(\phi(\beta)))^\top\right)$$

so if we let $\eta = \tau(\phi(\beta))$, we get by Fisher consistency that $g(\eta) = \beta$, which yields the asymptotic mean. Using (5.15) for the derivative Dg, we get for the asymptotic covariance with $\eta = \tau(\phi(\beta))$ and thus $g(\eta) = \beta$

$$\begin{aligned} Dg(\eta)\kappa(\phi(\beta))Dg(\eta)^\top &= i(g(\eta))^{-1} J(g(\eta))^\top \kappa(\phi(\beta)) J(g(\eta)) i(g(\eta))^{-1} \\ &= i(\beta)^{-1}\, i(\beta)\, i(\beta)^{-1} = i(\beta)^{-1}, \end{aligned}$$

and the result follows. □

For illustration, we again consider the bivariate normal distribution with mean on a semi-circle.

Example 5.30. [Continuation of Example 5.27] Consider again the case when the mean of a bivariate normal distribution is assumed to be located on a semi-circle in the right half-plane. We found the information function to be

$$I_n(\beta) = n(\bar{X}_1 \cos\beta + \bar{X}_2 \sin\beta)$$

yielding the Fisher information

$$i_n(\beta) = \mathbf{E}_\beta(I_n(\beta)) = n(\cos^2\beta + \sin^2\beta) = n$$

so we have

$$\hat{\beta}_n = \tan^{-1}(\bar{X}_2/\bar{X}_1) \overset{\text{as}}{\sim} \mathcal{N}(\beta, 1/n).$$

Note in particular that the asymptotic variance is constant in β. □

Example 5.31. [Continuation of Example 5.28] For the model of constant coefficient of variation in the normal distribution, we had

$$I_n(\beta) = 3\frac{SS_n}{\beta^4} - 2\frac{S}{\beta^3} - \frac{n}{\beta^2}$$

and by taking expectations we get

$$i_n(\beta) = ni(\beta) = \mathbf{E}\{I_n(\bar{T}_n, \beta)\} = \frac{6n\beta^2}{\beta^4} - \frac{2n\beta}{\beta^3} - \frac{n}{\beta^2} = \frac{3n}{\beta^2}$$

so we conclude that $\hat{\beta}_n \overset{\text{as}}{\sim} \mathcal{N}(\beta, \beta^2/3n)$. This could, of course, have been derived directly, using the delta method on expression (5.18); but the point is that we can avoid this work by using Theorem 5.29. □

5.4.3 *Geometric interpretation of the score equation*

As for the case of an affine subfamily, it is worth interpreting the score equation geometrically. By composite differentiation, we have

$$\frac{\partial \tau(\phi(\beta))}{\partial \beta} = \kappa(\phi(\beta))J(\beta)$$

and thus

$$J(\beta) = \kappa(\phi(\beta))^{-1}\frac{\partial \tau(\phi(\beta))}{\partial \beta} = \Sigma^{-1}_{\phi(\beta)}\frac{\partial \tau(\phi(\beta))}{\partial \beta}$$

where $\Sigma_\theta = \mathbf{V}_\theta\{t(X)\}$. Hence we can rewrite the score equation for $\hat{\eta}_n = \tau(\phi(\hat{\beta}_n))$ as

$$J(\hat{\beta}_n)^\top(\bar{t}_n - \hat{\eta}_n) = 0 \iff \left.\frac{\partial \tau(\phi(\beta))}{\partial \beta}^\top\right|_{\beta=\hat{\beta}_n} \Sigma^{-1}_{\phi(\hat{\beta}_n)}(\bar{t}_n - \hat{\eta}_n) = 0 \quad (5.19)$$

which expresses that the *residual* $\bar{t}_n - \hat{\eta}_n$ is *orthogonal* to the *tangent space* at $\hat{\eta}$ with respect to the *inner product* determined by $\Sigma^{-1}_{\phi(\hat{\beta}_n)}$.

The tangent space $\mathcal{T}(\beta)$ is the affine space around $\tau(\phi(\beta))$ spanned by the vectors

$$T_j(\phi(\beta)) = \left(\frac{\partial \tau_1(\phi(\beta))}{\partial \beta_j}, \ldots, \frac{\partial \tau_k(\phi(\beta))}{\partial \beta_j}\right)^\top, \; j = 1, \ldots, m$$

i.e. all vectors of the form

$$T(v, \beta) = \tau(\phi(\beta)) + \sum_j v_j T_j(\phi(\beta)), \; v \in \mathbb{R}^m.$$

For later use we also consider the function $h : O \to \tau(\Theta)$, where O is the open set in Lemma 5.25 and

$$h(\eta) = \tau(\phi(g(\eta))).$$

Note that then

$$\hat{\eta}_n = \tau(\phi(\hat{\beta}_n)) = h(\bar{t}_n)$$

is the MLE of $\tau(\phi(\beta)) = \mathbf{E}_{\phi(\beta)}(t(X))$. By composite differentiation, we find

$$T(\phi(\beta)) = \frac{\partial \tau(\phi(\beta))}{\partial \beta} = \kappa(\phi(\beta))J(\beta) \tag{5.20}$$

and further, using (5.15):

$$Dh(\eta) = \kappa(\phi(g(\eta)))J(g(\eta))i(g(\eta))^{-1}J(g(\eta))^\top = H(\eta). \tag{5.21}$$

The matrix $H(\eta)$ is often referred to as the *hat-matrix*, because of the following:

Lemma 5.32. *For $\eta = \tau(\phi(\beta))$, the hat-matrix $H(\eta) = Dh(\eta)$ represents the orthogonal projection onto $\mathcal{T}^*(\beta) = \mathcal{T}(\beta) - \eta$ with respect to the inner product determined by $\kappa(\phi(\beta))^{-1} = \Sigma_{\phi(\beta)}^{-1}$.*

Proof. We shall write H for $H(\eta)$, J for $J(g(\eta))$, etc., and let $\Sigma = \kappa$. Now we use Proposition 2.21 with $A = \Sigma J$ by (5.20), $K = \Sigma^{-1}$, and $i = J^\top \Sigma J$ to get

$$\begin{aligned} H &= \Sigma J i^{-1} J^\top = \Sigma J (J^\top \Sigma J)^{-1} J^\top = A((\Sigma J)^\top K (\Sigma J))^{-1} J^\top \\ &= A(A^\top K A)^{-1} J^\top \Sigma K = A(A^\top K A)^{-1} (\Sigma J)^\top K = A(A^\top K A)^{-1} A^\top K \end{aligned}$$

so H is indeed the matrix for the orthogonal projection. □

We shall illustrate this geometric result in the model of fixed coefficient of variation.

Example 5.33. [Continuation of Example 5.28] Assume that we have observed $\hat{t}_n = (0.5, 3)$ based on $n = 10$ observations leading by (5.18) to the MLE

$$\hat{\beta}_n = \frac{-1/2 + \sqrt{1/4 + 12}}{2} = 3/2 = 1.5.$$

The estimation is illustrated in Fig. 5.3. The picture does not immediate show orthogonality, but the inner product is also not the standard Euclidean one. The covariance of the canonical statistic on a point of the curve is

$$\Sigma_\beta = \mathbf{V}_\beta \left\{ \begin{pmatrix} X \\ X^2 \end{pmatrix} \right\} = \begin{pmatrix} \beta^2 & 2\beta^3 \\ 2\beta^3 & 6\beta^4 \end{pmatrix}$$

So the estimated precision at $\hat{\eta}_n = (1.5, 4.5)^\top$ is

$$\Sigma_{\hat{\beta}}^{-1} = \begin{pmatrix} \hat{\beta}^2 & 2\hat{\beta}^3 \\ 2\hat{\beta}^3 & 6\hat{\beta}^4 \end{pmatrix}^{-1} = \begin{pmatrix} \frac{3}{\hat{\beta}^2} & \frac{-1}{\hat{\beta}^3} \\ \frac{-1}{\hat{\beta}^3} & \frac{1}{2\hat{\beta}^4} \end{pmatrix} = \frac{1}{81} \begin{pmatrix} 108 & -24 \\ -24 & 18 \end{pmatrix}.$$

The tangent direction is obtained by differentiating $(\beta, 2\beta^2)^\top$ to yield $(1, 4\hat{\beta})^\top = (1, 6)^\top$ so the inner product between the residual and the tangent direction is

$$\frac{1}{81}(-1, -1.5) \begin{pmatrix} 108 & -24 \\ -24 & 18 \end{pmatrix} \begin{pmatrix} 1 \\ 6 \end{pmatrix} = (-108 + 36 + 144 - 72)/81 = 0$$

as it should be. □

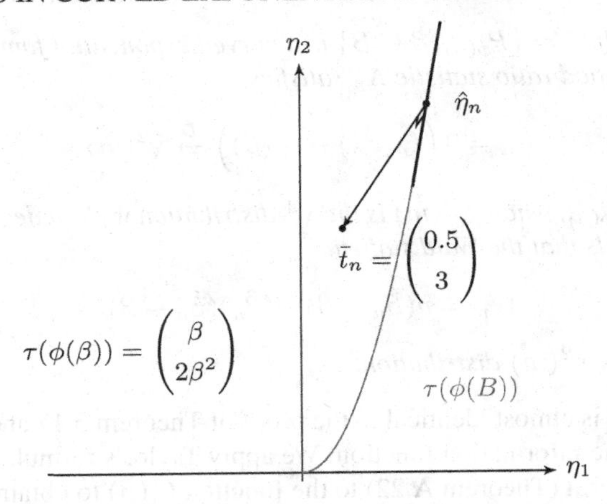

Figure 5.3 – Geometric illustration of maximum-likelihood estimation in the model with constant coefficient of variation. We have $\tau(\phi(\beta)) = (\beta, 2\beta^2)^\top$ and assume $\bar{t}_n = (0.5, 3)^\top$, leading to $\hat{\beta} = 1.5$. The tangent to the curve at the MLE is indicated with a thick line. The residual is orthogonal to the tangent with respect to the inner product determined by the inverse covariance.

5.4.4 *Likelihood ratios and quadratic forms*

Based on Lemma 5.32, we obtain the following for the joint distribution of the residual and estimate:

Corollary 5.34. *Let $(X_1, \ldots, X_n)$ be a sample from a curved exponential family of dimension m and order k with parameter $\beta \in B$. Then the Wald statistic W_n is the squared norm of the residual with respect to the inner product determined by $\kappa(\phi(\beta))^{-1} = \Sigma^{-1}_{\phi(\beta)}$*

$$W_n = n(\bar{T}_n - \hat{\eta}_n)^\top \kappa(\phi(\beta))^{-1}(\bar{T}_n - \hat{\eta}_n).$$

Also, W_n is asymptotically independent of $\hat{\beta}_n$ and $W_n \xrightarrow{\mathcal{D}} \chi^2(k-m)$ for $n \to \infty$.

Proof. The delta method (Theorem A.19) yields, using the abbreviations above:

$$\sqrt{n}\begin{pmatrix} \bar{T}_n - \hat{\eta}_n \\ \hat{\eta}_n - \eta \end{pmatrix} \stackrel{\text{as}}{=} \begin{pmatrix} (I-H)Z \\ HZ \end{pmatrix}$$

where $Z \sim \mathcal{N}_k(\eta, \kappa(\phi(\beta)))$, and the normal decomposition theorem — Theorem 2.24 — thus yields that $\hat{\eta}_n$ is asymptotically independent of the difference $\bar{T}_n - \hat{\eta}_n$ and $W_n \xrightarrow{\mathcal{D}} \chi^2(k-m)$. As $\hat{\beta}_n = g(\hat{\eta}_n)$, the result follows. □

Compare also to Corollary 5.20 to see that W_n here is closely related to the quadratic score statistics.

As in the case of a regular exponential family, we can readily extend Theorem 5.29 to a result on quadratic forms.

Theorem 5.35. *If $\mathcal{P} = \{P_{\phi(\beta)}, \beta \in B\}$ is a curved exponential family of dimension m, the log-likelihood ratio statistic Λ_n satisfies*

$$\Lambda_n = 2\left(\ell_n(\hat{\beta}_n) - \ell_n(\beta)\right) \xrightarrow{\mathcal{D}} \chi^2(m)$$

with respect to $P_{\phi(\beta)}$, where $\chi^2(m)$ is the χ^2-distribution with m degrees of freedom. In addition it holds that the Wald statistic

$$W_n = n(\hat{\beta}_n - \beta)^\top i(\hat{\beta}_n)(\hat{\beta}_n - \beta)$$

has an asymptotic $\chi^2(m)$-distribution.

Proof. The proof is almost identical to the proof of Theorem 5.17 and differs only in the handling of the information function. We apply Taylor's formula with remainder term in integral form (Theorem A.22) to the function $\ell_n(\beta)$ to obtain

$$\ell_n(\beta) = \ell_n(\hat{\beta}_n) + D\ell_n(\hat{\beta}_n)(\beta - \hat{\beta}_n) + \frac{n}{2}(\beta - \hat{\beta}_n)^\top \tilde{K}(\beta, \hat{\beta}_n)(\beta - \hat{\beta}_n),$$

where

$$\begin{aligned} \tilde{K}(\beta, \hat{\beta}_n) &= \frac{2}{n}\int_0^1 (1-t)D^2\ell_n(\beta + t(\hat{\beta}_n - \beta))\,dt \\ &= 2\int_0^1 (t-1)\bar{I}_n(\beta + t(\hat{\beta}_n - \beta))\,dt. \end{aligned}$$

where $D^2\ell_n(\beta) = -I_n(\beta)$ where I_n is the information function. Since $\hat{\beta}_n$ satisfies the score equation, we have $D\ell_n(\hat{\beta}_n) = 0$ and we thus get

$$\Lambda_n = 2\left(\ell_n(\hat{\beta}_n) - \ell_n(\beta)\right) = n(\hat{\beta}_n - \beta)^\top \tilde{K}(\beta, \hat{\beta}_n)(\hat{\beta}_n - \beta).$$

In contrast to the proof of Theorem 5.17, the average information function may not be equal to the Fisher information so we may have $\bar{I}_n \neq i$. However, from (B.4) we have the following explicit form of the information function:

$$\bar{I}_n(\beta) = I_n(\beta)/n = i(\beta) - \sum_u D^2\phi_u(\beta)(\bar{t}_{un} - \tau_u(\phi(\beta))). \tag{5.22}$$

But $\hat{\beta}_n$ is consistent so $\hat{\beta}_n \xrightarrow{P} \beta$ and as $\bar{t}_n \xrightarrow{P} \tau(\phi(\beta))$, the second term in (5.22) converges in probability to zero, so

$$\operatorname{plim}_{n\to\infty} \tilde{K}(\beta, \hat{\beta}_n) = \operatorname{plim}_{n\to\infty} 2\int_0^1 (t-1)i(\beta + t(\hat{\beta}_n - \beta))\,dt = 0.$$

Since $\hat{\beta}_n \overset{\text{as}}{\sim} \mathcal{N}_m(\beta, i(\beta)^{-1}/n)$, it holds that

$$n(\hat{\beta}_n - \beta)^\top i(\beta)(\hat{\beta}_n - \beta) \xrightarrow{\mathcal{D}} \chi^2(m).$$

Corollary A.12 in combination with Lemma A.18 yields the conclusion. □

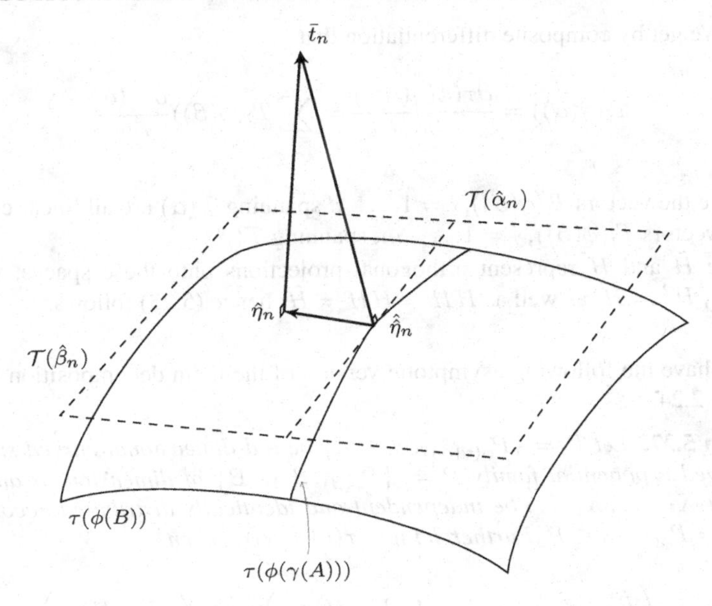

Figure 5.4 – Visualization of maximum-likelihood estimation in the situation where a system of nested models is considered. The residuals $\bar{t}_n - \hat{\eta}_n$ and $\bar{t}_n - \hat{\hat{\eta}}_n$ are orthogonal to the tangent spaces $\mathcal{T}(\hat{\beta}_n)$ and $\mathcal{T}(\hat{\alpha}_n)$ with respect to the inner products determined by the Fisher information at $\hat{\eta}_n$ and $\hat{\hat{\eta}}_n$.

We next consider the case where the likelihood ratio statistic is comparing a given family to a *curved subfamily*; more precisely, a submodel of the form

$$H_0 : \beta \in \gamma(A)$$

where $A \subseteq \mathbb{R}^d$ is open and $\gamma : A \to B$ is a smooth function satisfying the conditions ensuring $\tilde{\mathcal{P}} = \{P_{\phi(\gamma(\alpha))}, \alpha \in \alpha\}$ to be a curved exponential subfamily of a larger curved exponential family

$$\tilde{\mathcal{P}} = \{P_{\phi(\gamma(\alpha))}, \alpha \in A\} \subseteq \mathcal{P} = \{P_{\phi(\beta)}, \beta \in B\} \subseteq \{P_\theta, \theta \in \Theta\},$$

Note the larger family $\mathcal{P}$ may itself be a regular exponential family and the representation as a curved family only due to reparametrization.

We now let $\hat{\eta}_n = h(\bar{t}_n) = \tau(\phi(\hat{\beta}_n))$ and $\hat{\hat{\eta}}_n = \tilde{h}(\bar{t}_n) = \tau(\phi(\gamma(\hat{\alpha}_n)))$ denote the MLE of the mean value parameter under $\mathcal{P}$ and $\tilde{\mathcal{P}}$, respectively. The situation is described in Figure 5.4. Compare also with the linear normal case, as illustrated in Figure 2.2. Further, we let $H(\eta) = Dh(\eta)$ and $\tilde{H}(\eta) = D\tilde{h}(\eta)$ for $\eta = \tau(\phi(\gamma(\alpha)))$ be the corresponding hat-matrices as in Lemma 5.32. We have the following lemma:

Lemma 5.36. *The tangent space $\tilde{\mathcal{T}}(\alpha)$ at $\eta = \tau(\phi(\gamma(\alpha)))$ is a subspace of $\mathcal{T}(\beta)$ and we have*

$$(I - H)(H - \tilde{H}) = (I - \tilde{H})\tilde{H} = 0 \tag{5.23}$$

Proof. We get by composite differentiation that

$$T_i(\gamma(\alpha)) = \frac{\partial \tau(\phi(\gamma(\alpha)))}{\partial \alpha_i} = \sum_j T_j(\phi(\beta)) \frac{\partial \gamma_j(\alpha)}{\partial \alpha_i}$$

and hence the vectors $T_i(\gamma(\alpha)), i = 1, \ldots, d$ spanning $\tilde{\mathcal{T}}(\alpha)$ are all linear combinations of vectors $T_j(\phi(\beta)), j = 1, \ldots, m$, spanning $\mathcal{T}(\beta)$.

Since $\tilde{H}$ and H represent orthogonal projections onto these spaces, we have $\tilde{H}^2 = \tilde{H}$, $H^2 = H$, as well as $H\tilde{H} = \tilde{H}H = \tilde{H}$; hence (5.23) follows. □

We then have the following asymptotic version of the main decomposition theorem Theorem 2.24:

Theorem 5.37. *Let $\tilde{\mathcal{P}} = \{P_{\phi(\gamma(\alpha))}, \alpha \in A\}$ be a d-dimensional curved subfamily of a curved exponential family $\mathcal{P} = \{P_{\phi(\beta)}, \beta \in B\}$ of dimension m and order k and let $X_1, \ldots, X_n, \ldots$ be independent and identically distributed according to some $P = P_{\phi(\gamma(\alpha))} \in \tilde{\mathcal{P}}$. Further, let $\eta = \tau(\phi(\gamma(\alpha)))$. Then*

$$\sqrt{n}\begin{pmatrix} \bar{T}_n - \hat{\eta}_n \\ \hat{\eta}_n - \hat{\hat{\eta}}_n \\ \hat{\hat{\eta}}_n - \eta \end{pmatrix} \stackrel{\text{as}}{=} \sqrt{n}\begin{pmatrix} (I-H)\bar{T}_n \\ (H-\tilde{H})\bar{T}_n \\ \tilde{H}\bar{T}_n - \eta \end{pmatrix} \xrightarrow{\mathcal{D}} \begin{pmatrix} (I-H)Z \\ (H-\tilde{H})Z \\ \tilde{H}Z \end{pmatrix},$$

where $Z \sim \mathcal{N}_k(0, \Sigma)$ for $\Sigma = \kappa(\phi(\gamma(\alpha)))$. It holds in particular that the quantities $T_n - \hat{\eta}_n, \hat{\eta}_n - \hat{\hat{\eta}}_n, \hat{\hat{\eta}}_n$ are asymptotically mutually independent and normally distributed on appropriate subspaces of $\mathbb{R}^k$ with concentrations Σ^{-1}.

Proof. The statement about asymptotic equivalence follows directly from the delta method. The remaining part of the conclusion follows from the normal decomposition theorem, using Lemma 5.36. □

We have a convenient corollary:

Corollary 5.38. *Under the assumptions in Theorem 5.37, the Wald statistics are asymptotically equivalent and satisfy*

$$W_n = n(\hat{\eta}_n - \hat{\hat{\eta}}_n)^\top \hat{\hat{\Sigma}}_n^{-1}(\hat{\eta}_n - \hat{\hat{\eta}}_n) \xrightarrow{\mathcal{D}} \chi^2(m-d)$$
$$\tilde{W}_n = n(\hat{\eta}_n - \hat{\hat{\eta}}_n)^\top \hat{\Sigma}_n^{-1}(\hat{\eta}_n - \hat{\hat{\eta}}_n) \xrightarrow{\mathcal{D}} \chi^2(m-d),$$

where $\hat{\hat{\Sigma}}_n = \kappa(\phi(\gamma(\hat{\alpha}_n)))$ and $\hat{\Sigma}_n = \kappa(\phi(\hat{\beta}_n))$

Proof. This follows from Theorem 5.37 and the normal decomposition theorem combined with Lemma A.18, since the variance estimates are consistent. □

We are now ready to show the main result concerning the asymptotic distribution of the likelihood ratio. We define the maximized *log-likelihood ratio statistic* as

$$\begin{aligned}\Lambda_n = &= -2\log\frac{\sup_{\beta\in\gamma(A)} L(\beta; X_1,\ldots,X_n)}{\sup_{\beta\in B} L(\beta; X_1,\ldots,X_n)}\\ &= -2\log\frac{L(\gamma(\hat\alpha_n); X_1,\ldots,X_n)}{L(\hat\beta_n; X_1,\ldots,X_n)}\end{aligned}$$

where $\hat\alpha_n$ is the MLE of α in $\mathcal{P}_0$ and $\hat\beta_n$ the MLE of β in $\mathcal{P}$. We then have

Theorem 5.39 (Wilks). *Let $\tilde{\mathcal{P}} = \{P_{\phi(\gamma(\alpha))}, \alpha \in A\}$ be a d-dimensional curved subfamily of a curved exponential family $\mathcal{P} = \{P_{\phi(\beta)}, \beta \in B\}$ of dimension m. Then the log-likelihood ratio statistic Λ_n for the subfamily $\gamma(A)$ satisfies*

$$\Lambda_n \xrightarrow{\mathcal{D}} \chi^2(m-d)$$

with respect to any $P \in \tilde{\mathcal{P}}$.

Proof. As usual, the proof is based on Taylor expansion. We let $\hat\theta_n = \phi(\hat\beta_n)$ and $\hat{\hat\theta}_n = \phi(\gamma(\hat\alpha_n))$. We then have

$$\Lambda_n = 2\left(\ell(\hat\theta_n) - \ell(\hat{\hat\theta}_n)\right) = 2n\left(\psi(\hat{\hat\theta}_n) - \psi(\hat\theta_n)\right) + 2n(\hat{\hat\theta}_n - \hat\theta_n)^\top \bar t_n.$$

Next we apply Taylor's formula with remainder in integral form (Theorem A.22) on ψ to get

$$\begin{aligned}\psi(\hat{\hat\theta}_n) - \psi(\hat\theta_n) &= \tau(\hat\theta_n)^\top(\hat{\hat\theta}_n - \hat\theta_n) + \frac{1}{2}(\hat{\hat\theta}_n - \hat\theta_n)^\top K(\hat{\hat\theta}_n, \hat\theta_n)(\hat{\hat\theta}_n - \hat\theta_n)\\ &= (\hat{\hat\theta}_n - \hat\theta_n)^\top \hat\eta_n + \frac{1}{2}(\hat{\hat\theta}_n - \hat\theta_n)^\top K(\hat{\hat\theta}_n, \hat\theta_n)(\hat{\hat\theta}_n - \hat\theta_n)\end{aligned}$$

where

$$K(\hat{\hat\theta}_n, \hat\theta_n) = 2\int_0^1 (t-1)\kappa(\hat\theta + t(\hat{\hat\theta}_n - \hat\theta))\,dt.$$

Inserting this into the expression above yields

$$\Lambda_n = 2n(\hat{\hat\theta}_n - \hat\theta_n)^\top(\bar t_n - \hat\eta_n) + n(\hat{\hat\theta}_n - \hat\theta_n)^\top K(\hat{\hat\theta}_n, \hat\theta_n)(\hat{\hat\theta}_n - \hat\theta_n).$$

Now since $\hat\theta_n = \tau^{-1}(\hat\eta_n)$, $\hat{\hat\theta}_n = \tau^{-1}(\hat{\hat\eta}_n)$ are both asymptotically normally distributed and $D\tau^{-1} = \kappa^{-1}$, the delta method (Theorem A.19) gives

$$\sqrt{n}(\hat\theta_n - \hat{\hat\theta}_n) \overset{\text{as}}{=} \sqrt{n}\,\kappa_0^{-1}(\hat\eta_n - \hat{\hat\eta}_n)$$

where $\kappa_0 = \kappa(\phi(\gamma(\alpha)))$; thus we can express everything in η-terms as

$$\Lambda_n \overset{\text{as}}{=} 2n(\hat{\hat\eta}_n - \hat\eta_n)^\top \kappa_0^{-1}(\bar t_n - \hat\eta_n) + n(\hat{\hat\eta}_n - \hat\eta_n)^\top \kappa_0^{-1} K(\hat{\hat\theta}_n, \hat\theta_n)\kappa_0^{-1}(\hat{\hat\eta}_n - \hat\eta_n).$$

But we now get from Theorem 5.37 that

$$\begin{aligned} 2n(\hat{\eta}_n - \hat{\eta}_n)^\top \kappa_0^{-1}(\bar{t}_n - \hat{\eta}_n) &\overset{\text{as}}{=} 2n\bar{T}_n^\top(\tilde{H} - H)^\top \Sigma^{-1}(I - H)\bar{T}_n \\ &= 2n\bar{T}_n^\top \Sigma^{-1}(\tilde{H} - H)(I - H)\bar{T}_n = 0 \end{aligned}$$

where we have used that $\tilde{H}$ and H are self-adjoint and $(\tilde{H} - H)(I - H) = 0$ by Lemma 5.36, so we conclude

$$\Lambda_n \overset{\text{as}}{=} n(\hat{\hat{\eta}}_n - \hat{\eta}_n)^\top \kappa_0^{-1} K(\hat{\hat{\theta}}_n, \hat{\theta}_n)\kappa_0^{-1}(\hat{\hat{\eta}}_n - \hat{\eta}_n).$$

Since both of $\hat{\theta}_n$ and $\hat{\hat{\theta}}_n$ are consistent, we conclude that

$$\kappa_0^{-1} K(\hat{\hat{\theta}}_n, \hat{\theta}_n)\kappa_0^{-1} \overset{P}{\to} \kappa_0^{-1} = \Sigma^{-1}.$$

Now Lemma A.18 and Corollary A.12 imply

$$\Lambda_n \overset{\mathcal{D}}{\to} \chi^2(m - d),$$

as desired. □

We emphasize that although the limiting distribution of Λ_n is $\chi^2(m - d)$ for all $P \in \tilde{\mathcal{P}}$, the quality of approximating the true distribution with the asymptotic distribution can be very different for different $P \in \tilde{\mathcal{P}}$.

5.5 More about asymptotics

The classical asymptotic results for MLE were established by Cramér (1946) and Wald (1949). Cramér (1946) shows under regularity conditions — which include smoothness assumptions and majorized boundedness of third-order derivatives of the likelihood function — that there is a consistent solution to the score equation and that this solution is asymptotically normally distributed with the inverse Fisher information as its asymptotic covariance. For the sake of completeness, we shall here state Cramér's theorem in full but without proof.

Theorem 5.40 (Cramér). *Consider a sample $X_1, \dots, X_n$ from a smooth and locally stable family with parameter space $\Theta \subseteq \mathbb{R}^k$. Assume further that for every $\theta \in \Theta$ there is a neighbourhood U_θ so the family of densities satisfy*

$$\left| \frac{\partial^3}{\partial \eta_i \eta_j \eta_k} f_\eta(x) \right| \leq H_\theta(x) \text{ for all } \eta \in U_\theta$$

where $\sup_{\eta \in U_\theta} \mathbf{E}_\eta\{H_\theta(X)\} \leq M_\theta < \infty$. Then it holds for every neighbourhood V_θ of θ that

$$\lim_{n \to \infty} P_\theta\{\exists \hat{\theta}_n \in V_\theta : S(X_1, \dots, X_n, \hat{\theta}_n) = 0\} = 1.$$

In other words, the probability that the score equation has a root $\hat{\theta}_n$ near θ tends to unity. Further, this root is asymptotically distributed as $\hat{\theta}_n \overset{\text{as}}{\sim} \mathcal{N}_k(\theta, (ni(\theta))^{-1})$.

Note that there is no guarantee that the solution to the score equation maximizes the likelihood — see Example 5.27 — and in specific cases the regularity conditions may either be hard to verify or they may be violated. However, it does follow from Lemma B.2 that regular exponential families satisfy the conditions in the theorem and this is also easy to show for curved exponential families. So *the conclusion in Cramér's theorem holds for curved exponential families.*

Wald (1949) shows under a number of regularity conditions—involving uniform convergence of log-likelihood functions to their expectation and other (non-smooth) conditions on the parameter spaces—that the MLE is asymptotically well-defined and consistent, without smoothness assumptions, nor smoothness results. And again this applies to curved exponential families, but we have established this specifically, in Theorem 5.26.

The asymptotic theory for curved exponential families was originally developed by Andersen (1969) and Berk (1972). As the authors above, we have only considered the case where the sample consists of independent and identically distributed observations; but the results turn out to be applicable well beyond that case although it is hard to give a precise general set of conditions that is easy to verify in a given case.

Indeed, the important condition for the results to hold seems to be—apart from smoothness conditions—that *the Fisher information $i_n(\beta)$ tends to infinity in a way that is reasonably uniform in β*; this holds in the i.i.d. case because $i_n(\beta) = ni(\beta)$, where $i(\beta)$ is the Fisher information in a single observation. The reason for this is that we need a central limit theorem for the score statistic of the form

$$\sqrt{i_n(\theta)}\bar{S}_n(X_n, \theta) \xrightarrow{\mathcal{D}} \mathcal{N}_k(0, I_k).$$

There are many variants of the central limit theorem without assuming that the score statistic is the sum of i.i.d. random variables. In the non i.i.d. case, the asymptotic results may often be true as well and may occasionally be confirmed by simulation.

Example 5.41. [Poisson asymptotics] As an example of this, let us consider a Poisson random variable X with unknown mean $\lambda \in \mathbb{R}_+$. We have seen in Example 4.19 that the MLE of λ is $\hat{\lambda} = X$ and in Example 1.26 we found that the Fisher information in the Poisson model was $i(\lambda) = \lambda^{-1}$. Now let us assume that λ varies with n so that $\lambda_n = n\mu$ and reparametrize with $\mu = \lambda_n/n$ so that $\hat{\mu}_n = X/n$ and the Fisher information about μ becomes $\tilde{i}_n(\mu) = n\mu^{-1}$ and we note that $\tilde{i}_n(\mu) \to \infty$ for $n \to \infty$.

But X has the same distribution as $X = Y_1 + \cdots + Y_n$, where $Y_1, \ldots, Y_n$ are independent and identically Poisson distributed with parameter $\mu \in \mathbb{R}_+$. Thus we conclude from the asymptotic results that

$$\hat{\mu}_n = X/n \overset{\text{as}}{\sim} \mathcal{N}\left(\mu, \frac{1}{n}\mu\right) \text{ for } n \to \infty.$$

or, equivalently, that

$$\hat{\lambda}_n = X \overset{\text{as}}{\sim} \mathcal{N}(\lambda_n, \lambda_n) \text{ for } n \to \infty.$$

We may now phrase this as

$$\hat{\lambda} = X \overset{\text{as}}{\sim} \mathcal{N}(\lambda, \lambda) \text{ for } \lambda \to \infty$$

with the formal meaning that

$$\frac{1}{\sqrt{\lambda}}(\hat{\lambda} - \lambda) \overset{\mathcal{D}}{\to} \mathcal{N}(0, 1) \text{ for } \lambda \to \infty$$

thus establishing asymptotic normality of the MLE for large λ. Hence, if λ is large, we may in effect treat the case as if we have a large number of observations. □

For state of the art in asymptotic statistical theory, the reader is referred to van der Vaart (2012) who also develops the asymptotic theory for general M-estimators, including the MLE.

5.6 Exercises

Exercise 5.1. Let $X_1, \ldots, X_n, \ldots$ and $Y_1, \ldots, Y_n, \ldots$ be a sequence of real-valued random variables so that $\text{plim}_{n\to\infty} X_n = 0$ and $|Y_n| \leq C$, where C is a positive constant. Show that $\text{plim}_{n\to\infty} X_n Y_n = 0$.

Exercise 5.2. Consider $X_n = (Y_1 + \cdots + Y_n)/n$, where $Y_1, \ldots, Y_n$ are independent and identically Poisson distributed with parameter $\mu \in \mathbb{R}_+$ so that

$$X_n \overset{\text{as}}{\sim} \mathcal{N}\left(\mu, \frac{\mu}{n}\right).$$

Use the delta method to show that

$$X_n^{-1} \overset{\text{as}}{\sim} \mathcal{N}\left(\frac{1}{\mu}, \frac{1}{n\mu^3}\right)$$

and contrast this with the fact that X_n^{-1} does not have finite variance. Explain what is going on here. This is an example where the asymptotic variance is not a limit of variances, but the variance in the limiting distribution.

Exercise 5.3. Consider the estimator

$$S_n^2 = \frac{1}{n-1} \sum_{i=1}^{n} (X_i - \bar{X})^2$$

of the variance σ^2 in the simple standard normal model, so that S_n^2 is distributed as $S_n^2 \sim \sigma^2 \chi^2(n-1)/(n-1)$.

a) Show that S_n^2 is asymptotically normally distributed with

$$S_n^2 \overset{\text{as}}{\sim} \mathcal{N}\left(\sigma^2, \frac{2\sigma^4}{n}\right).$$

b) Consider the transformation $Y_n = \log S_n^2$. Use the delta method to find the asymptotic distribution of Y_n.

Exercise 5.4. Consider the simple Bernoulli model where $X_1, \ldots, X_n$ are independent and identically distributed with

$$P_\mu(X_i = 1) = 1 - P_\mu(X_i = 0) = \mu$$

with $\mu \in (0, 1)$ being unknown.

a) Show that the MLE of μ is $\hat{\mu}_n = \bar{X}_n$ and that this is asymptotically normally distributed with

$$\hat{\mu}_n \overset{\text{as}}{\sim} \mathcal{N}\left(\mu, \frac{\mu(1-\mu)}{n}\right).$$

b) Consider the transformation $Y_n = \sin^{-1}(\sqrt{\bar{X}_n})$. Use the delta method to find the asymptotic distribution of Y_n.

Exercise 5.5. Consider a sample $Z_1, \ldots, Z_n$ of independent and identically distributed random variables, where $Z_i = X_i - Y_i$ with X_i and Y_i independent and exponentially distributed with mean θ.

a) Show that $\mathbf{E}_\theta(Z_i^4) = 24\theta^4$;

b) Find the asymptotic distribution of the moment estimator $\tilde{\theta}_n$ of θ based on the statistic $t(z) = z^2$;

c) Compare the asymptotic distribution of $\tilde{\theta}_n$ to that of

$$\hat{\theta}_n = \frac{1}{2n}\sum_{i=1}^{n}(X_i + Y_i)$$

which would be the MLE of θ had $X_1, \ldots, X_n$ and $Y_1, \ldots, Y_n$ also been observed.

Exercise 5.6. Consider the family of Pareto distributions with densities

$$f_\alpha(x) = \alpha x^{-\alpha-1}\mathbf{1}_{(1,\infty)}(x)$$

with respect to standard Lebesgue measure on $\mathbb{R}$ and assume $\alpha > 2$ but unknown. Consider a sample $X_1, \ldots, X_n$ from this distribution.

a) Determine a moment estimator $\tilde{\alpha}_n$ for α based on the statistic $t(x) = x$ and the sample $X_1, \ldots, X_n$;

b) Find the asymptotic distribution of this moment estimator;

c) Show that the MLE $\hat{\alpha}_n$ of α based on the sample $X_1, \ldots, X_n$ is given as

$$\hat{\alpha}_n = \frac{n}{\sum_{i=1}^{n}\log X_i};$$

d) Find the asymptotic distribution of $\hat{\alpha}_n$;

e) Compare the estimators $\tilde{\alpha}_n$ and $\hat{\alpha}_n$.

Exercise 5.7. Consider the family of distributions with densities

$$f_\theta(x) = \theta x^{\theta-1}\mathbf{1}_{(0,1)}(x), \quad \theta \in \Theta = \mathbb{R}_+$$

with respect to standard Lebesgue measure on $\mathbb{R}$ and consider a sample $X_1, \ldots, X_n$ from this distribution.

a) Determine a moment estimator $\tilde{\theta}_n$ for θ based on the statistic $t(x) = x$;

b) Find the asymptotic distribution of this moment estimator;

c) Show that the family is a regular one-dimensional exponential family and identify the canonical parameter, canonical statistic, and cumulant function;

d) Show that the MLE of θ based on a sample $X_1, \ldots, X_n$ is given as

$$\hat{\theta}_n = \frac{-n}{\sum_{i=1}^n \log X_i};$$

e) Find the asymptotic distribution of $\hat{\theta}_n$;

f) Compare the estimators $\tilde{\theta}_n$ and $\hat{\theta}_n$.

Exercise 5.8. Consider the family of gamma distributions with identical parameters for shape and scale, i.e. with densities

$$f_\beta(x) = \frac{x^{\beta-1}e^{-x/\beta}}{\Gamma(\beta)\beta^\beta}, \quad \beta \in \mathbb{R}_+$$

with respect to standard Lebesgue measure on $\mathbb{R}_+$, and let $X_1, \ldots, X_n$ be a sample from this distribution.

a) Determine a moment estimator $\tilde{\beta}_n$ for β based on the statistic $t(x) = x$ and the sample $X_1, \ldots, X_n$ and find its asymptotic distribution.

b) Show that a moment estimator $\check{\beta}_n$ for β based on the statistic $t(x) = x^2$ and the sample $X_1, \ldots, X_n$ is well-defined and find its asymptotic distribution.

c) Find the asymptotic distribution of MLE $\hat{\beta}_n$ of β based on the sample $X_1, \ldots, X_n$.

d) Compare the asymptotic distributions of the estimators $\tilde{\beta}_n$, $\check{\beta}_n$, and $\hat{\beta}_n$.

Exercise 5.9. Let X and Y be independent random variables with X Poissons distributed with mean λ and Y exponentially distributed with rate λ, where $\lambda > 0$, and let $(X_1, Y_1), \ldots, (X_n, Y_n)$ be a sample from this distribution.

a) Determine a moment estimator $\tilde{\lambda}_n$ for λ based on the statistic $t(x, y) = x - y$ and the sample $(X_1, Y_1), \ldots, (X_n, Y_n)$ and find its asymptotic distribution.

b) Find the asymptotic distribution of the MLE $\hat{\lambda}_n$ of λ.

c) Compare the asymptotic distributions of $\hat{\lambda}_n$ and $\tilde{\lambda}_n$.

Exercise 5.10. Let X and Y be independent and exponentially distributed random variables with $\mathbf{E}(X) = \beta$ and $\mathbf{E}(Y) = 1/\beta$ where $\beta > 0$ as in Exercise 3.9 and Exercise 4.10, and let $(X_1, Y_1), \ldots, (X_n, Y_n)$ be a sample from this distribution.

a) Determine a moment estimator $\tilde{\beta}_n$ for β based on the statistic $t(x, y) = x - y$ and the sample $(X_1, Y_1), \ldots, (X_n, Y_n)$ and find its asymptotic distribution;

b) Find the asymptotic distribution of the MLE $\hat{\beta}_n$ of β based on the sample $(X_1, Y_1), \ldots, (X_n, Y_n)$;

c) Compare the asymptotic distributions of the estimators $\tilde{\beta}_n$ and $\tilde{\beta}_n$.

Exercise 5.11. Consider the situation in Exercise 3.6, where (X, Y) are random variables taking values in $\mathbb{R}_+ \times \mathbb{N}_0$ with density

$$f_{(\lambda,\beta)}(x, y) = \frac{1}{\lambda} e^{-x/\lambda} \frac{(\beta x)^y}{y!} e^{-\beta x} \quad x > 0, \; y = 0, 1, \ldots \tag{5.24}$$

with respect to $\nu \times m$, where ν is the standard Lebesgue measure on $\mathbb{R}_+$ and m is counting measure on $\mathbb{N}_0$. Consider now n independent and identically distributed observations $(X_1, Y_1), \ldots, (X_n, Y_n)$, where (X_i, Y_i)has density formen (5.24) and $(\lambda, \beta) \in \mathbb{R}_+^2$ are unknown.

a) Determine the MLE for (λ, β) and identify when it is well-defined;
b) Find the asymptotic distribution of the MLE $(\hat{\lambda}_n, \hat{\beta}_n)$;
c) Determine the moment estimator $(\tilde{\lambda}_n, \tilde{\beta}_n)$ for (λ, β) based on the function $s(X, Y) = (X, XY)$;
d) Find the asymptotic distribution of the moment estimator $\tilde{\beta}_n$ and compare it to the asymptotic distribution of $\hat{\beta}_n$.

Chapter 6

Set Estimation

6.1 Basic issues and definition

Estimation, as described in Chapter 4, yields methods for 'guessing' the unknown parameter θ involved in generating the observation x. But it fails to give an indication of the *precision* of that estimate. One way of indicating precision is by giving an *set estimate* rather than a *point estimate*.

More precisely, if $\Theta \subseteq \mathbb{R}^k$, we let $\mathcal{C}$ be a collection of subsets of $\mathbb{R}^k$. For example, $\mathcal{C}$ could be the set of spheres in $\mathbb{R}^k$. For $k = 1$ we would typically let $\mathcal{C}$ be the set of open intervals, indicated by their endpoints.

So consider as usual a parametrized statistical model with representation space $(\mathcal{X}, \mathbb{E})$ and associated family $\mathcal{P} = \{P_\theta \mid \theta \in \Theta\}$. We define

Definition 6.1. A *set estimator* is a map $C : \mathcal{X} \mapsto \mathcal{C}$ with the property that the induced maps

$$x \mapsto \mathbf{1}_{C(x)}(\theta)$$

are measurable for all $\theta \in \Theta$. Its *coverage* c_θ is given as

$$c_\theta = \mathbf{E}_\theta\{\mathbf{1}_{C(X)}(\theta)\} = P_\theta\{C(X) \ni \theta\}.$$

We say that $C(X)$ is an $1 - \alpha$ *confidence set* or *confidence region* if the coverage is at least $1 - \alpha$, i.e. if $c_\theta \geq 1 - \alpha$ for all $\theta \in \Theta$.

Note that it is the set $C(X)$ that is random and not θ. The 'confidence' does not say anything about how likely θ is once x is observed, but if $\alpha = 0.05$, say, it says that the method for producing our confidence set will include the true value 95% of the time. Or rather about 5% of our confidence sets will be wrong, i.e. not include the true value. The distinction is subtle and it is quite common to get this wrong when referring to statistical investigations, for example in the popular press.

6.2 Exact confidence regions by pivots

Consider a parametrized statistical model $\mathcal{P} = \{P_\theta \mid \theta \in \Theta\}$ with representation space $(\mathcal{X}, \mathbb{E})$ and let $(\mathcal{Y}, \mathbb{F})$ be a measurable space. We shall consider functions of the form $R : \mathcal{X} \times \Theta \mapsto \mathcal{Y}$ satisfying

$$R(\cdot, \theta) : \mathcal{X} \mapsto \mathcal{Y} \text{ is measurable for all } \theta \in \Theta.$$

DOI: 10.1201/9781003272359-6

Definition 6.2. A function $R : \mathcal{X} \times \Theta \mapsto \mathcal{Y}$ as above is a *pivot* if it holds for all $B \in \mathbb{F}$ that $P_\theta\{R(X,\theta) \in B\}$ does not depend on θ.

Pivots are plenty, and the pivot below is in some sense universal:

Example 6.3. [The universal pivot] If $Y = t(X)$ is a real-valued random variable with distribution function F_θ, i.e.

$$P_\theta\{Y \leq y\} = F_\theta(y)$$

that is continuous on the support of $t(P_\theta)$, the function

$$R_0(X,\theta) = F_\theta(t(X))$$

is indeed such a pivot, since

$$P_\theta\{R_0(t(X),\theta) \leq t\} = P_\theta\left\{Y \leq \inf\{y \,|\, F_\theta(y) \geq t\}\right\} = t$$

so $F_\theta(Y)$ is uniformly distributed on the unit interval for all $\theta \in \Theta$. □

The universal pivot in Example 6.3 can in principle be used to construct a set estimate as follows. Choose any $B \subseteq (0,1)$ with Lebesgue measure $\lambda(B) = 1 - \alpha$. Next, let

$$C_B(x) = \{\theta \,|\, F_\theta(t(x)) \in B\}.$$

Then, clearly, $C_B(X)$ is a set estimator with coverage $1 - \alpha$. For some B this might be sensible, but in general there are too many different possibilities for B, and it might be impossible or difficult to compute $C_B(x)$ for a given x and B.

Example 6.4. [Universal pivot for the normal distribution] Let $X = x$ be an observation from a normal distribution on $\mathbb{R}$ with mean $\xi \in \mathbb{R}$ and variance $\sigma^2 > 0$. The distribution function is thus

$$R(x;\xi,\sigma^2) = F_{\xi,\sigma^2}(x) = \Phi\left(\frac{x-\xi}{\sigma}\right),$$

where Φ is the distribution function for $\mathcal{N}(0,1)$. To ensure coverage $1-\alpha$, we may for example choose

$$B = B_\delta = (\delta\alpha, 1-(1-\delta)\alpha)$$

for any $\delta \in (0,1)$, since then

$$\lambda(B_\delta) = 1 - (1-\delta)\alpha - \delta\alpha = 1 - \alpha.$$

This now leads to the set estimate $C_\delta(x)$ determined as

$$(\xi,\sigma^2) \in C_\delta(x) \iff \alpha\delta < \Phi\left(\frac{x-\xi}{\sigma}\right) < 1-(1-\delta)\alpha.$$

Taking inverses and moving things around, we see that this is equivalent to

$$(\xi,\sigma^2) \in C_\delta(x) \iff x + \sigma z_{1-(1-\delta)\alpha} < \xi < x + \sigma z_{\delta\alpha}$$

where $z_y = \Phi^{-1}(y)$ is the normal quantile function. Now exploiting symmetry in the normal distribution and hence in the quantile function we note that $z_{1-y} = -z_y$ and thus obtain the set

$$C_\delta(x) = \{(\xi, \sigma^2) : x - \sigma z_{(1-\delta)\alpha} < \xi < x + \sigma z_{\delta\alpha}\}.$$

For fixed value of σ^2, this is a *confidence interval*, but generally, it is a rather large region in $\mathbb{R} \times \mathbb{R}_+$. If $\delta = 0$ or $\delta = 1$, we say the confidence interval is *one-sided*. If $\delta = 1/2$, we say that the interval is *two-sided* and *symmetric*. □

Example 6.5. [Student's T as pivot] Example 6.4 gives a formally correct region for the pair of parameters (ξ, σ^2) jointly, but this is huge and not really very helpful. Typically, we are interested in ξ only, whereas σ^2 is a nuisance parameter. So consider independent and identically distributed observations $X_1, \ldots, X_n$, where $X_i \sim \mathcal{N}(\xi, \sigma^2)$ with $\xi \in \mathbb{R}$ and $\sigma^2 > 0$ both unknown. We may now use these repeated observations to obtain information about σ^2 and form the quantity

$$T_n = T(\xi, X_1, \ldots, X_n) = \sqrt{n}\frac{\bar{X}_n - \xi}{S}$$

where as usual $\bar{X}_n = (X_1 + \cdots + X_n)/n$ and $S^2 = \sum_i (X_i - \bar{X}_n)^2/(n-1)$. For all values of (ξ, σ^2), T_n follows a Student's t-distribution with $n-1$ degrees of freedom and hence T_n is a pivot. We thus get a $1-\alpha$ confidence region for ξ as

$$C_T(X) = \left\{\xi \in \mathbb{R} \mid t^{n-1}_{\alpha/2} < T(\xi, X_1, \ldots, X_n) < t^{n-1}_{1-\alpha/2}\right\},$$

where t^{n-1}_β is the β-quantile in the t-distribution with $n-1$ degrees of freedom. Equivalently we may write

$$C_T(X) = \left(\bar{X}_n - t^{n-1}_{1-\alpha/2}\frac{S}{\sqrt{n}}, \bar{X}_n + t^{n-1}_{1-\alpha/2}\frac{S}{\sqrt{n}}\right).$$

Note also that we in fact have $T_n^2 = W_n$, where W_n is the Wald statistic in (5.7). In this case, W_n is distributed as $F(1, n-1)$ which tends to $\chi^2(1)$ for $n \to \infty$. □

Example 6.6. [Linear parameter functions in the normal model] An analogous but slightly more sophisticated example is the following. Let us again consider the linear normal model , so that $X \sim \mathcal{N}_V(\xi, \sigma^2 I_V)$, where $(V, \langle\cdot,\cdot\rangle)$ is a d-dimensional Euclidean space, $\xi \in L$ with $L \subseteq V$ an m-dimensional linear subspace of V, and $\sigma^2 > 0$ with both of ξ and σ^2 unknown.

Suppose our parameter of interest is a real-valued linear function of ξ, i.e. we have $\eta = \langle u, \xi\rangle$, for some $u \in V$. From Example 4.22, we know that the MLE of ξ is $\hat{\xi} = \Pi_L(X)$ and hence

$$\hat{\eta} = \langle u, \hat{\xi}\rangle = \langle u, \Pi_L(X)\rangle = \langle \Pi_L(u), X\rangle$$

and this is distributed as

$$\hat{\eta} \sim \mathcal{N}(\langle u, \xi\rangle, \sigma^2\langle u, \Pi_L(u)\rangle) = \mathcal{N}(\eta, \sigma^2\|\Pi_L(u)\|^2).$$

As in the previous example we have information about σ^2 from $X - \Pi_L X$ which is independent of $\hat{\xi}$ and yields an unbiased estimator of σ^2 as

$$\tilde{\sigma}^2 = \frac{\|X - \Pi_L(X)\|^2}{d-m} \sim \sigma^2 \chi^2(d-m).$$

We may now construct a pivot as

$$T(\eta, X) = \frac{\hat{\eta} - \eta}{\tilde{\sigma}\|\Pi_L(u)\|} = \sqrt{d-m}\frac{\langle u, \Pi_L(X) - \xi\rangle}{\|X - \Pi_L(X)\|\|\Pi_L(u)\|}$$

which follows Student's t-distribution with $d - m$ degrees of freedom. This leads to the following confidence interval for η

$$C(X) = \left(\langle u, \Pi_L(X)\rangle - t^{d-m}_{1-\alpha/2}\widehat{\mathbf{SE}}, \langle u, \Pi_L(X)\rangle + t^{d-m}_{1-\alpha/2}\widehat{\mathbf{SE}}\right)$$

where

$$\widehat{\mathbf{SE}} = \frac{\|\Pi_L(u)\|\|X - \Pi_L(X)\|}{\sqrt{d-m}}$$

is the estimated *standard error* of the estimate $\hat{\eta}$. Note the complete analogy to the confidence region found in Example 6.5, where we had $d = n$, $m = 1$, $u = (1, \ldots, 1)/n$, $\langle u, \Pi_L(X)\rangle = \bar{X}_n$, and $\|\Pi_L(u)\| = 1/n$. □

Example 6.7. [Pivots for the exponential distribution] Let us consider a sample $X_1, \ldots, X_n$ of independent and identically exponentially distributed with unknown expectation $\theta \in \Theta = \mathbb{R}_+$. Since the exponential distribution is also a gamma distribution with shape parameter $\alpha = 1$, we deduce that $\bar{X}_n \sim \Gamma(n, \theta/n)$ and hence

$$R_1(X_1, \ldots, X_n; \theta) = \frac{\bar{X}_n}{\theta} \sim \Gamma(n, 1/n) \tag{6.1}$$

is a pivot. So if we let $g^n_{0.025}$ and $g^n_{0.075}$ denote the corresponding quantiles in the $\Gamma(n, 1/n)$ distribution, it holds for all $\theta \in \Theta$ that

$$P_\theta\left\{g^n_{0.025} < \frac{\bar{X}_n}{\theta} < g^n_{0.975}\right\} = 0.95,$$

or, equivalently, by taking reciprocals and multiplying with $\bar{X}_n$

$$P_\theta\left\{\frac{\bar{X}_n}{g^n_{0.975}} < \theta < \frac{\bar{X}_n}{g^n_{0.025}}\right\} = 0.95$$

and thus

$$C_1 = C_1(X_1, \ldots, X_n) = \left(\frac{\bar{X}_n}{g^n_{0.975}}, \frac{\bar{X}_n}{g^n_{0.025}}\right)$$

is a 95% confidence interval for θ. For illustration throughout this chapter, we consider a sample of $n = 8$ observations with values

```
0.581 0.621 3.739 4.354 0.409 1.843 1.705 0.312
```

here leading to the confidence interval $C_1 = (0.94, 3.93)$. A universal pivot would exploit that $Y = X_1 + \cdots + X_n \sim \Gamma(n, \theta)$ so

$$R_2(X_1, \ldots, X_n; \theta) = G_{(n,\theta)}(Y) = G_{(n,1/n)}(Y/(n\theta)) = G_{(n,1/n)}(\bar{X}_n/\theta)$$

where $G_{(\alpha,\beta)}$ is the Gamma distribution function. Then for $B = (0.025, 0.975)$, the confidence interval C_B based on R_2 would be identical to the interval C_1 found above. □

In general there is a strong element of arbitrariness in constructing set estimators, both in terms of the choice of pivot R and the choice of the set B used in the construction; still the computation of $C_B(x)$ might be challenging. In the following we shall give some general methods for constructing set estimators.

6.3 Likelihood-based regions

It seems consequent to attempt to base a set estimate on the likelihood function or, equivalently, on the log-likelihood function. More precisely a *likelihood-based region* has the form

$$C^a(x) = \{\theta : \Lambda(x, \theta) \leq a\} \tag{6.2}$$

for some $a \geq 0$, where $\Lambda = \Lambda(x, \theta)$ is the *log-likelihood ratio*

$$\Lambda(x, \theta) = 2\left(\ell_x(\hat{\theta}) - \ell_x(\theta)\right) = -2 \log \frac{L_x(\theta)}{L_x(\hat{\theta})} \tag{6.3}$$

with $\hat{\theta} = \hat{\theta}(x)$ denoting an MLE of the unknown parameter θ. Note that we use the term log-likelihood ratio even though we have a factor -2 in front and here we explicitly need to consider the dependence of this statistic on data and parameters.

This type of set appears particularly natural since it necessarily contains the MLE, i.e. it holds that $\hat{\theta} \in C^a(x)$ for all $a > 0$. This is fine if the construction works. However, the problem here is that the coverage $c_\theta(a)$

$$c_\theta(a) = P_\theta\{C^a(X) \ni \theta\}$$

in general depends on the unknown parameter θ, and hence a cannot be determined to achieve a specific level of confidence.

However, in some examples we are lucky and the log-likelihood ratio is actually a pivot. In such cases the coverage can be determined by Monte Carlo methods even though the distribution of $\Lambda(X, \theta)$ cannot easily be expressed in analytic form.

Example 6.8. We consider again a sample $X = (X_1, \ldots, X_n)$ from an exponential distribution with unknown mean $\theta \in \Theta = \mathbb{R}_+$. We have previously—in Example 4.20—derived the log-likelihood function and MLE to be

$$\ell_X(\theta) = -n \log \theta - \frac{\sum_i X_i}{\theta}, \quad \hat{\theta}_n = \bar{X}_n = \frac{X_1 + \cdots + X_n}{n}$$

and thus the maximized log-likelihood function is

$$\ell_X(\hat{\theta}) = -n \log \bar{X}_n - n.$$

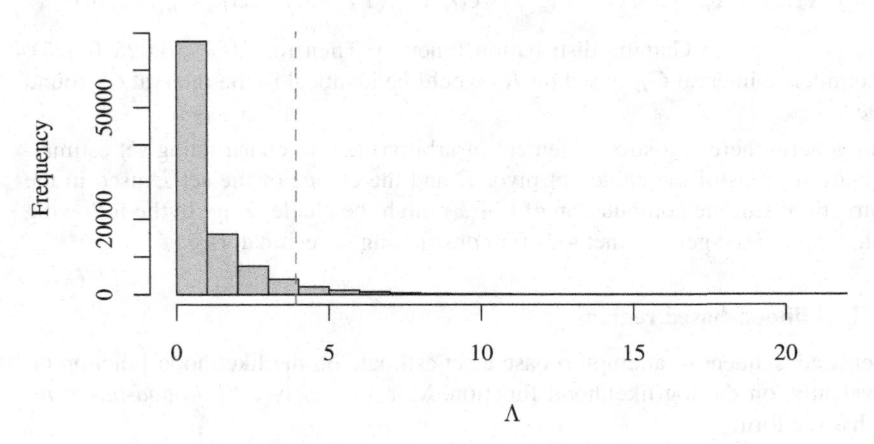

Figure 6.1 – The figure displays a histogram of 100000 simulated values of $\Lambda_n(X, \theta)$ in the exponential distribution for $n = 8$. Since Λ_n is a pivot, it is sufficient to simulate for $\theta = 1$. The 95% quantile in the empirical distribution is $\lambda_{0.95} = 3.91$ and is indicated by a vertical dotted line.

Hence, the log-likelihood ratio statistic becomes

$$\Lambda_n(X, \theta) = 2\left(n \log \theta + \frac{\sum_i X_i}{\theta} - n \log \bar{X}_n - n\right) = 2n\left(\frac{\bar{X}_n}{\theta} - \log \frac{\bar{X}_n}{\theta} - 1\right).$$

Since $\bar{X}_n/\theta$ is a pivot, the same holds for $\Lambda_n(X, \theta)$; indeed the likelilhood ratio statistic has the same distribution as $2n(Y - \log Y - 1)$, where $Y \sim \Gamma(n, 1/n)$.

We may now determine, say, the 95% quantile $\lambda_{0.95}$ in the distribution of Λ by simulation. Fig. 6.1 shows the result of such a simulation for the case of $n = 8$, leading to the quantile $\lambda_{0.95} = 3.91$. The 95% confidence interval is thus

$$C_2(X) = \{\theta \in \Theta \mid \Lambda_n(X, \theta) < 3.91\}$$

which yet again demands a numerical solution. Fig. 6.2 shows the numerical calculation for data in Example 6.7 with $n = 8$ observations, leading to the confidence interval $C_2 = (0.91, 3.74)$. □

6.4 Confidence regions by asymptotic pivots

Although confidence regions constructed via pivots represent some elegance, they are not widely available and it is mostly necessary to rely on approximate methods. Likelihood-based regions may be difficult to handle exactly, but when the number n of observations is large we note that it follows from Theorem 5.17 and Theorem 5.35

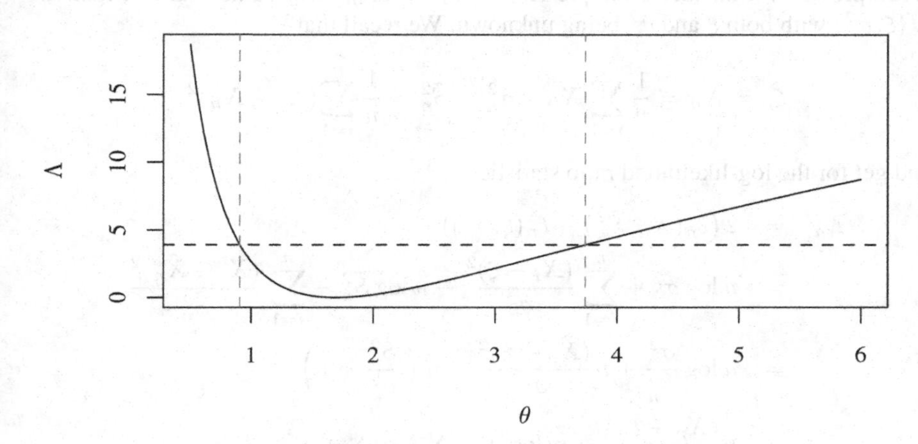

Figure 6.2 – Determination of likelihood ratio confidence interval for the unknown mean. The horizontal line in the diagram is placed at the 95% quantile $y_{0.95} = 3.91$ in the empirical distribution as determined in Fig. 6.1. The endpoints of the interval are given by the intersection of this line with the curve of log-likelihood ratio values.

that *the likelihood ratio statistic* Λ_n *is an asymptotic pivot* as is true also for the *Wald statistics* and the *quadratic score statistic*, since all of these have an asymptotic χ^2-distributions, i.e. an asymptotic distribution that does not depend on the unknown parameter θ.

6.4.1 Asymptotic likelihood-based regions

Consider a curved or regular exponential family so that the log-likelihood ratio statistic $\Lambda_n(\theta) = \Lambda(X_1, \ldots, X_n, \theta)$ approximately follows a $\chi^2(m)$ distribution where m is the dimension of the family. Further, let $\gamma_{1-\alpha}(m)$ denote the $1 - \alpha$ quantile in this distribution. The *asymptotic likelihood region* is given as

$$C_{1-\alpha}(x) = C^{\gamma_{1-\alpha}(m)}(x) = \{\theta : \Lambda_n(\theta) \leq \gamma_{1-\alpha}(m).\} \qquad (6.4)$$

Even though we may confide in the asymptotic results for determining the coverage, we would normally still have to solve the equation by numerical means, which can be quite involved in many dimensions.

Example 6.9. Let us again consider the exponential distribution as in Example 6.8. We calculated the 95% quantile by simulation to be $\lambda_{0.95} = 3.91$, whereas the asymptotic 95% quantile for a $\chi^2(1)$ distribution is $\gamma_{0.95}(1) = 3.84$. The asymptotic confidence interval for the data example is now determined in the same way as described in Fig. 6.2 to be $C_3 = (0.91, 3.68)$. □

The next example shows that issues may be somewhat more involved when the parameter space is more than one-dimensional.

Example 6.10. Consider a sample $X = (X_1, \ldots, X_n)$ from a normal distribution $\mathcal{N}(\xi, \sigma^2)$ with both ξ and σ^2 being unknown. We recall that

$$\hat{\xi}_n = \bar{X}_n = \frac{1}{n}\sum_{i=1}^{n} X_i, \quad \hat{\sigma}_n^2 = \tilde{S}_n^2 = \frac{1}{n}\sum_{i=1}^{n}(X_i - \bar{X}_n)^2$$

and get for the log-likelihood ratio statistic

$$\begin{aligned}
\Lambda_n &= 2\left(\ell_n(\bar{X}_n, \tilde{S}_n^2) - \ell_n(\xi, \sigma^2)\right) \\
&= n \log \sigma^2 + \sum_{i=1}^{n} \frac{(X_i - \xi)^2}{\sigma^2} - n \log \tilde{S}_n^2 - \sum_{i=1}^{n} \frac{(X_i - \bar{X}_n)^2}{\tilde{S}_n^2} \\
&= n \log \frac{\sigma^2}{\tilde{S}_n^2} + n\frac{(\bar{X}_n - \xi)^2}{\sigma^2} + n\left(\frac{\tilde{S}_n^2}{\sigma^2} - 1\right) \\
&= n\frac{(\bar{X}_n - \xi)^2}{\sigma^2} - n(\log(1 + \Delta_n) - \Delta_n) \\
&= n\frac{(\bar{X}_n - \xi)^2}{\sigma^2} + n\frac{\Delta_n^2}{2} + \epsilon(\Delta_n)||\sqrt{n}\Delta_n||^2 \\
&\overset{\text{as}}{=} n\frac{(\bar{X}_n - \xi)^2}{\sigma^2} + n\frac{\Delta_n^2}{2}.
\end{aligned}$$

where we have let $\Delta_n = \tilde{S}_n^2/\sigma^2 - 1$, used Taylor's formula with Peano's remainder term (Theorem A.24), and the fact that $\sqrt{n}\Delta_n \overset{\mathcal{D}}{\to} \mathcal{N}(0, 2)$ to obtain the last two equations.

Although we shall not use this directly, the calculation illustrates how the approximate $\chi^2(2)$-distribution appears as the sum of the term $n(\bar{X}_n - \xi)^2/\sigma^2$—which has an exact $\chi^2(1)$-distribution—and the independent term

$$-n(\log(1 + \Delta_n) - \Delta_n) \overset{\text{as}}{=} n\frac{\Delta_n^2}{2},$$

which has an approximate $\chi^2(1)$-distribution. Even in this simple case, the asymptotic confidence region

$$C^{\gamma_{1-\alpha}(2)}(X) = \left\{(\xi, \sigma^2) : n \log \frac{\sigma^2}{\tilde{S}_n^2} + n\frac{(\bar{X}_n - \xi)^2}{\sigma^2} + n\left(\frac{\tilde{S}_n^2}{\sigma^2} - 1\right) \leq \gamma_{1-\alpha}(2)\right\}$$

can only be calculated numerically. □

6.4.2 *Quadratic score regions*

An alternative form of set estimators is based on the *quadratic score* statistic from Definition 1.24

$$Q_n(X, \theta) = n\bar{S}_n(X, \theta)i(\theta)^{-1}\bar{S}_n(X, \theta)^\top$$

which has an asymptotic $\chi^2(m)$-distribution if the family is smooth and stable and of dimension m by Corollary 1.35 and hence also the quadratic score Q_n is an asymptotic pivot. The corresponding confidence set is then

$$C = \{\theta \mid n\bar{S}_n(X,\theta)i(\theta)^{-1}\bar{S}_n(X,\theta)^\top \leq \gamma_{1-\alpha}(m)\}$$

where $\bar{S}_n$ is the average score statistic and $\gamma_{1-\alpha}(m)$ is the $1-\alpha$-quantile in the $\chi^2(m)$-distribution. Clearly, this demands that the score statistic S_n and Fisher information is available and the set itself would typically need to be determined numerically.

Example 6.11. [Quadratic score intervals for the exponential distribution] For the exponential distribution with mean θ, we calculated the score statistic, Fisher information, and quadratic score in Example 1.32 for the mean θ to yield

$$Q_n = n\theta^2\left(\frac{\bar{x}_n}{\theta^2} - \frac{1}{\theta}\right)^2 = n\left(\frac{\bar{x}_n}{\theta} - 1\right)^2.$$

Since the quantiles in the normal distribution satisfy $z_{1-\alpha/2} = \sqrt{\gamma_{1-\alpha}}$ for the $\chi^2(1)$-distribution, we get the confidence interval

$$C = \bar{x}_n\left(\frac{1}{1+z_{1-\alpha}/\sqrt{n}}, \frac{1}{1-z_{1-\alpha}/\sqrt{n}}\right).$$

For the data in Example 6.7, we have $n = 8$ and $z_{0.975} = 1.96$, leading to the confidence interval $C_4 = (1.00, 5.52)$. □

6.4.3 *Wald regions*

When likelihood ratio regions are difficult to calculate, an alternative is to use the quadratic approximation to the log-likelihood function as described in Theorem 5.17 or Theorem 5.35 using the Wald statistic

$$W_n = n(\hat{\beta}_n - \beta)^\top i(\hat{\beta}_n)(\hat{\beta}_n - \beta)$$

which is asymptotically distributed as $\chi^2(m)$ where m is the dimension of the model under investigation so the corresponding set estimator is

$$C = \left\{\beta \,\middle|\, n(\hat{\beta}_n - \beta)^\top i(\hat{\beta}_n)(\hat{\beta}_n - \beta) \leq \gamma_{1-\alpha}(m)\right\},$$

where $\gamma_{1-\alpha}(m)$ is the $1-\alpha$ quantile in the $\chi^2(m)$-distribution.

The confidence region is an ellipse centred around the MLE $\hat{\beta}_n$. If the unknown parameter β is one-dimensional, the ellipse becomes an interval and may alternatively be calculated as

$$C = \left(\hat{\beta}_n - \frac{z_{(1-\alpha/2)}}{\sqrt{ni(\hat{\beta}_n)}}, \hat{\beta}_n + \frac{z_{(1-\alpha/2)}}{\sqrt{ni(\hat{\beta}_n)}}\right) = \hat{\beta}_n \pm z_{(1-\alpha/2)}\widehat{\mathbf{SE}},$$

where $z_{(1-\alpha/2)}$ is the $1-\alpha/2$ quantile in the standard normal distribution and $\widehat{\mathbf{SE}}$ is the estimated standard error $(ni(\hat{\beta}_n))^{-1/2}$ of the estimate.

Example 6.12. We consider again the exponential distribution where we previously have calculated the information for a single observation to be $i(\theta) = \theta^{-2}$ and the MLE as $\hat{\theta}_n = \bar{x}_n$ and thus the Wald region is given as

$$C = \left(\hat{\theta}_n - \frac{z_{(1-\alpha/2)}}{\sqrt{n\hat{\theta}^{-2}}}, \hat{\theta}_n + \frac{z_{(1-\alpha/2)}}{\sqrt{n\hat{\theta}^{-2}}}\right) = \bar{x}_n\left(1 - \frac{z_{(1-\alpha/2)}}{\sqrt{n}}, 1 + \frac{z_{(1-\alpha/2)}}{\sqrt{n}}\right),$$

which for the example data yields the interval $C_5 = (0.52, 2.87)$. □

Example 6.13. [Continuation of Example 5.33] Consider again the model with constant coefficient of variation, i.e. we assume $X_i \sim \mathcal{N}(\beta, \beta^2)$, where $\beta > 0$ is unknown and assume that we have $n = 10$ observations with $\bar{t}_n = (0.5, 3)^\top$ leading to $\hat{\beta}_n = 1.5$, as calculated in Example 5.33.

To obtain a likelihood-based interval, we recall the log-likelihood function was derived in Example 5.28 as

$$\ell_n(\beta) = -\frac{SS_n}{2\beta^2} + \frac{S_n}{\beta} - n\log\beta$$

and since a solution to the score equation satisfied $SS_n = S_n\hat{\beta}_n + n\hat{\beta}_n^2$, we have for the maximized log-likelihood function

$$\ell_n(\hat{\beta}_n) = -\frac{S_n\hat{\beta}_n + n\hat{\beta}_n^2}{2\hat{\beta}_n^2} + \frac{S_n}{\hat{\beta}_n} - n\log\hat{\beta}_n = -\frac{n}{2} + \frac{S_n}{2\hat{\beta}_n} - n\log\hat{\beta}_n.$$

We thus get for the log-likelihood ratio statistic

$$\begin{aligned}
\Lambda_n(\beta) &= 2(\ell_n(\hat{\beta}_n) - \ell_n(\beta)) \\
&= -n + \frac{S_n}{\hat{\beta}_n} - 2n\log\hat{\beta}_n + \frac{SS_n}{\beta^2} - \frac{2S_n}{\beta} + 2n\log\beta \\
&= -10 + \frac{5}{1.5} - 20\log 1.5 + \frac{30}{\beta^2} - \frac{10}{\beta} + 20\log\beta \\
&= \frac{30}{\beta^2} - \frac{10}{\beta} + 20\log\beta - 14.776. \qquad (6.5)
\end{aligned}$$

An asymptotic likelihood-based interval is now obtained by finding the roots of the equation $\Lambda_n(\beta) = 3.84$, since 3.84 is the .95 quantile in the χ^2-distribution with one degree of freedom. The equation must be solved numerically and yields the 95% confidence interval $C_\Lambda = (1.05, 2.41)$.

In Example 5.31 we established that the asymptotic distribution of $\hat{\beta}_n$ was $\mathcal{N}(\beta, \beta^2/3n)$, and hence, the Wald statistic using the theoretical variance is

$$W_n(\beta) = \frac{3n}{\beta^2}(\hat{\beta}_n - \beta)^2 = 30\left(\frac{1.5}{\beta} - 1\right)^2. \qquad (6.6)$$

Solving the equation $W_n(\beta) < 3.84$ yields the interval

$$C_W = \hat{\beta}_n\left(\frac{1}{1 + 1.96\sqrt{1/30}}, \frac{1}{1 - 1.96\sqrt{1/30}}\right) = (1.10, 2.34).$$

Finally, a Wald-based confidence interval using the estimated asymptotic variance $\hat{\beta}_n^2/3n$ would be

$$C_{\tilde{W}} = 1.5 \pm 1.96\sqrt{1.5^2/30} = 1.5(1 \pm 1.96\sqrt{1/30}) = (0.96, 2.04).$$

Again this is somewhat shorter than the other intervals as we often see using this type of Wald statistic. □

We conclude with a multivariate example.

Example 6.14. [Set estimates for gamma distribution] Since an exponential distribution is also a gamma distribution, it makes sense to analyse the data from Example 6.7 using the gamma distribution with unknown shape α and scale β where $(\alpha, \beta) \in \mathbb{R}_+^2$.

The maximum-likelihood estimates of the unknown parameters are

$$\hat{\alpha} = 1.330, \quad \hat{\beta} = 1.275$$

and the Fisher information was calculated in Example 5.16 to

$$i(\alpha, \beta) = \begin{pmatrix} \Psi_1(\alpha) & \frac{1}{\beta} \\ \frac{1}{\beta} & \frac{\alpha}{\beta^2} \end{pmatrix}.$$

We shall calculate the Wald statistic $\tilde{W}$ so the estimated information matrix becomes

$$i(\hat{\alpha}, \hat{\beta}) = \begin{pmatrix} 1.099 & 0.784 \\ 0.784 & 0.818 \end{pmatrix}.$$

The 95% quantile in the $\chi^2(2)$-distribution is $\gamma_{0.95} = 5.99$ so the corresponding Wald region based on these $n = 8$ observations is the ellipse

$$8(1.099(\alpha - 1.33)^2 + 1.57(\alpha - 1.33)(\beta - 1.275) + 0.818(\beta - 1.275)^2) < 5.99$$

which is displayed in Figure 6.3.

The confidence ellipse may be compared to the asymptotic likelihood region determined as

$$\Lambda(x_1, \ldots, x_8; \alpha, \beta) < 5.99.$$

This likelihood regions can only be determined numerically and are displayed in Fig. 6.4. We note that the likelihood regions are not so well approximated with ellipses and have more the shape of a banana; the likelihood-based regions also appear to be larger than the Wald regions, suggesting that the Wald regions may not give the right coverage. □

6.4.4 Confidence regions for parameter functions

Here we describe a simple way of constructing approximate confidence regions for general smooth parameter functions. Suppose we are particularly interested in estimating a smooth parameter function $\lambda = \phi(\theta)$, where $\phi : \Theta \mapsto \mathbb{R}^m$ is not necessarily

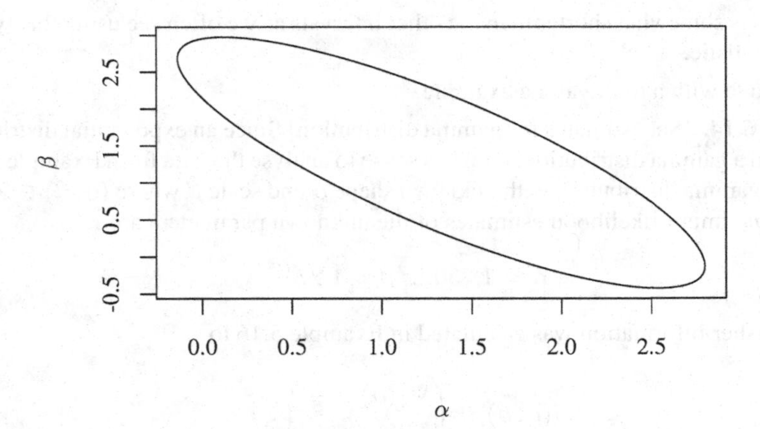

Figure 6.3 – The 95% confidence ellipse for the parameters of a gamma distribution based on the Wald statistic.

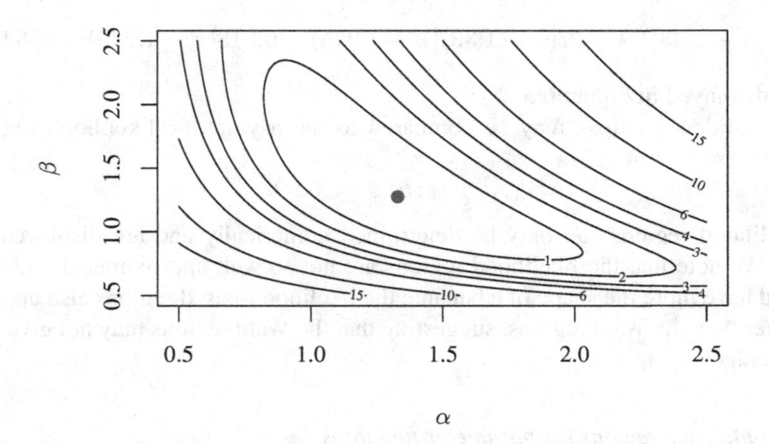

Figure 6.4 – Level curves of the log-likelihood ratio, and the corresponding likelihood ratio set is determined by the contour for $\Lambda = 6$. The dot represents the MLE $(\hat{\alpha}, \hat{\beta})$.

injective, but the Jacobian $D\phi(\theta)$ has full rank m. It is then fairly easy to construct confidence sets for λ based on Wald statistics.

Indeed, if we are in a situation where $\hat{\theta}_n \overset{\text{as}}{\sim} \mathcal{N}(\theta, i(\theta)^{-1}/n)$ we may use the delta method (Theorem A.19) to deduce that

$$\hat{\lambda}_n = \phi(\hat{\theta}_n) \overset{\text{as}}{\sim} \mathcal{N}\left(\lambda, \frac{1}{n} D\phi(\theta) i(\theta)^{-1} D\phi(\theta)^\top\right)$$

and by the usual arguments we conclude that the Wald statistic

$$W_n = n(\hat{\lambda}_n - \lambda)^\top \left\{D\phi(\hat{\theta}_n) i(\hat{\theta}_n)^{-1} D\phi(\hat{\theta}_n)^\top\right\}^{-1} (\hat{\lambda}_n - \lambda) \overset{\mathcal{D}}{\to} \chi^2(m)$$

which now gives a sound basis for making confidence sets. The most common application of this is to construct confidence intervals for single coordinates, corresponding to the case with $m = 1$. In that case, the regions become intervals and will all have the form

$$C = \hat{\lambda}_n \pm z_{1-\alpha/2} \widehat{\mathbf{SE}}$$

where $\widehat{\mathbf{SE}}$ is the estimated standard error

$$\widehat{\mathbf{SE}} = \sqrt{\left\{D\phi(\hat{\theta}_n) i(\hat{\theta}_n)^{-1} D\phi(\hat{\theta}_n)^\top / n\right\}}.$$

Example 6.15. We illustrate this procedure in the gamma example. Suppose we specifically wish to construct a confidence interval just for the shape parameter α. In Example 6.14 we calculated the information matrix to

$$i(\hat{\alpha}, \hat{\beta}) = \begin{pmatrix} 1.099 & 0.784 \\ 0.784 & 0.818 \end{pmatrix}$$

with scaled inverse

$$i(\hat{\alpha}, \hat{\beta})^{-1}/8 = \begin{pmatrix} 0.36 & -0.34 \\ -0.34 & 0.48 \end{pmatrix}$$

so the 95% Wald interval for α becomes

$$C = \hat{\alpha}_n \pm 1.96\sqrt{0.36} = 1.33 \pm 1.18 = (0.15.2.51).$$

If we instead consider a confidence interval for the mean of the distribution, i.e. $\phi(\alpha, \beta) = \alpha\beta = \mathbf{E}_{\alpha,\beta}\{X\}$, we get

$$D\phi(\alpha, \beta) = (\beta, \alpha)$$

implying that the asymptotic variance is

$$\frac{(\beta, \alpha)}{8(\alpha\Psi_1(\alpha) - 1)} \begin{pmatrix} \alpha & -\beta \\ -\beta & \Psi_1(\alpha)\beta^2 \end{pmatrix} (\beta, \alpha)^\top = \frac{\alpha\beta^2}{8}$$

and since $\hat{\alpha}\hat{\beta}^2/8 = 0.27$, this leads to the interval

$$C_6 = 1.6955 \pm 1.96\sqrt{0.27} = (1.17, 2.23)$$

which is comparable to the intervals we earlier found for the mean, using the exponential distribution.

This could also have been derived directly by realizing that the MLE in an exponential family is a moment estimator, and hence, the mean $\mathbf{E}_{\alpha,\beta}(X) = \alpha\beta$ is estimated by

$$\hat{\alpha}_n\hat{\beta}_n = \bar{X}_n \overset{\text{as}}{\sim} \mathcal{N}\left(\alpha\beta, \frac{1}{n}\mathbf{V}_{\alpha,\beta}(X)\right) = \mathcal{N}\left(\alpha\beta, \frac{\alpha\beta^2}{8}\right)$$

leading to the same result. □

6.5 Properties of set estimators

We have seen a number of methods for constructing set estimators but have not paid much attention to discussing what are good properties. We first consider how the set estimators behave under reparametrization.

6.5.1 Reparametrization

We consider a diffeomorphism $\phi : \Theta \mapsto \Lambda$ and recall from Theorem 1.30 that the log-likelihood function and information function in the two parametrizations satisfy

$$\tilde{\ell}_x(\theta) = \ell_x(\lambda), \quad \tilde{i}(\theta) = D\phi(\theta)^\top i(\lambda) D\phi(\theta)$$

for $\lambda = \phi(\theta)$. Since the log-likelihood function is equivariant, the likelihood ratio based regions are as well, since for any a

$$\theta \in \tilde{C}^a = \{\theta \mid \tilde{\Lambda}(x,\theta) < a\} \iff \lambda \in C^a = \{\lambda \mid \Lambda(x,\lambda) < a\}$$

and thus we have

$$C^a = \phi(\tilde{C}^q).$$

The same holds for sets based on pivots, since if $R(X,\lambda)$ is a pivot, we have $\tilde{R}(X,\theta) = R(X,\lambda)$ for $\lambda = \phi(\theta)$ so such sets are also equivariant.

As the quadratic score statistic is equivariant so are the associated confidence regions. However, this is *not true for the sets based on the Wald statistics* since, for example,

$$W'(\theta) = n(\hat{\theta}_n - \theta)^\top \tilde{i}(\theta)(\hat{\theta}_n - \theta)$$

whereas in the λ parametrization we have for $\lambda = \phi(\theta)$

$$W(\lambda) = n(\hat{\lambda}_n - \lambda)^\top D\phi(\theta)^\top i(\lambda) D\phi(\theta)(\hat{\lambda}_n - \lambda)$$

and similarly for other types of Wald statistics.

As a consequence, we shall be careful with the choice of parametrization when calculating Wald intervals and often we shall first choose a suitable parametrization and then transform the relevant interval back to the scale we want. We illustrate this in an example

Example 6.16. [Exponential rates] Let us illustrate the above considerations for the simple model with an exponential distribution, assuming that we are interested in confidence intervals for the *rate* $\lambda = 1/\theta$. We previously found that

$$C_1 = C_1(X_1, \ldots, X_n) = \left(\frac{\bar{X}_n}{g^n_{0.975}}, \frac{\bar{X}_n}{g^n_{0.025}}\right)$$

is a 95% confidence interval for θ and the similar confidence interval for the rate is now

$$C_1' = C_1'(X_1, \ldots, X_n) = \left(\frac{g^n_{0.025}}{\bar{X}_n}, \frac{g^n_{0.975}}{\bar{X}_n},\right).$$

The same is true for the likelihood-based intervals.

If we, however, consider the Wald intervals, we have a different story. In Example 6.12 we found the following confidence interval for the mean θ

$$C = \bar{x}_n \left(1 - \frac{z_{(1-\alpha/2)}}{\sqrt{n}}, 1 + \frac{z_{(1-\alpha/2)}}{\sqrt{n}}\right).$$

We have previously calculated the Fisher information in the λ parametrization to be $i(\lambda) = 1/\lambda^2$ and thus the Wald set for the rate becomes

$$C' = \left\{\lambda \;\middle|\; n\frac{(\lambda - \bar{x}_n^{-1})^2}{\lambda^2} < \gamma_{1-\alpha}(1)\right\},$$

which again leads to a quadratic equation but now the interval becomes

$$C' = \left(\frac{\bar{x}_n^{-1}}{1 - z_{1-\alpha}(1)/\sqrt{n}}, \frac{\bar{x}_n^{-1}}{1 + z_{1-\alpha}(1)/\sqrt{n}}\right) \neq 1/C.$$

The absence of equivariance may in certain circumstances be an advantage. Suppose we consider the parameter $\eta = \log\theta$. Then we get that the Fisher information for η satisfies

$$i(\theta) = \frac{1}{\theta}\tilde{i}(\eta)\frac{1}{\theta}$$

and since we have $i(\theta) = \theta^{-2}$, the Fisher information for η is constant: $i(\eta) = 1$. This implies that all variants of Wald intervals for η are identical and given as

$$C'' = \left(\log\bar{x}_n - \frac{z_{0.975}}{\sqrt{n}}, \log\bar{x}_n - \frac{z_{0.975}}{\sqrt{n}}\right).$$

A transformation of this type is known as a *variance stabilizing transformation*. We may now choose to transform this interval to the original scale and use

$$C^* = \bar{x}_n\left(e^{-z_{0.975}/\sqrt{n}}, e^{z_{0.975}/\sqrt{n}}\right)$$

as a confidence interval for θ yielding $C_7 = (0.85, 3.39)$ for our data example. □

Table 6.1 – Empirical coverage and average length (in parenthesis) of 95% confidence intervals for θ based on 5000 repeated samples of size n from an exponential distribution with mean $\theta = 5$.

	Pivot	Λ	Q	W	W^*
$n = 10$	0.946	0.947	0.949	0.901	0.939
	(7.49)	(7.20)	(10.05)	(6.19)	(6.59)
$n = 50$	0.955	0.956	0.956	0.941	0.952
	(2.87)	(2.87)	(2.99)	(2.76)	(2.80)
$n = 500$	0.950	0.951	0.951	0.948	0.949
	(0.88)	(0.89)	(0.88)	(0.88)	(0.88)

6.5.2 *Coverage and length*

We have now several possibilities for calculating a set stimate, and for the simple exponential model, it leads to seven different intervals:

a) Based on the exact pivot $R(x, \theta) = \bar{X}_n/\theta$: $C_1(x) = (0.94, 3.93)$;

b) Based on $\Lambda_n(x, \theta)$ with exact cutoff: $C_2(x) = (0.91, 3.74)$;

c) Based on the likelihood ratio with asymptotic cutoff: $C_3(x) = (0.91, 3.68)$;

d) Based on the quadratic score: $C_4(x) = (1.00, 5.52)$;

e) Based on the Wald statistic: $C_5(x) = (0.52, 2.87)$

f) Based on the Wald statistic in the gamma model: $C_6(x) = (1.17, 2.23)$;

g) Stabilized Wald $W_n^* = n(\log\hat{\theta} - \log\theta)^2$: $C_7(x) = (0.85, 3.39)$

They all have either exact or approximate coverage equal to 95% but the approximations may be of different quality. They have quite different lenghts, and if we can keep the coverage, we would rather have short intervals than long intervals, to give as precise an estimate as possible.

Table 6.1 displays the result of a simulation experiment based on 5000 repetitions of samples of size n for $n = 10, 50, 500$ from an exponential distribution with mean $\theta = 5$. In each case five different intervals, corresponding to a), b), d), e), and g) have been calculated. We have also calculated the average length of the intervals and the empirical coverage, i.e. the number of intervals containing the true value $\theta = 5$.

We first note that the empirical coverage follows a binomial distribution with success parameter being the true coverage and $N = 5000$. If we let $1 - \alpha = 0.95$, confidence intervals for the empirical coverage are obtained by adding

$$\pm 1.96\sqrt{0.95 \times 0.05/5000} = \pm 0.006$$

to the numbers in the table.

The coverage is mostly as it should be, save for the case $n = 10$, where W and W^* have coverages that are too small, the stabilized W^* performing better than W. This phenomenon is still visible for $n = 50$.

The shortest intervals are those given by W and W^*, but the price is paid in terms of failing coverage. The quadratic score intervals Q appear to have excessively long intervals for $n = 10$ and still rather long intervals for $n = 50$. Among the intervals that give correct coverage, it appears that the two exact intervals based on the exact pivot and Λ are shortest and therefore preferable.

For $n = 500$ it really does not matter and all types of set estimators perform equally well.

6.6 Credibility regions

Confidence sets are often misinterpreted, and it is a common mistake to say that a 95% confidence interval, C say, contains the true parameter with 95% probability or—rephrasing—that the probability that $\theta \in C$ is 95%. This is incorrect and indeed the statement makes little sense as θ is not random, but rather the set $C = C(X)$ is. In fact, *the coverage of 95% is a property of the procedure used to construct the interval, not a property of the interval itself.*

However, there is an alternative statistical paradigm—*the Bayesian paradigm*—where statements of this kind do make sense and we shall briefly sketch the arguments within this paradigm without going too much into detail.

Consider a (Fisherian) statistical model $\mathcal{P} = \{P_\theta, \theta \in \Theta\}$ with an associated family $\mathcal{F} = \{f_\theta \mid \theta \in \Theta\}$ of densities with respect to a base measure μ on the representation space $(\mathcal{X}, \mathbb{E})$. A *Bayesian statistical model* adjoins what is known as a *prior distribution* π on the parameter space $(\Theta, \mathbb{T})$ which then must have an associated σ-algebra $\mathbb{T}$. The prior distribution reflects what is believed about θ prior to observing $X = x$ and in combination with the Fisherian model, it specifies a joint distribution P over $(\Theta \times \mathcal{X}, \mathbb{T} \times \mathbb{E})$ through the relation

$$P(A \times B) = \int_A \int_B f_\theta(x)\, \mu(dx)\, \pi(d\theta), \quad A \in \mathbb{T}, B \in \mathbb{E}$$

Within the Bayesian paradigm, a Fisherian model is simply incompletely specified, as the prior knowledge about θ fails to be represented in the model, as it only specifies the distribution of X for fixed values of θ.

Having observed the outcome $X = x$, the information about θ is updated using *Bayes' formula* to yield the *posterior distribution* π^* given as

$$\pi^*(\theta \in A) = \frac{\int_A f_\theta(x)\, d\pi(\theta)}{\int_\Theta f_\eta(x)\, d\pi(\eta)} = k(x)^{-1} \int_A L_x(\theta)\, d\pi(\theta),$$

where $k(x)$ is a normalizing constant ensuring that the integral is equal to one:

$$k(x) = \int_\Theta L_x(\eta)\, d\pi(\eta).$$

In other words, *the likelihood function L_x is the density of the posterior distribution π^* with respect to the prior distribution π*, sometimes written as

$$\textit{posterior} \propto \textit{likelihood} \times \textit{prior}.$$

Note that any arbitrary multiplicative constant in the likelihood function cancels in the normalization process so the posterior distribution π^* only depends on the shape of L_x and not its absolute size. Note that this makes a lot of sense, since likelihood functions are only well-defined up to arbitrary multiplicative constants; see Theorem 1.16.

A set $A(x) \subseteq \Theta$ with $A(x) \in \mathbb{T}$ is a $1 - \alpha$ *credibility region* for θ if it holds that

$$\pi^*\{\theta \in A(x)\} = k(x)^{-1} \int_{A(x)} L_x(\theta)\, d\pi(\theta) = 1 - \alpha.$$

Note here that θ *is random* (unknown) while $A(x)$ *is fixed and known;* so in this paradigm, it makes sense to say that the probability that θ is in the region is equal to $1 - \alpha$, but it demands an alternative interpretation of the notion of probability, interpreting probabilities via betting, so $P(A) = p$ means that the *odds*

$$\text{odds} = \frac{p}{1 - p}$$

would be fair odds in a bet on the event A occurring. Such an interpretation of probability is known as *subjective probability*.

Example 6.17. [Exponential distribution] Let us again consider the model for the exponential distribution. If we say that the rate $\lambda = 1/\theta$ has an exponential prior distribution with mean 1, i.e.

$$\pi(\lambda) = e^{-\lambda}, \quad \lambda > 0$$

we find the posterior distribution to be

$$\pi^*(\lambda) \propto \lambda^n e^{-\lambda \sum_i x_i} e^{-\lambda} = \lambda^n e^{-\lambda(1 + \sum_i x_i)}, \quad \lambda > 0$$

which we recognize as a gamma distribution with scale parameter $(1 + \sum_i x_i)^{-1}$ and shape parameter $n + 1$.

In our standard data example, we have $n = 8$ and $\sum_i x_i = 13.564$, meaning that the posterior distribution of λ is $\Gamma(9, 1/14.564)$. Thus we have the following 95% credibility interval for λ:

$$g_{0.025}(9, 14.564) < \lambda < g_{0.975}(9, 14.564)$$

where $g_\gamma(\alpha, \delta)$ is the γ quantile in the gamma distribution with shape α and rate δ. This gives the following interval for the mean $\theta = 1/\lambda$

$$\frac{1}{g_{0.975}(9, 14.564)} < \theta < \frac{1}{g_{0.025}(9, 14.564)}$$

which yields the credibility interval $A(x) = (1.070, 3.539)$. □

The credibility interval $A(x)$ found above is quite similar to the confidence intervals we found using other methods, even though the interpretation is different. This is not a coincidence. Indeed one can show (we refrain from doing this here) that it holds asymptotically *in the posterior distribution* for large n that

$$\theta \overset{\text{as}}{\sim} \mathcal{N}_k(\hat{\theta}_n, i(\hat{\theta}_n)^{-1}/n).$$

Here in the Bayesian paradigm θ is random (since it is unknown), whereas $\hat{\theta}_n$ is fixed (since it has been observed). The consequence is that it holds—also in the posterior distribution—that the Wald statistic W_n satisfies

$$W_n(x,\theta) = n(\theta - \hat{\theta}_n)^\top i(\hat{\theta}_n)(\theta - \hat{\theta}_n) \overset{\text{as}}{\sim} \chi^2(k).$$

This means that the 95% confidence interval

$$C(x) = \{\theta \mid W_n(x,\theta) < \gamma_k\}$$

is also a credibility interval. So in such cases, there is some justification in (mis)interpreting the confidence interval as a credibility interval.

6.7 Exercises

Exercise 6.1. The *Weibull distribution* with shape parameter $\alpha > 0$ has density function

$$f_\alpha(x) = \alpha x^{\alpha-1} e^{-x^\alpha}$$

with respect to standard Lebesgue measure on $\mathbb{R}_+$. The Weibull distribution is, for example, much used in Reliability Theory as a distribution of the time to failure of a component in a complex system. Assume that X follows such a Weibull distribution. Construct a 95% two-sided confidence interval $C(X)$ for α based on the universal pivot.

Exercise 6.2. The *von Mises distribution* is a distribution on the unit circle or, equivalently, the interval $(-\pi, \pi]$ with density

$$f_{\kappa,\theta}(x) = \frac{e^{\kappa\cos(x-\theta)}}{2\pi I_o(\kappa)}$$

with respect to standard Lebesque measure on this interval. Here $\kappa > 0$ is the *precision* of the distribution and $\theta \in \Theta = (-\pi, \pi]$ is the *principal direction*; the normalizing constant $I_0(\kappa)$ is known as the modified Bessel function of order 0. This distribution is important for analyzing directional data.

Now, consider κ to be fixed and known and θ unknown and assume that a random variable X with this distribution is observed. We are interested in constructing a confidence set for the principal direction θ.

a) Show that the likelihood ratio statistic is

$$\Lambda(X,\theta) = 2\kappa(1 - \cos(X - \theta)).$$

b) Show that $\Lambda(X,\theta)$ is a pivot.

c) Determine 95% confidence sets $C^\kappa(X)$ for θ based on observation of X for a range of values of κ, for example by determining the relevant quantiles by Monte Carlo simulation.

Exercise 6.3. Consider again the inverse normal distribution as in Exercise 3.4 with density

$$f_{\mu,\lambda}(x) = \sqrt{\frac{\lambda}{2\pi x^3}} \exp\left\{\frac{-\lambda(x-\mu)^2}{2\mu^2 x}\right\}$$

with respect to standard Lebesgue measure on $\mathbb{R}_+$ and consider the subfamily $\mathcal{P}_0$ determined by the restriction $\mu = 1$, parametrized with λ. Let $X_1, \ldots, X_n$ be a sample from this family.

a) Show that the MLE $\hat{\lambda}_n$ for λ is determined as

$$\hat{\lambda}_n = \frac{1}{\bar{X}_n + \bar{Y}_n - 2}, \text{ where } \bar{Y}_n = \frac{1}{n}\sum_{i=1}^{n} \frac{1}{X_i}.$$

b) Determine an asymptotic confidence interval for λ based on the quadratic score statistic for the sample.

Exercise 6.4. Let X and Y be independent and exponentially distributed random variables with $\mathbf{E}(X) = \beta$ and $\mathbf{E}(Y) = 1/\beta$ where $\beta > 0$ as in Exercise 5.10 and let $(X_1, Y_1), \ldots, (X_n, Y_n)$ be a sample from this distribution.

a) Determine an asymptotic confidence interval for β based on the quadratic score statistic using the sample $(X_1, Y_1), \ldots, (X_n, Y_n)$.

b) Determine an asymptotic confidence interval for β from the moment estimator $\tilde{\beta}_n$ for β based on $(X_1, Y_1), \ldots, (X_n, Y_n)$ and the statistic $t(x, y) = x - y$.

c) Make a simulation study to compare the coverage and length for the two types of interval.

Exercise 6.5. Let X follow a log-normal distribution as in Exercise 1.3 with parameters $(\xi, \sigma^2) \in \mathbb{R} \times \mathbb{R}_+$ which are both considered unknown. In other words, $Y = \log X$ with $Y \sim \mathcal{N}(\xi, \sigma^2)$. Consider a sample $X_1, \ldots, X_n$ from this distribution.

a) Determine an asymptotic Wald confidence interval for the *median* λ of the distribution

$$\lambda = \phi_1(\xi, \sigma^2) = \mathrm{med}_{\xi,\sigma^2}(X) = e^{\xi}.$$

b) Determine an asymptotic Wald confidence interval for the *mean* μ of the distribution

$$\mu = \phi_2(\xi, \sigma^2) = \mathbf{E}_{\xi,\sigma^2}(X) = e^{\xi + \sigma^2/2}.$$

c) Determine an asymptotic Wald confidence interval for the *coefficient of variation* δ of the distribution

$$\delta = \phi_3(\xi, \sigma^2) = \mathbf{C}_{\xi,\sigma^2}(X) = \sqrt{e^{\sigma^2} - 1}.$$

d) Investigate the coverage of these intervals for various values of n and σ^2 by simulation.

Exercise 6.6. Let $X \sim \text{binom}(n, \mu), \mu \in (0, 1)$. Determine the following approximate confidence intervals for the unknown probability of success μ:

a) A confidence interval based on the quadratic score statistic $Q(X, \mu)$;

b) A confidence interval based on a Wald statistic for the log-odds ratio

$$\theta = \log \frac{\mu}{1 - \mu},$$

transformed back to μ.

c) A confidence interval based on a Wald statistic for the parameter $\gamma = \sin^{-1}(\sqrt{\mu})$, transformed back to μ.

d) Make a simulation study to compare these intervals with respect to coverage and length.

Note that the length n in the binomial distribution of X can both be considered just as a label, but also representing that $X = Y_1 + \cdots + Y_n$ where $Y_1, \ldots, Y_n$ are independent and identically Bernoulli distributed. In this way, the use of asymptotic results is justified, simply because the Fisher information tends to infinity. Compare this to the discussion in Example 5.41.

Chapter 7

Significance Testing

7.1 The problem

We consider a statistical model with associated family $\mathcal{P} = \{P_\theta \mid \theta \in \Theta\}$ on the representation space $(\mathcal{X}, \mathbb{E})$ and an observation $X = x$. We are interested in determining whether this observation supports specific statements about the unknown parameter, for example that the parameter has a specific value $\theta_0 \in \Theta$, or the parameter is inside a specific subset $\Theta_0 \subseteq \Theta$, known as a *hypothesis*.

Although hypothesis testing is very much at the core of many applications of statistics, the subject is quite controversal, partly because hypothesis testing is used in a huge variety of contexts and some considerations may not appear equally relevant in all contexts.

7.2 Hypotheses and test statistics

7.2.1 Formal concepts

Formally we consider a *null hypothesis*

$$H_0 : \theta \in \Theta_0,$$

where $\Theta_0 \subset \Theta$. Generally we do not distinguish between the hypothesis H_0 and the representing subset Θ_0. We use the term *alternative hypothesis* for the complement

$$H_A : \theta \in \Theta \setminus \Theta_0.$$

We may equivalently represent a hypothesis as a *subfamily* $\mathcal{P}_0 = \{P_\theta \mid \theta \in \Theta_0\} \subset \mathcal{P}$ of the family associated with the basic statistical model.

When testing a statistical hypothesis, the first problem is to choose a suitable *test statistic* which is a map $d : \mathcal{X} \mapsto \mathbb{R}$. Without loss of generality, we assume here and later that the test statistic is chosen so that large values of d indicate that the hypothesis is unlikely or, in other words, d measures a *deviation* from the hypothesis. A suitable transformation of any chosen statistic will always have this property. We express this by saying that *large values of d are critical*.

We have already seen such statistics in the previous chapter, as they have been used to construct set estimators. A canonical choice of test statistic is the *maximized*

DOI: 10.1201/9781003272359-7

likelihood ratio test statistic or, in short, the *likelihood ratio*

$$\Lambda = \Lambda(x) = -2 \log \frac{\sup_{\theta \in \Theta_0} L_x(\theta)}{\sup_{\theta \in \Theta} L_x(\theta)}$$

obtained by comparing the highest possible values of the likelihood function assuming the null hypothesis with the highest possible value without this restriction on the parameter.

Indeed, if we let $\hat{\hat{\theta}}$ and $\hat{\theta}$ denote the maximum-likelihood estimates (MLE) assuming the smaller model and the model without this assumption

$$\hat{\hat{\theta}} = \arg\max_{\theta \in \Theta_0} L_x(\theta), \quad \hat{\theta} = \arg\max_{\theta \in \Theta} L_x(\theta),$$

we have

$$\Lambda(x) = 2\left(\ell_x(\hat{\theta}) - \ell_x(\hat{\hat{\theta}})\right). \tag{7.1}$$

Alternatives include statistics of the Wald type, and others, as we shall see in the following.

7.2.2 *Classifying hypotheses by purpose*

The hypothesis might have its own substantive interest, such as whether a treatment is ineffective or not, or it might just represent a desirable simplification of the model. Below we shall highlight some of the relevant possibilities.

Simplifying hypotheses: We wish to investigate whether a simplification of the model represented by the smaller family $\mathcal{P}_0$ could be suitable; here the issue is to avoid models that are unnecessarily complex, for example because our estimates of the unknown parameters then become more precise and reliable. Or we might wish to use the scientific principle sometimes known as *Occam's razor* that would always prefer a simple model to a complex one, if the simpler model is still satisfactory.

Confirmatory hypotheses: We may have a theoretical reason that predicts a specific value of a parameter function, but wish to confirm the theoretical value by an empirical investigation.

Reverse hypotheses: Consider an experiment that is made with the purpose of establishing that a certain treatment, medical or other, definitely has an effect, say of magnitude $\theta \in \mathbb{R}$. It is then customary to formulate the null hypothesis in a reverse manner, as $H_0 : \theta = 0$, i.e. the formal statistical hypothesis says that there is no effect. This is one of many examples of statisticians using the term 'hypothesis' in a manner different from most scientists. The hope is that the statistical test can *falsify* the hypothesis and thus establish beyond reasonable doubt that the treatment actually has an effect.

Model criticism: Occasionally we perform statistical hypothesis testing as an activity of the Devil's advocate; we wish to subject our statistical model to criticism and make sure the model survives that criticism, thus increasing our confidence in its validity. Such hypotheses are often tested by graphical methods, say in residual analysis, rather than formal quantitative tests. Another term that is often used for this type of tests is *goodness-of-fit*. It is characteristic for this type of hypothesis that the alternative hypothesis H_A may not be fully specified, or not specified at all, and certainly does not play an an important role: we only specify the test statistic $D = d(X)$ and investigate whether its value conforms with H_0.

The list of possibilities given above is far from exhaustive, as the terminology and methodology of hypothesis testing is used in an almost infinite variety of ways. As a consequence, it is hard to present a single formal theory of hypothesis testing that seems suitable for all these different purposes, and the theory will from time to time seem awkward in specific situations.

7.2.3 *Mathematical classification of hypotheses*

Another dimension for the classification of statistical hypothesis is through the mathematical properties of the subset $\Theta_0 \subset \Theta$.

Simple or composite: We say that the hypothesis is *simple* if $\Theta_0 = \{\theta\}$ consists of a single point. If the hypothesis is not simple, we say that it is *composite*.

Linear hypotheses are hypotheses given as $\Theta_0 = L \cap \Theta$, where $L \subseteq V$ is a linear subspace of a vector space V with $\Theta \subseteq V$. The subspace L can either be given as an image or inverse image of a linear map:

$$\Theta_0 = \{\theta \in \Theta \mid \theta = A\beta, \beta \in \mathbb{R}^m\}, \quad \Theta_0 = \{\theta \in \Theta \mid H\theta = 0\}.$$

Affine hypotheses are hypotheses given as $\Theta_0 = L \cap \Theta$, where $L \subseteq V$ is an affine subspace, given as an image or inverse image of an affine map:

$$\Theta_0 = \{\theta \in \Theta \mid \theta = A\beta + b, \beta \in \mathbb{R}^m\}, \quad \Theta_0 = \{\theta \in \Theta \mid H\theta = \beta_0\}.$$

Smooth hypotheses are given as an image or inverse image of a smooth map for Θ being an open subset of $\mathbb{R}^k$:

$$\Theta_0 = \{\theta \in \Theta \mid \theta = \phi(\beta), \beta \in B \subseteq \mathbb{R}^m\}, \quad \Theta_0 = \{\theta \in \Theta \mid h(\theta) = \beta_0\}$$

where maps ϕ, h are smooth and have Jacobi matrices with full rank. For exponential families, such hypotheses correspond to curved subfamilies.

7.3 Significance and p-values

To judge whether a given hypothesis is reasonable, we may (at least in principle) calculate the *p-value*

$$p = \sup_{\theta \in \Theta_0} P_\theta\{D \geq d(x)\}$$

where $D = d(X)$ is the random variable corresponding to our test statistic and $d(x)$ the observed value. The p-value is the highest probability that can be achieved for the event $\{D \geq d(x)\}$ while maintaining the hypothesis that $\theta \in \Theta_0$. If this is very small, say $p < \varepsilon$, we evoke *Borel's single law of chance*, also sometimes referred to a *Cournot's principle*:

IMPROBABLE EVENTS DO NOT OCCUR.

We then conclude that the hypothesis cannot be maintained. We also say that there is *significant* evidence against the hypothesis. With some imprecision, we may also say that the *test is significant*.

Note that the process of *significance testing* as described above fits well into Karl Popper's theory of scientific progress through *falsification* (Popper, 1959), reflecting the time period when it was conceived, early in the 20th century while *Karl Popper* (1902–1994) also developed his philosophical theory of scientific evidence.

It remains to be quantified what 'very small' is. This appears to be mostly culturally defined and depends often on the context. Émile Borel (1943) set the following scales for probabilities to be small when evoking his single law of chance:

- *l'échelle humaine* (human scale): $\varepsilon \sim 10^{-6}$
- *l'échelle terrestre* (earthly scale): $\varepsilon \sim 10^{-15}$
- *l'échelle cosmique* (cosmic scale): $\varepsilon \sim 10^{-50}$.

Modern statistical practice mostly uses $\varepsilon \in \{0.05, 0.01, 0.001\}$, and it is common to speak about 'one-, two-, or three-starred significance', but in Particle Physics, for example, an ε between 10^9 and 10^{12} is used for a scientific discovery to be acknowledged. In general we need different scales in different areas to allow for scientific progress and simultaneously prevent too many false conclusions.

If we have decided in advance what 'very small' means, we speak of a *level of significance* $\alpha \in (0, 1)$ and we would then reject a hypothesis if $p < \alpha$. We then say that the test is *significant at level* α. Clearly, if $\alpha_1 > \alpha$, any test that is significant at level α is also significant at level α_1.

We emphasize that statistical significance is not the same as importance. Consider the following simple example.

Example 7.1. [Male births] In the year 1998 there were a total of 66170 live births in Denmark, of which 34055 were boys and 32115 girls. We wish to investigate whether the probability of a random child being born as a boy is the same as that of being born as a girl.

We let X denote a random variable corresponding to the number of male births and specify a corresponding statistical model $\mathcal{P} = \{P_\theta \mid \theta \in \Theta = (0, 1)\}$ where P_θ is the binomial distribution of length $n = 66170$ with parameter θ. Our null hypothesis is $H_0 : \theta = 1/2$. Under the hypothesis, we should expect around $66170/2 = 33085$ male births. We choose to use the test statistic

$$d(x) = |x - 33085|$$

measuring the deviation from this expectation and calculate the p-value approximately using the central limit theorem yielding a normal approximation to the binomial distribution

$$\begin{aligned} p &= P_{1/2}\{D \geq |34055 - 33085| = 970\} \\ &\approx 2\left(1 - \Phi\left(\frac{970}{\sqrt{66170/4}}\right)\right) = 4.64 \times 10^{-14} \end{aligned}$$

so this is extremely significant, even at a level close to Borel's earthly scale and certainly at a level corresponding to what is used in modern Particle Physics. Nevertheless, the frequency of male births is

$$x/n = 34055/66170 = 0.514659\ldots$$

so for most practical purposes, this is a neglible deviation from the equiprobable and therefore typically *unimportant*; see also Example 8.2 for a similar situation. □

The fact that a statistical test is significant just means that it would *not be reasonable to attribute the observed deviation to chance,* however small and unimportant the deviation may be.

7.4 Critical regions, power, and error types

An alternative formulation of hypothesis testing is to see this as a decision problem and an associated partition of the representation space into two regions:

- A *critical region* $\mathcal{K} \subseteq \mathcal{X}$.
- An *acceptance region* $\mathcal{A} = \mathcal{X} \setminus \mathcal{K}$.

The interpretation is then that the hypothesis is *rejected* if $x \in \mathcal{K}$ and *accepted* if $x \in \mathcal{A}$. Again we can without loss of generality assume that these regions are determined by a test statistic d as before with

$$\mathcal{K} = \{x : d(x) > d_{\text{crit}}\}, \quad \mathcal{A} = \{x : d(x) \leq d_{\text{crit}}\}$$

where d_{crit} is the *critical value* for the test. Thus we can identify the test with its critical region, or with the pair (d, d_{crit}) of the test statistic and its associated critical value.

To further investigate the behaviour of such a test, we introduce the *power function* $\gamma : \Theta \to [0, 1]$:

$$\gamma_{\mathcal{K}}(\theta) = P_\theta\{\mathcal{K}\} = P_\theta\{D > d_{\text{crit}}\} = 1 - P_\theta\{\mathcal{A}\}$$

giving the rejection probability as a function of the unknown parameter θ.

When accepting or rejecting a hypothesis, we may commit two types of error. We say that we commit an *error of type I* if we reject a hypothesis that is true, i.e. if

$$x \in \mathcal{K} \text{ and } \theta \in \Theta_0.$$

Similarly, we say that we commit an *error of type II* if we accept a hypothesis that is false, i.e. if

$$x \in \mathcal{A} \text{ and } \theta \notin \Theta_0.$$

The power function determines the probability of committing these errors although they in general depend on the unknown value of θ; indeed, the probability of a type I error is

$$\gamma_{\mathcal{K}}(\theta) = P_\theta\{\mathcal{K}\}, \quad \theta \in \Theta_0$$

whereas the probability of a type II error is

$$\beta_{\mathcal{A}}(\theta) = 1 - \gamma_{\mathcal{K}}(\theta), \quad \theta \in \Theta \setminus \Theta_0.$$

We would ideally like both of these errors to be small. The *size* $\delta_{\mathcal{K}}$ of the test is the lowest upper bound for the probability of a type I error:

$$\delta_{\mathcal{K}} = \sup_{\theta \in \Theta_0} \gamma_{\mathcal{K}}(\theta) = \sup_{\theta \in \Theta_0} P_\theta\{\mathcal{K}\}.$$

It might be difficult to determine the exact size of a test so we also say that the test has *level* $\alpha \in [0, 1]$ if its size is at most α

$$\delta_{\mathcal{K}} \leq \alpha.$$

When constructing a test, we shall attempt to maximize its power outside the hypothesis, i.e. minimize the type II error probability, while ensuring a given level α, i.e. controlling the probability of a type I error.

We emphasize that power considerations are only relevant in the planning phase, when constructing tests and test statistics, or deciding on the sample size for an experiment yet to be conducted. Once a test has been designed and the test statistic calculated, it is formally the associated p-value that carries the relevant information. However, it is worth knowing about the power of the test, simply because a test with low power may not be able to falsify a hypothesis and therefore provides only weak evidence.

Example 7.2. [Multiple choice examinations] We consider the design of a set of exam questions for a multiple choice examination. The idea is to have three choices for every n questions in the exam. So a student who knows about 50% of the material for the course would then give correct answers to half of the questions and guess the rest, thus having $1/2 + 1/6 = 2/3$ of the n questions right in expectation. Motivated by this, we shall demand that 65% of the questions are correctly answered for the student to pass the exam.

Now we are interested in designing the examination in such a way that students who have not followed the course at all and simply guess all answers will fail with high probability. So how many questions do we need to achieve that such students will fail with probability .999?

We may see the examination as a statistical test based on observing the number of correct answers X taking values is the representation space $\mathcal{X} = \{0, \ldots, n\}$ with critical region

$$\mathcal{K}_n = \{x \in \mathcal{X} \mid x \leq 65n/100\}.$$

We also assume that $X \sim \text{binom}(n, \theta)$, where $\theta \in \Theta = [1/3, 1]$ and wish to determine n such that the power

$$\gamma_n(\theta) = P_\theta(\mathcal{K}_n)$$

is at least .999 at $\theta = 1/3$. We have

$$\gamma_n(1/3) = F_{n,1/3}(.65n)$$

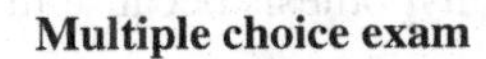

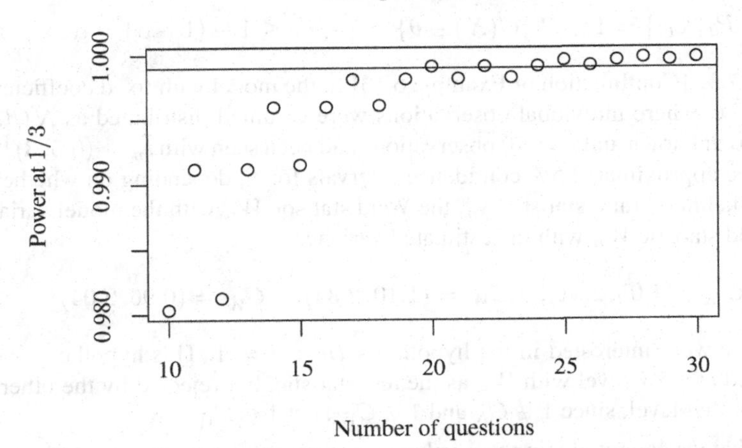

Figure 7.1– The power of a multiple choice examination with three choices in each question as a function of the total number of questions. The horizontal line is at .999.

where $F_{n,\theta}$ is the cumulative distribution function for a binomial distribution of length n and success probability $1/3$. This function is plotted in Figure 7.1, and we conclude that $n = 20$ questions are necessary for us to achieve this goal. Then the student must answer $x = 13$ questions right to pass the exam. Note that the power decreases from $n = 20$ to $n = 21$ as the pass boundary remains at $x = 13$. The discreteness of the space is the reason that the power is not a monotone function of the number of questions n. □

7.5 Set estimation and testing

There is a fundamental logical relation between set estimation and testing: Given acceptance regions $\mathcal{A}(\theta)$ for tests of simple hypotheses of the form $\Theta_0 = \{\theta\}$, the associated confidence set is the set of θ-values for which the hypothesis would be accepted, i.e.

$$C(x) = \{\theta \mid x \in \mathcal{A}(\theta)\}.$$

If the tests all have level α, we get for the coverage

$$q_\theta = P_\theta\{C(X) \ni \theta\} = P_\theta\{\mathcal{A}(\theta)\} = 1 - P_\theta\{\mathcal{K}(\theta)\} \geq 1 - \alpha$$

so $C(X)$ is an $(1-\alpha)$-confidence set for θ. Similarly, if $C = C(x)$ is a confidence set with coverage at least $1-\alpha$, the critical region for its associated test of the simple hypothesis $\Theta_0 = \{\theta\}$ is simply the set of observations for which the confidence set does not include θ:

$$x \in \mathcal{K}_C \iff \theta \notin C(x).$$

Then, if $C(X)$ is a $(1-\alpha)$-confidence set for θ, $\mathcal{K}_C$ becomes the critical region of a test of level α for the simple hypothesis because then

$$P_\theta\{\mathcal{K}_C\} = 1 - P_\theta\{C(X) \ni \theta\} = 1 - d_\theta \leq 1 - (1-\alpha) = \alpha.$$

Example 7.3. [Continuation of Example 6.13] In the model with fixed coefficient of variation, i.e. where individual observations were assumed distributed as $\mathcal{N}(\beta, \beta^2)$ with $\beta > 0$ unknown and $n = 10$ observations had been seen with $\bar{t}_n = (0.5, 3)^\top$, we found three approximate 95% confidence intervals for β, depending on whether we used the likelihood ratio statistic Λ_n, the Wald statistic W_n with the model variance, or the Wald statistic $\tilde{W}_n$ with the estimated variance:

$$C_\Lambda = (1.05, 2.41), \quad C_W = (1.10, 2.34), \quad C_{\tilde{W}} = (0.96, 2.04)$$

Suppose we were interested in the hypothesis $H_0 : \beta = 1$. This hypothesis would be accepted at a 5% level with $\tilde{W}_n$ as the test statistic, but rejected by the other test statistics at that level, since $1 \notin C_\Lambda$ and $1 \notin C_W$, but $1 \in C_{\tilde{W}}$. □

Note that we cannot obtain the p-value from the confidence interval, and we cannot obtain the confidence interval from the observed p-value. To get the p-value from the confidence intervals, we need the full system of intervals for different degrees α of confidence. Similarly, to get the confidence intervals from a system of p-values, we need to consider the entire family of tests for hypotheses of the form $H_0 : \beta = \beta_0$ for $\beta_0 \in \mathbb{R}_+$.

7.6 Test in linear normal models

7.6.1 *The general case*

We consider a linear normal model with $X \sim \mathcal{N}_V(\xi, \sigma^2 I_V)$ where $(V, \langle\cdot,\cdot\rangle)$ is a d-dimensional Euclidean vector space. Different tests appear as we vary specifications of restrictions on ξ and σ^2. Below we shall derive likelihood ratio test statistics for some of these.

7.6.1.1 *Linear hypothesis, variance known*

We first consider the situation where σ^2 is known and $\xi \in V$ is completely unknown. We wish to test the hypothesis $H_0 : \xi \in L$, where $L \subseteq V$ is an m-dimensional linear subspace of V. The MLE of ξ in the unrestricted model is $\hat{\xi} = X$ and the maximized log-likelihood is

$$\ell(\hat{\xi}) = -\frac{\|X - \hat{\xi}\|^2}{2\sigma^2} = 0$$

where we have ignored additive constant terms involving 2π and σ^2. Similarly, under the hypothesis, the MLE is $\hat{\hat{\xi}} = \Pi_L(X)$ leading to the maximized log-likelihood

$$\ell(\hat{\hat{\xi}}) = -\frac{\|X - \hat{\hat{\xi}}\|^2}{2\sigma^2} = -\frac{\|X - \Pi_L(X)\|^2}{2\sigma^2}$$

We thus get for the likelihood ratio test statistic

$$\Lambda(X) = 2(\ell(\hat{\xi}) - \ell(\hat{\hat{\xi}})) = \frac{\|X - \Pi_L(X)\|^2}{\sigma^2}.$$

From Theorem 2.24, we get that $\Lambda(X)$ follows a $\chi^2(d-m)$ distribution leading to the p-value

$$p = P\{\Lambda(X) \geq \Lambda(x)\} = 1 - F^{d-m}(\Lambda(x))$$

where F^{d-m} is the distribution function for the $\chi^2(d-m)$ distribution.

7.6.1.2 Linear subhypothesis, variance unknown

A more common model has both of $\sigma^2 > 0$ and ξ unknown, assuming $\xi \in L$, where L is an m-dimensional linear subspace of V. We then wish to test the hypothesis $H_1 : \xi \in L_1$, where $L_1 \subseteq V$ is a k-dimensional linear subspace of L. The MLE of ξ in the larger model is $\hat{\xi} = \Pi_L(X)$ leading to the profile log-likelihood, maximized over $\xi \in L$

$$\ell(\hat{\xi}, \sigma^2) = -\frac{d}{2}\log\sigma^2 - \frac{\|X - \hat{\xi}\|^2}{2\sigma^2} = -\frac{d}{2}\log\sigma^2 - \frac{\|X - \Pi_L(X)\|^2}{2\sigma^2}$$

where we have ignored additive constants involving π, but this time we need the terms involving σ^2 as these are not constant. Maximizing again over σ^2 yields

$$\hat{\sigma}^2 = \|X - \Pi_L(X)\|^2/d$$

leading to the full maximized log-likelihood

$$\ell(\hat{\xi}, \hat{\sigma}^2) = -\frac{d}{2}\log\hat{\sigma}^2 - \frac{\|X - \Pi_L(X)\|^2}{2\hat{\sigma}^2} = -\frac{d}{2}\log\frac{\|X - \Pi_L(X)\|^2}{d} - \frac{d}{2}.$$

Similarly, the maximized likelihood under the hypothesis H_1 becomes

$$\ell(\hat{\hat{\xi}}, \hat{\hat{\sigma}}^2) = -\frac{d}{2}\log\hat{\hat{\sigma}}^2 - \frac{\|X - \Pi_{L_1}(X)\|^2}{2\hat{\hat{\sigma}}^2} = -\frac{d}{2}\log\frac{\|X - \Pi_{L_1}(X)\|^2}{d} - \frac{d}{2}$$

leading to the likelihood ratio test statistic

$$\Lambda(X) = 2(\ell(\hat{\xi}, \hat{\sigma}^2) - 2(\ell(\hat{\hat{\xi}}, \hat{\hat{\sigma}}^2)) = d\log\frac{\|X - \Pi_{L_1}(X)\|^2}{\|X - \Pi_L(X)\|^2}.$$

Since we have $\|X - \Pi_{L_1}(X)\|^2 = \|X - \Pi_L(X)\|^2 + \|\Pi_L(X) - \Pi_{L_1}(X)\|^2$, we may rewrite this as

$$\Lambda(X) = d\log\left(1 + \frac{\|\Pi_L(X) - \Pi_{L_1}(X)\|^2}{\|X - \Pi_L(X)\|^2}\right)$$

and thus we have $\Lambda(X) > \Lambda(x)$ if and only if $F(X) > f(x)$ where F is the normalized ratio of squared distances

$$F = F(X) = \frac{\|\hat{\xi} - \hat{\hat{\xi}}\|^2/(m-k)}{\|X - \Pi_L(X)\|^2/(d-m)} = \frac{\|\Pi_L(X) - \Pi_{L_1}(X)\|^2/(m-k)}{\|X - \Pi_L(X)\|^2/(d-m)}.$$

In other words, large values of the squared distance between the estimates are compared to the squared length of the residual indicated deviations from the hypothesis.

By Theorem 2.27, $\|\Pi_L(X) - \Pi_{L_1}(X)\|^2/\sigma^2$ and $\|X - \Pi_L(X)\|^2/\sigma^2$ are independent and χ^2-distributed with $m-k$ and $d-m$ degrees of freedom under the hypothesis, so F then follows an $F_{m-k,d-m}$-distribution. We may therefore calculate the p-value for the test as

$$p = P\{\Lambda(X) > \Lambda(x)\} = P\{F(X) > f(x)\} = 1 - F_{m-k,d-m}(f(x))$$

where $F_{m-k,d-m}$ is the distribution function for an F distribution with the relevant degrees of freedom.

7.6.2 *Some standard tests*

We shall here first mention a number of standard tests of which most may be seen as special instances of those derived in Section 7.6.1. These are then all equivalent to likelihood ratio tests in the sense that, for each of them, it holds that there is a function h such that

$$\Lambda(x) \geq \lambda_0 \iff d(x) \geq h(\lambda_0)$$

where $\Lambda(x)$ is the likelihood ratio statistic as defined in (7.1) and $d(x)$ is the test statistic in the specific example discussed. In other words, the system of critical regions defined by Λ and D are identical.

All of these standard tests are of *constant level*, meaning that the distribution of the test statistic is the same for all parameter values conforming with the hypothesis. In other words the test statistic is a *pivot* under the hypothesis, see also the next section for further discussion.

7.6.2.1 *Z-test for a given mean, variance known*

We consider a sample $X = (X_1, \ldots, X_n)$ from a normal distribution $\mathcal{N}(\mu, \sigma^2)$ where $\sigma^2 > 0$ is known and $\mu \in \mathbb{R}$ is unknown. The null hypothesis is $H_0 : \mu = \mu_0$ where μ_0 is a specific value. The test statistic is $d(x) = |Z(x)|$ where

$$Z = Z(X) = \frac{\bar{X}_n - \mu_0}{\sigma/\sqrt{n}}$$

which for all μ_0 follows a standard $\mathcal{N}(0,1)$ distribution so

$$p(x) = P_{\mu_0}\{|Z| > z(x)\} = 2(1 - \Phi(z(x)))$$

where Φ is the standard normal distribution function. This is an instance of the situation in Section 7.6.1.1 applied to $Y = X - \mu_0\mathbf{1}$.

7.6.2.2 *T-test for a given mean, variance unknown*

We consider a sample $X = (X_1, \ldots, X_n)$ from a normal distribution $\mathcal{N}(\mu, \sigma^2)$ where $\sigma^2 > 0$ and $\mu \in \mathbb{R}$ are both unknown. The null hypothesis is composite and given as $H_0 : \mu = \mu_0$ where μ_0 is a specific value, whereas $\sigma^2 > 0$ is also considered unknown under the hypothesis.

When the variance σ^2 is not known, the value of Z cannot be calculated and we use instead the test statistic $d(X) = T = |t(X)|$ where

$$T = t(X) = \frac{\bar{X}_n - \mu_0}{S_n/\sqrt{n}},$$

where

$$S_n = \sqrt{\frac{\sum_{i=1}^{n}(X_i - \bar{X}_n)^2}{n-1}}.$$

The distribution of T is for all (μ_0, σ^2) a Student's t-distribution so

$$p = p(x) = P_{\mu_0,\sigma^2}\{|T| > t(x)\} = 2(1 - F^T_{n-1}(t(x)))$$

where F^T_{n-1} is the distribution function for Student's T with $n-1$ degrees of freedom. This is again a special instance of the situation in Section 7.6.1.2 applied to $Y = X - \mu_0\mathbf{1}$. Then $L = \text{span}\{\mathbf{1}\}$ and $L_1 = \{0\}$ and we have $T^2 = F \sim F_{1,n-1}$.

7.6.2.3 T-test for comparing means

We consider two independent samples $X = (X_1, \dots, X_m)$ and $Y = (Y_1, \dots, Y_n)$ where $X_i \sim \mathcal{N}(\mu^X, \sigma^2)$ and $Y_i \sim \mathcal{N}(\mu^X, \sigma^2)$ are all mutually independent. We consider $(\mu^X, \mu^Y, \sigma^2) \in \mathbb{R} \times \mathbb{R} \times \mathbb{R}_+$ to be unknown and we are interested in the composite hypothesis $H_0 : \mu^X = \mu^Y$, whereas σ^2 is considered unknown. Then we use the test statistic $q(X, Y) = |t(X, Y)|$:

$$T = t(X, Y) = \frac{\bar{X}_m - \bar{Y}_n}{\bar{S}\sqrt{\frac{1}{m} + \frac{1}{n}}},$$

where

$$\bar{S}^2 = \frac{\sum_{i=1}^{m}(X_i - \bar{X}_m)^2 + \sum_{i=1}^{n}(Y_i - \bar{Y}_n)^2}{m+n-2}.$$

This test statistic follows for all values of $\mu = \mu^X = \mu_Y$ a Student's t-distribution with $m+n-2$ degrees of freedom, so again the relevant p-value is

$$p = p(x, y) = 2(1 - F^T_{m+n-1}(t(x, y))).$$

This is again special instance of the situation in Section 7.6.1.2. Here $V = \mathbb{R}^{m+n}$, $L = \text{span}\{(\mathbf{1}, 0), (0, \mathbf{1})\}$ and $L_1 = \text{span}\{(\mathbf{1}, \mathbf{1})\}$. Since $\dim L - \dim L_1 = 1$, we get $T^2 = F$.

It is important that the variances for the two samples are assumed identical. If this is not the case, no test of constant level exists and other *ad hoc* or approximate methods must be used. This situation is known as the *Behrens–Fisher problem.*

7.6.2.4 T-test for paired comparisons

Here we consider a *paired sample* $(X, Y) = (X_1, Y_1, \ldots, X_n, Y_n)$ where all single observations are independent and normally distributed as

$$X_i \sim \mathcal{N}(\mu_i^X, \sigma_X^2), \quad Y_i \sim \mathcal{N}(\mu_i^Y, \sigma_Y^2)$$

where $(\mu_i^X, \mu_i^Y, i = 1, \ldots, n; \sigma_X^2, \sigma_Y^2)$ are all unknown. We further assume that

$$\mu_i^Y = \mu_i^X + \delta, \tag{7.2}$$

i.e. that the difference of means in the two groups is the same for all i. Note that this is not an instance of the standard test in linear normal models unless we further assume $\sigma_X^2 = \sigma_Y^2$. For then the family of concentration matrices is not proportional to the identity.

But we may transform the data and let $D_i = X_i - Y_i$, noticing that then $D_i \sim \mathcal{N}(\delta, \sigma_X^2 + \sigma_Y^2)$. The problem of unequal variances has now disappeared after transforming the data to the set of differences. We are interested in the hypothesis $H_0 : \delta = 0$, corresponding to the situation where the means in the two samples are pairwise identical. Note that by considering the differences rather than the original observation has implied that the composite hypothesis in the original problem has become a simple hypothesis in terms of the differences and thus been reduced to the simple T-test in Section 7.6.2.2, so we use the test statistic $d(X) = |t(X, Y)|$ where

$$T = t(X, Y) = \frac{\bar{D}_n}{S_d \sqrt{n}}, \quad S_d^2 = \frac{1}{n-1} \sum_{i=1}^{n} (D_i - \bar{D})^2,$$

which under the hypothesis follows a Student's t-distribution with $n - 1$ degrees of freedom so the p-value becomes

$$p = p(x, y) = 2(1 - F_{n-1}^T(t(x, y))),$$

where F_{n-1}^T is the distribution function of Student's T with $n-1$ degrees of freedom.

7.7 Determining p-values

For any testing procedure to be operational, we need to be able to calculate the associated p-value and this section is devoted to methods for doing so. The simplest case is when the distribution of the test statistic $D = d(X)$ is the same for all $\theta \in \Theta_0$ or, in other words, $d(X)$ is a *pivot* under the hypothesis. We then say that the test has *constant level*, i.e. if it holds that

$$\gamma_\theta(d) = P_\theta\{D \geq d\} = \gamma(d), \quad \text{for all } \theta \in \Theta_0 \text{ and all } d \in \mathbb{R}.$$

Then the p-value for an outcome x is just

$$p = p(x) = \gamma(d(x)).$$

As we have seen in Section 7.6, a number of classical tests satisfy this property, including most tests associated with the general linear model, typically leading to test statistics following a normal distribution, Student's T, the χ^2-distribution, or the F-distribution.

7.7.1 Monte Carlo p-values

Above we have seen a number of examples where we have been able to calculate the p-values exactly, as the test statistics happened to equivalent to quantities having standard distributions with well-known properties. This is not always the case. However, often we are able to *simulate* from the distribution of the test statistic $D = d(X)$ and are therefore able to get a Monte Carlo estimate of the p value.

7.7.1.1 Simple hypotheses

Consider first a simple hypothesis $H_0 : \theta = \theta_0$ and assume that we have observed $Q = d(x)$ in our original sample. We may then generate a new and artificial sample $X^* = (x_1^*, \ldots, x_N^*)$ of size N from the distribution P_{θ_0} of X and then estimate the p-value by the relevant frequency:

$$\hat{p}_N = p_{\text{MC}}(d(x)) = \frac{1}{N}\sum_{i=1}^{N} \mathbf{1}_{(d(x),\infty)}(d(X_i^*)) = \frac{1}{N}\sum_{i=1}^{N} Y_i$$

where we have let $Y_i = \mathbf{1}_{(d(x),\infty)}(d(X_i^*))$. Then Y_i are independent Bernoulli random variable with success probability equal to the p-value $p = p(d(x))$ we are looking for, so we obtain an approximate $(1-\alpha)$-confidence interval for p using the Wald interval

$$\hat{p}_N \pm z_{1-\alpha/2} \times \sqrt{\hat{p}_N(1-\hat{p}_N)/N}.$$

7.7.1.2 Composite hypotheses

In the case of a composite hypothesis, we may be lucky and have a test statistic with constant level so that the distribution of D under the hypothesis is known. If this is the case, we essentially proceed as above for an arbitrary choice of $\theta_0 \in \Theta_0$. However, if this is not the case, we may use the following procedure which is known as *parametric bootstrap*. First, estimate θ under the hypothesis, say by maximum likelihood:

$$\hat{\hat{\theta}} = \text{argmax}_{\theta\in\Theta_0}\, \ell_x(\theta).$$

Then proceed as above, just simulating the artificial sample $X^* = (x_1^*, \ldots, x_N^*)$ from the estimated distribution $P_{\hat{\hat{\theta}}}$.

7.7.2 Asymptotic p-values

As an alternative to exact and Monte Carlo methods, we may use the results derived in Chapter 5 to enable us to calculate approximate p-values in a wide range of problems.

7.7.2.1 Simple hypotheses

For the case of a simple hypothesis in a model given by a curved exponential family, Theorem 5.35 yields exactly what we need for the approximate calculation of p-values for the likelihood ratio and Wald type statistics.

Example 7.4. [Continuation of Example 6.13] In the model with fixed coefficient of variation, i.e. where individual observations were assumed distributed as $\mathcal{N}(\beta, \beta^2)$ with $\beta > 0$ unknown and $n = 10$ observations had been seen with $\bar{t}_n = (0.5, 3)^\top$, we calculated the log-likelihood ratio statistic in (6.5) to

$$\Lambda_n(\beta) = \frac{30}{\beta^2} - \frac{10}{\beta} + 20 \log \beta - 14.776,$$

so the value of the test statistic for the hypothesis $H_0 : \beta = 1$ is

$$\Lambda_n(1) = 30 - 10 + 0 - 14.776 = 5.224.$$

We thus get the approximate p-value from a $\chi^2(1)$ distribution as

$$p = P\{\Lambda_n(1) \geq 5.22\} \approx 0.022$$

corresponding to the fact that the 95% confidence interval based on this statistic did not contain the value 1, see Example 7.3

Similarly, using the Wald statistic with model variance we get from (6.6) that

$$W_n(1) = 30/4 = 7.5$$

leading to the asymptotic p-value $p = .006$, yielding the same conclusion. Finally, we have

$$\tilde{W}_n(\beta) = \frac{3n}{\hat{\beta}_n^2}(\hat{\beta}_n - \beta)^2 = \frac{40}{3}(1.5 - \beta)^2$$

so $\tilde{W}_n(1) = 10/3 = 3.33$, leading to an asymptotic p-value of 0.068, corresponding to the fact that $\beta = 1$ was included in the last 95% Wald confidence interval. □

Example 7.5. [Continuation of Example 5.27] We consider again the curved family, where the mean of a bivariate normal distribution is assumed to be located on a semi-circle in the right half-plane and recall from Example 5.27 that the MLE based on n observations is

$$\hat{\beta}_n = \tan^{-1}(\bar{x}_{2n}/\bar{x}_{1n})$$

provided $\bar{x}_{1n} > 0$. From (3.10) the maximized log-likelihood function based on n observations then becomes

$$\ell_n(\hat{\beta}_n) = n\phi(\hat{\beta}_n)^\top \bar{x}_n = n(\bar{x}_{1n}/R, \bar{x}_{2n}/R)^\top \bar{x}_n = n\|\bar{x}_n\|^2/R = nR_n,$$

where $R = \|\bar{x}_n\| = \sqrt{\bar{x}_{1n}^2 + \bar{x}_{2n}^2}$.

Suppose now that we wish to test the hypothesis $H_0 : \beta = 0$. The log likelihood ratio statistic becomes

$$\Lambda_n = 2\left(\ell_n(\hat{\beta}_n) - \ell_n(0)\right) = 2nR - 2n\bar{x}_{1n} = 2n(R - \bar{x}_{1n}).$$

Thus this test statistic has an asymptotic $\chi^2(1)$-distribution and compares the length of the observation to the length of the first coordinate (recall that we assume $\bar{x}_{1n} > 0$ to ensure existence of the MLE). □

7.7.2.2 Composite hypotheses

Example 7.6. [Continuation of Example 5.30 and Example 7.5] We may wish to make a statistical test for the hypothesis that the mean is actually on the given semi-circle.

In the larger exponential family we assume no restrictions on the mean $\theta \in \mathbb{R}^2$, so we have the usual MLE $\hat{\theta} = \bar{x} = (\bar{x}_1, \bar{x}_2)^\top$ with the maximized log-likelihood function being

$$2\ell_n(\hat{\theta}_n) = -\sum_{i=1}^{n}(x_{i1} - \bar{x}_1)^2 - \sum_{i=1}^{n}(x_{i2} - \bar{x}_2)^2,$$

ignoring terms that are constant in θ. If we estimate in the model determined by

$$\theta = \phi(\beta) = (\sin\beta, \cos\beta)^\top, \beta \in (0, \pi)$$

we have seen that

$$\hat{\beta}_n = \tan^{-1}(\bar{x}_2/\bar{x}_1),$$

provided that $\bar{x}_1 > 0$. Thus, the log-likelihood function maximized under the hypothesis $H_0 : P \in \mathcal{P}_0$ is

$$\begin{aligned}2\ell_n(\hat{\beta}_n) &= -\sum_{i=1}^{n}(x_{i1} - \cos\hat{\beta}_n)^2 - \sum_{i=1}^{n}(x_{i2} - \sin\hat{\beta}_n)^2 = \\ &-\sum_{i=1}^{n}(x_{i1} - \bar{x}_1)^2 - \sum_{i=1}^{n}(x_{i2} - \bar{x}_2)^2 - n(\bar{x}_1 - \cos\hat{\beta}_n)^2 - n(\bar{x}_2 - \sin\hat{\beta}_n)^2 \\ &= 2\ell_n(\hat{\theta}_n) - n(\bar{x}_1 - \cos\hat{\beta}_n)^2 - n(\bar{x}_2 - \sin\hat{\beta}_n)^2\end{aligned}$$

Using now the expressions for $\cos\hat{\beta}_n, \sin\hat{\beta}_n$ in (5.17) from Example 5.27 we get if $\bar{x}_1 > 0$

$$\begin{aligned}\Lambda'_n &= 2(\ell_n(\hat{\theta}_n) - \ell_n(\hat{\beta}_n)) = n(\bar{x}_1 - \cos\hat{\beta}_n)^2 + n(\bar{x}_2 - \sin\hat{\beta}_n)^2 \\ &= n\left(\bar{x}_1 - \frac{\bar{x}_1}{R}\right)^2 + n\left(\bar{x}_2 - \frac{\bar{x}_2}{R}\right)^2 = nR^2\left(\frac{(1-R)}{R}\right)^2 = n(1-R)^2\end{aligned}$$

and this is asymptotically distributed as $\chi^2(2-1) = \chi^2(1)$, i.e. with 1 degree of freedom. The test statistic R measures how much the length R of the observed average differs from 1. Note however also that Λ_n only makes proper sense if $\bar{x}_1 > 0$ since otherwise the MLE under the hypothesis is not well-defined.

If we consider the simple hypothesis $H_0 : \theta = \theta_0 = (1, 0)^\top$ without assuming the semi-circle model, we would get the test statistic

$$\Lambda''_n = 2\left(\ell_n(\hat{\theta}_n) - \ell_n(\theta_0)\right) = n\|\bar{x}_n - \theta_0\|^2$$

and here the asymptotic $\chi^2(2)$-distribution would actually be exact. However, we also have

$$\begin{aligned}\Lambda''_n &= 2\left(\ell_n(\hat\theta_n) - \ell_n(\theta_0)\right)\\ &= 2\left(\ell_n(\hat\theta_n) - \ell_n(\hat\beta_n)\right) + 2\left(\ell_n(\hat\beta_n) - \ell_n(\theta_0)\right) = \Lambda'_n + \Lambda_n\end{aligned}$$

where Λ_n was calculated in Example 7.5. This may also be verified by the calculation

$$\begin{aligned}\Lambda'_n + \Lambda_n &= n(1-R)^2 + 2n(R - \bar x_{1n})\\ &= n(1 + R^2 - 2R + 2R - 2\bar x_1 n)\\ &= n(1 + \bar x_{1n}^2 + \bar x_{2n}^2 - 2\bar x_{1n})\\ &= n\left((\bar x_{1n} - 1)^2 + \bar x_{2n}^2\right) = n\|\bar x_n - \theta_0\|^2 = \Lambda''_n.\end{aligned}$$

Whereas Λ''_n has an exact $\chi^2(2)$-distribution, the individual constituents Λ_n and Λ'_n are only asymptotically distributed as $\chi^2(1)$. □

We shall also illustrate the testing of composite hypotheses in the case of constant coefficient of variation.

Example 7.7. [Continuation of Example 5.33] Assume that we in this example observed $\bar t_n = (.5, 3)^\top$ based on $n = 10$ observations and we wish to investigate whether the model is correct, i.e. whether the coefficient of variation is actually equal to 1, formally formulated as $H_0 : \sigma^2 = \xi^2$. Without this restriction we have $\hat\eta_n = \bar T_n$ so—ignoring the constant 2π—the log-likelihood becomes

$$2\ell_n(\hat\eta_n) = -\frac{\sum(X_i - \bar X_n)^2}{\hat\sigma^2} - n\log\hat\sigma^2 = -n(1 + \log\hat\sigma^2)$$

For the specific example, we get

$$\hat\sigma^2 = (SS/n - S^2/n^2) = 3 - 1/4 = 2.75, \quad 2\ell_n(\hat\eta_n) = -20.12.$$

Further, the maximized log-likelihood assuming the model is correct is

$$2\ell_n(\hat{\hat\eta}_n) = 2\ell_n(\hat\beta_n) = -\frac{SS_n}{\hat\beta_n^2} + \frac{2S_n}{\hat\beta_n} = n\log\hat\beta_n = -14.78.$$

So we get $\Lambda_n = 20.12 - 14.78 = 5.34$ with an asymptotic p-value at $p = 0.021$ judged in a χ^2-distribution with $2 - 1 = 1$ degrees of freedom. So the hypothesis cannot reasonably be maintained.

Alternatively we could consider the Wald statistic, and we shall first look at the version where the covariance is estimated under the hypothesis. We get

$$\begin{aligned}W_n &= n(\hat\eta_n - \hat{\hat\eta}_n)^\top \hat{\hat\Sigma}_n^{-1}(\hat\eta_n - \hat{\hat\eta}_n)\\ &= \frac{10}{81}(-1, -1.5)\begin{pmatrix}108 & -24\\ -24 & 18\end{pmatrix}\begin{pmatrix}-1\\ -1.5\end{pmatrix}\\ &= 10(108 - 36 - 36 + 81/2)/81 = 85/9 = 9.44.\end{aligned}$$

Corresponding to a p-value of $p = 0.002$ as we also here use a χ^2-distribution with $2 - 1 = 1$ degrees of freedom, leading to the same conclusion as above.

If we estimate outside the hypothesis we get $\hat{\sigma}^2 = 11/4$, $\hat{\xi} = 1/2$ and thus using expressions derived in Example 3.21, the estimated covariance becomes

$$\Sigma_{\hat{\xi},\hat{\sigma}^2} = \mathbf{V}_{\hat{\xi},\hat{\sigma}^2}\left\{\begin{pmatrix} X \\ X^2 \end{pmatrix}\right\} = \begin{pmatrix} \hat{\sigma}^2 & 2\hat{\sigma}^2\hat{\xi} \\ 2\hat{\sigma}^2\hat{\xi} & 2\hat{\sigma}^4 + 4\hat{\sigma}^2\hat{\xi}^2 \end{pmatrix} = \frac{11}{8}\begin{pmatrix} 2 & 2 \\ 2 & 13 \end{pmatrix}$$

which has inverse

$$\hat{\Sigma}_n^{-1} = \frac{4}{121}\begin{pmatrix} 13 & -2 \\ -2 & 2 \end{pmatrix}$$

so

$$\begin{aligned} \tilde{W}_n &= \frac{40}{121}(-1, -1.5)\begin{pmatrix} 13 & -2 \\ -2 & 2 \end{pmatrix}\begin{pmatrix} -1 \\ 1.5 \end{pmatrix} \\ &= 20(26 - 12 + 9)/121 = 430/121 = 3.55. \end{aligned}$$

This yields a p-value of $p = 0.059$ using a χ^2-distribution with $2 - 1 = 1$ degrees of freedom, so here we do not clearly reject the hypothesis. It is typically true that the Wald statistic with covariance estimated outside the hypothesis loses power compared to the other test statistics due to the extra variability in the estimate of the covariance. □

7.7.2.3 Smooth hypotheses

The theorem identifying the χ^2-distribution of the log-likelihood ratio statistic can be generalized to the slightly wider case of a *smooth hypothesis* of order d:

$$H_0 : h(\beta) = 0 \tag{7.3}$$

where $h : B \to \mathbb{R}^{m-d}$ is a smooth map with Jacobian having full rank $m - d$ for all $\beta \in B$ with $h(\beta) = 0$. This fact is sometimes also referred to as *Wilks' theorem* although Wilks (1938) actually was giving the slightly weaker version in Theorem 5.39.

The hypothesis H_0 may, for example, be of the form that certain parameters in the larger family have the value 0:

$$H_0 : \beta_{d+1} = 0, \ldots, \beta_m = 0, \tag{7.4}$$

in which case we have $\alpha = \Pi_d(B)$ where Π_d is the coordinate projection onto the first d coordinates of $\beta \in B$. Then the map γ is simply

$$\gamma(\alpha_1, \ldots, \alpha_d) = (\alpha_1, \ldots, \alpha_d, 0, \ldots, 0)$$

in which case the smooth family is also a curved family. If the hypothesis has the form (7.4) we say the hypothesis *has standard form*.

Theorem 7.8. *Let $H_0 : h(\beta) = 0$ be a smooth hypothesis of order d in a curved exponential family $\mathcal{P} = \{P_{\phi(\beta)}, \beta \in B\}$ of dimension m. Then the maximized log-likelihood ratio statistic Λ_n for the hypothesis satisfies*

$$\Lambda_n \overset{\mathcal{D}}{\to} \chi^2(m-d)$$

with respect to any $P \in \tilde{\mathcal{P}} = \{P_{\phi(\beta)} \mid \beta \in B, h(\beta) = 0\}$. As before, $\chi^2(m-d)$ denotes the χ^2-distribution with $m-d$ degrees of freedom.

Proof. Clearly, a curved composite hypothesis as in Theorem 5.39 is a special case of a smooth hypothesis, since a curved hypothesis can always be represented in standard form (7.4).

But, in fact, the implicit function theorem exactly says that the converse is true, although only locally. If β_0 satisfies the hypothesis, i.e. if $h(\beta_0) = 0$, there is a neighbourhood around $\phi(\beta_0)$ which has a smooth parametrization as a curved subfamily. Since, asympotically, only the local structure around $\phi(\beta_0)$ matters, Theorem 5.39 therefore essentially covers the smooth case as well. We refrain from giving the full technical details. □

Example 7.9. [Bivariate normal with mean on circle] A simple example of smooth hypothesis is a modification of Example 3.28. We again assume

$$X \sim \mathcal{N}_2(\theta, I_2)$$

for $\Theta = \mathbb{R}^2$ but assume now further that the mean θ lies on the unit circle, i.e.

$$H_0 : \theta_1^2 + \theta_2^2 = 1.$$

Note that this differs from Example 3.28 where we only considered a half-circle. This subfamily does *not* correspond to a curved exponential family since there is no smooth homeomorphism from an open interval in $\mathbb{R}$ to the unit circle. However, as we shall see, it does not really matter.

The Jacobian of the map $(x, y) \to x^2 + y^2$ is $(2x, 2y)$ and hence this has full rank everywhere except at $(x, y) = (0, 0)$. Going through the same mimimization exercise as in Example 5.27, we get that under H_0, the MLE is uniquely determined if and only if $\bar{x} \neq 0$ and then

$$\hat{\hat{\theta}} = \frac{1}{\sqrt{\bar{x}_1^2 + \bar{x}_2^2}} \begin{pmatrix} \bar{x}_1 \\ \bar{x}_2 \end{pmatrix} = \bar{x}/R.$$

The log-likelihood ratio test statistic here is exactly the same as in Example 7.6:

$$\Lambda_n = n(1-R)^2 \overset{\text{as}}{\sim} \chi^2(1),$$

we leave the details to the reader as an exercise. The difference is now that *the test statistic Λ_n now makes sense if only $(\bar{x}_1, \bar{x}_2) \neq (0, 0)$*, since the MLE always exists and is well defined. □

7.8 Exercises

Exercise 7.1. Show that the simple T-test in Section 7.6.2.2 is equivalent to the likelihood ratio test in a suitable linear normal model.

Exercise 7.2. Consider comparing means as in Section 7.6.2.3 and show that the test described is equivalent to the likelihood ratio test when variances in the groups are assumed equal. Show that if the variances are not assumed equal, the projection onto the space of identical means depends on the ratio of variances.

Exercise 7.3. The data below are based on Fisher (1947) and represent the weight of the heart in 16 cats as a percentage of the weight of the body. Do you see any difference between male and female cats?

Gender	Percentage of heart weight								
Female	0.276	0.247	0.288	0.274	0.200	0.276	0.242		
Male	0.275	0.221	0.280	0.197	0.283	0.260	0.226	0.289	0.206

Exercise 7.4. Consider the situation with paired comparisons as in Section 7.6.2.4. Show that the suggested test based on differences is equivalent to the likelihood ratio test based on the original observations if and only if the variances are assumed equal.

Exercise 7.5. Consider the Pareto distribution with fixed threshold c ($c = 1$ in Exercise 3.3) and index parameter $\theta > 0$:

$$f_\theta^c(x) = \frac{\theta c^\theta}{x^{\theta+1}}, \qquad \text{for } x > c.$$

The journal *Forbes Magazine* lists every year the world's largest personal assets. The data below are from 2017 and 2018 and gives the values of personal assets above 50 billion U.S. dollars, as listed by Forbes Magazine.

	Assets in billions of US dollars								
2017	86	76	73	71	56	54	52		
2018	112	90	84	72	71	70	67	60	58

Assuming that these are Pareto distributed with threshold $c = 50$ and index parameter $\theta_i > 0, i = 2017, 2018$, we would like to test the hypothesis that the index parameter is unchanged from 2017 to 2018, i.e. $H_0 : \theta_{2017} = \theta_{2018}$.

a) Calculate the likelihood ratio test for this hypothesis, but calculate the p-value both via Monte Carlo methods and by asymptotic approximation.

b) Under the assumption that H_0 may be upheld, determine a 95% confidence interval for the index parameter θ.

c) Would the hypothesis $H_1 : \theta_{2017} = \theta_{2018} = 1$ be compatible with the data from Forbes magazine?

Exercise 7.6. Two methods for analysing starch content in potatoes are to be analysed and compared. Sixteen potatoes with varying content of starch are measured by the two methods with results displayed in Table 7.1. Are the measurement methods comparable?

Table 7.1 – Starch content in potatoes in % as measured in two different ways. Source: Hald (1952) based on von Scheele et al. (1935).

Sample	Method I	Method II
1	21.7	21.5
2	18.7	18.7
3	18.3	18.3
4	17.5	17.4
5	18.5	18.3
6	15.6	15.4
7	17.0	16.7
8	16.6	16.9
9	14.0	13.9
10	17.2	17.0
11	21.7	21.4
12	18.6	18.6
13	17.9	18.0
14	17.7	17.6
15	18.3	18.5
16	15.6	15.5

Exercise 7.7. Let X and Y be independent and exponentially distributed random variables with $\mathbf{E}(X) = \lambda_1$ and $\mathbf{E}(Y) = \lambda_2$ where $\lambda = (\lambda_1, \lambda_2) \in \mathbb{R}^2_+$. and let $(X_1, Y_1), \ldots, (X_n, Y_n)$ be a sample from this distribution. Consider the hypothesis $H_0 : \lambda_1\lambda_2 = 1$ as in Exercise 3.9, Exercise 4.10, and Exercise 5.10.

a) Determine the log-likelihood ratio statistic for the composite hypothesis H_0.

b) Determine a Wald test statistic for the composite hypothesis based on the fact that the hypothesis may be expressed through the parameter function

$$\phi(\lambda) = \log \lambda_1 + \log \lambda_2$$

as $H_0 : \phi(\lambda) = 0$.

c) Does the following (simulated) sample support the hypothesis H_0? Use both the likelihood ratio and Wald test statistics as derived above?

x	0.581	0.621	3.739	4.354	0.409	1.843	1.705	0.312
y	0.469	0.084	0.362	2.050	2.704	0.034	0.061	0.419

Now we shall investigate the simple hypothesis $H_1 : \lambda_1 = \lambda_2 = 1$ within the model determined by H_0.

d) Derive the likelihood ratio statistic for H_1 under the assumption of H_0.

e) Derive the quadratic score test statistic for H_1 under the assumption of H_0.

f) Are the data under c) compatible with H_1?

g) Determine the Monte Carlo p-value for the above tests.

Exercise 7.8. Let $X_1, \ldots, X_n$ be a sample from a gamma distribution with unknown shape and scale.

a) Derive the likelihood ratio and Wald test statistics for the hypothesis $H_0 : \alpha = 1$, i.e. the hypothesis that the observations are from an exponential distribution.

b) Calculate the asymptotic and Monte Carlo p-value for the hypothesis H_0 when the following (simulated) data have been observed

0.469	0.084	0.362	2.050	2.704	0.034	0.061	0.419

Exercise 7.9. Suppose a multiple choice exam is to be constructed, with four possible answers in each category. If we make it a pass criterion that 62.5% of the questions must be correctly answered, how many questions do we need to ensure that a student who is merely guessing is failing with probability 0.999?

Exercise 7.10. A clinical trial is to be conducted with a new medicine for controlling hypertension. A total of $2n$ patients who suffer from high blood pressure will be randomized into two groups of equal size, treatment and control. Patients in the treatment group are given the new medicine and patients in the control group are given a traditional medicine. The blood pressure of the patients is measured (in mm Hg) at the beginning of the trial, and three months later. The change in the average of systolic and diastolic blood pressure is recorded for each patient. The standard deviation σ for differences of measurements of this type is known to be around 10 mm Hg. How many patients are needed to detect a difference in effect of 10mm Hg with probability .99?

Chapter 8

Models for Tables of Counts

In this chapter we shall further investigate a number of specific and much used models for data given in the form of tables of counts. The models considered are all either curved or regular exponential models and the chapter therefore illustrates the use of the concepts developed in the previous chapters.

8.1 Multinomial exponential families

8.1.1 *The unrestricted multinomial family*

We consider n objects, classified into $k+1$ groups which we here shall considered labeled as $\mathcal{X} = (0, 1, \dots, k)$. We may thus think of corresponding random variables as $X_1, \dots, X_n$ with values in $\mathcal{X}$. We shall consider these independent and identically distributed with $\pi_x = P(X_i = x)$ denoting the probability that an object falls in the category $x \in \mathcal{X}$.

We further assume that $\pi_x > 0$ for all $x \in \mathcal{X}$ so $\pi = (\pi_0, \dots, \pi_k)$ is an element of the open *k-dimensional simplex* Δ^k:

$$\pi \in \Delta^k = \left(\pi \in \mathbb{R}^{k+1} \;\middle|\; \pi_x > 0, x \in \mathcal{X}, \sum_{x=0}^{k} \pi_x = 1 \right). \tag{8.1}$$

We note also that Δ^k is smoothly parametrized with $\pi_{\setminus 0} = (\pi_1, \dots, \pi_k)$ so that we may write

$$\Delta^k \sim \Delta^k_0 = \left(\pi_{\setminus 0} \in \mathbb{R}^k \;\middle|\; \pi_x > 0, x \in \mathcal{X}_{\setminus 0}, \sum_{x=1}^{k} \pi_x < 1 \right)$$

where $\mathcal{X}_{\setminus 0} = \mathcal{X} \setminus (0)$. To identify this as an exponential family, we may for $\pi \in \Delta^k$ write the density of a random variable X as

$$f_\pi(x) = \prod_{j=0}^{k} \pi_j^{\mathbf{1}_{(j)}(x)} = \exp\left(\sum_{j=1}^{k} \theta_j \mathbf{1}_{(j)}(x) - \psi(\theta) \right) = \exp\left(\theta^\top t(x) - \psi(\theta) \right)$$

DOI: 10.1201/9781003272359-8

with $\theta_j = \log(\pi_j/\pi_0), j = 1, \dots k, t(x) = (\mathbf{1}_{(1)}(x), \dots, \mathbf{1}_{(k)}(x))$, and

$$\psi(\theta) = \log\left(1 + \sum_{j=1}^{k} e^{\theta_j}\right) = -\log \pi_0.$$

This identifies the family of multinomial distributions on $\mathcal{X}$ as a k-dimensional regular exponential family with moment map

$$\tau_j(\theta) = \mathbf{E}_\theta(t_j(X)) = P_\theta(X = j) = \pi_j = \frac{e^{\theta_j}}{1 + \sum_{j=1}^k e^{\theta_j}}, \quad j \in \mathcal{X}_0$$

that is a diffeomorphism between $\Theta = \mathbb{R}^k$ and Δ_0^k with inverse

$$\tau^{-1}(\pi_{\setminus 0})_j = \theta_j = \log \frac{\pi_j}{1 - \sum_{j=1}^k \pi_j} = \log \frac{\pi_j}{\pi_0}.$$

We thus note that if we let $Y_j = \sum_{i=1}^n t_j(X_i)$ denote the number of objects in category j, we have the MLE determined by the corresponding moment equation and hence

$$\hat{\pi}_j = Y_j/n, \quad j = 1, \dots, k.$$

Since $\pi_0 = 1 - \sum_{j=1}^n \pi_j$ and $\sum_{j \in \mathcal{X}} Y_j = n$, we have $Y_0 = n - \sum_{j \in \mathcal{X}_{\setminus 0}} Y_j$ and therefore in fact also

$$\hat{\pi}_0 = 1 - \sum_{j=1}^k \hat{\pi}_j = Y_0/n.$$

The maximum-likelihood estimator is well-defined if all categories have been observed, i.e. if $Y_j > 0$ for all $j = 1, \dots, k$. The distribution of $Y = (Y_0, \dots, Y_k)$ is multinomial:

$$\begin{aligned} P_\pi(Y = y) &= \binom{n}{y_0, \dots y_k} \prod_{j=0}^k \pi_j^{y_j} = \frac{n!}{y_0! \cdots y_k!} \prod_{j=0}^k \pi_j^{y_j} \\ &= \exp\left(\theta^\top \sum_{i=1}^n t(x_i) - n\psi(\theta)\right) \frac{n!}{y_0! \cdots y_k!}. \end{aligned}$$

The covariance becomes

$$\kappa(\theta)_{rs} = \mathbf{V}_\theta(t_r(X) t_s(X)) = \frac{1}{n} \mathbf{V}_\theta(Y_r, Y_s) = \begin{cases} \pi_r(1 - \pi_r) & \text{if } r = s \\ -\pi_r \pi_s & \text{if } r \neq s \end{cases}$$

which can be obtained by differentiation of τ or directly, using that we in fact have $t_j(X)^2 = t_j(X)$ since t_j has values in $(0, 1)$. If we denote this covariance by $\Sigma_{\pi_{\setminus 0}}$, we may write

$$\mathbf{V}_\theta(t(X)) = \Sigma_{\pi_{\setminus 0}} = D(\pi_{\setminus 0}) - \pi_{\setminus 0} \pi_{\setminus 0}^\top$$

where $D(\pi_{\setminus 0})$ is the $k \times k$ diagonal matrix

$$D(\pi_{\setminus 0}) = \begin{pmatrix} \pi_1 & 0 & \cdots & 0 \\ 0 & \pi_2 & \cdots & 0 \\ \vdots & \vdots & \ddots & \vdots \\ 0 & 0 & \cdots & \pi_k \end{pmatrix}. \tag{8.2}$$

We then have

Lemma 8.1. *The inverse covariance matrix for $t(X)$ is given as*

$$\Sigma_{\pi_{\setminus 0}}^{-1} = D(1/\pi_{\setminus 0}) + \frac{1}{\pi_0}\mathbf{1}_k\mathbf{1}_k^\top.$$

Proof. This follows from direct matrix multiplication. Using that

$$\pi_{\setminus 0}^\top \mathbf{1}_k = \sum_1^k \pi_i = 1 - \pi_0$$

we have

$$\begin{aligned} \Sigma_{\pi_{\setminus 0}}\left(D(1/\pi_{\setminus 0}) + \frac{1}{\pi_0}\mathbf{1}_k\mathbf{1}_k^\top\right) &= \\ D(\pi_{\setminus 0})D(1/\pi_{\setminus 0}) - \pi_{\setminus 0}\pi_{\setminus 0}^\top D(1/\pi_{\setminus 0}) + \frac{1}{\pi_0}D(\pi_{\setminus 0})\mathbf{1}_k\mathbf{1}_k^\top - \frac{1}{\pi_0}\pi_{\setminus 0}\pi_{\setminus 0}^\top\mathbf{1}_k\mathbf{1}_k^\top \\ = I_k - \pi_{\setminus 0}\mathbf{1}_k^\top + \frac{1}{\pi_0}\pi_{\setminus 0}\mathbf{1}_k^\top - \frac{1-\pi_0}{\pi_0}\pi_{\setminus 0}\mathbf{1}_k^\top = I_k, \end{aligned}$$

as desired. □

8.1.2 Curved multinomial families

In many examples we may know or postulate more about the structure of the vector π of probabilities, then typically specified as an m-dimensional curved subfamily of the multinomial family, i.e. we have

$$\theta = \phi(\beta), \beta \in B \subseteq \mathbb{R}^m$$

where ϕ satisfies the conditions in Definition 3.24. It follows that the asymptotic theory for maximum-likelihood estimation (MLE) and testing developed previously applies, so that $\hat{\beta}_n$ is asymptotically well-defined and asymptotically distributed as $\mathcal{N}(\beta, i(\beta)^{-1}/n)$; we just have to solve the score equation (often numerically), establish whether the solution corresponds to a global maximum of the score equation, and calculate the Fisher information.

8.1.2.1 Score and information

We let

$$\pi_j(\beta) = \tau_j(\phi(\beta)) = \mathbf{E}_{\phi(\beta)}(t_j(X)), j \in \mathcal{X}_0, \quad \pi_0(\beta) = P_\beta(X = 0)$$

and get for the log-likelihood and score functions for a single observation

$$\ell(x, \beta) = \log \pi_x(\beta), \quad S(x, \beta) = \frac{D\pi_x(\beta)}{\pi_x(\beta)} \tag{8.3}$$

and further for the Fisher information, using the second Bartlett identity in Theorem 1.23:

$$i(\beta) = \mathbf{E}_\beta(S(X, \beta)^\top S(X, \beta)) = \sum_{j=0}^{k} \frac{D\pi_j(\beta)^\top D\pi_j(\beta)}{\pi_j(\beta)}. \tag{8.4}$$

For the case of n observations, we then get from (8.3) that the log-likelihood and score function is

$$\ell_n(\beta) = \sum_{j=0}^{k} Y_j \log \pi_j(\beta), \quad S_n(\beta) = \sum_{j=0}^{k} Y_j \frac{D\pi_j(\beta)}{\pi_j(\beta)}. \tag{8.5}$$

8.1.2.2 Likelihood ratio

If we let $\mathcal{P}$ denote the set of distributions in the unrestricted multinomial family and $\mathcal{P}_0$ the set of distributions specified by the curved multinomial subfamily, the likelihood ratio test for the composite hypothesis $H_0 : P \in \mathcal{P}_0$ becomes

$$\Lambda_n = 2\left(\ell_n(\hat{\pi}) - \ell_n(\hat{\beta})\right) = 2\sum_{j=0}^{k} Y_j \log \frac{Y_j/n}{\pi_j(\hat{\beta})} = 2\sum_{j=0}^{k} \mathrm{OBS}_j \log \frac{\mathrm{OBS}_j}{\mathrm{EXP}_j} \tag{8.6}$$

where OBS_j is the *observed* number of objects in group j, $\mathrm{EXP}_j = n\pi_j(\hat{\beta})$ is the *expected* number of objects in category j, and the log likelihood ratio Λ_n is asymptotically distributed as $\chi^2(k - m)$.

Note that, strictly speaking, this test statistic is not well-defined unless it holds that $\mathrm{OBS}_j = Y_j > 0$ for all j. However, if we use the convention that $0 \cdot \log 0 = 0$, the statistic makes sense just if $\mathrm{EXP}_j \neq 0$.

8.1.2.3 Wald statistics

The Wald statistic for the same hypothesis would be given as

$$W_n = n\left(\hat{\pi}_{\setminus 0} - \pi(\hat{\beta})_{\setminus 0}\right)^\top \Sigma^{-1}_{\pi(\hat{\beta})_{\setminus 0}} \left(\hat{\pi}_{\setminus 0} - \dot{\pi}(\hat{\beta})_{\setminus 0}\right).$$

Using Lemma 8.1 yields a simpler expression for W_n:

$$\begin{aligned} W_n &= n\left(\hat{\pi}_{\setminus 0}-\pi(\hat{\beta})_{\setminus 0}\right)^\top D(1/\pi(\hat{\beta})_{\setminus 0})\left(\hat{\pi}_{\setminus 0}-\pi(\hat{\beta})_{\setminus 0}\right) \\ &\quad + n\frac{\left((\hat{\pi}-\pi(\hat{\beta}))^\top \mathbf{1}_k\right)^2}{\hat{\pi}_0(\beta)} \\ &= n\sum_{j=1}^k \frac{(Y_j/n-\pi_j(\hat{\beta}))^2}{\pi_j(\hat{\beta})} + n\frac{(\hat{\pi}_0-\pi_0(\hat{\beta}))^2}{\hat{\pi}_0(\beta)} \\ &= \sum_{j=0}^k \frac{(Y_j-n\pi_j(\hat{\beta}))^2}{n\pi_j(\hat{\beta})} = \sum_{j=0}^k \frac{(\mathrm{OBS}_j-\mathrm{EXP}_j)^2}{\mathrm{EXP}_j}. \end{aligned} \tag{8.7}$$

The Wald statistic in (8.7) is also known as *Pearson's* χ^2 and has an asymptotic χ^2-distribution with $k-m$ degrees of freedom as was also true for the likelihood ratio Λ_n in (8.6). In fact, since this Wald statistic is based on the mean value parameter, it is also equal to the quadratic score statistic and the expression in (8.7) is therefore invariant under reparametrization. This may also be seen directly as the last expression is free from terms that depend on the specific parametrization of the two models involved.

In the statistical literature on the analysis of tables of counts, see for example Agresti (2002), *it is common to use the notation* G^2 *for* Λ *and* X^2 *for* W which we shall also do here. Also, we shall not in any detail discuss issues relating to the quality of the χ^2-approximation to the test statistics, but just mention that experience has shown that the approximation is adequate whenever $\mathrm{EXP}_j \geq 5$ for all j. If this is not the case, Monte Carlo methods may be used. Sometimes the condition $\mathrm{EXP}_j \geq 5$ may be achieved by joining categories with small expected numbers, see Example 8.2.

For the sake of completeness, we display the alternative Wald statistic where the variance is estimated in the larger model:

$$\begin{aligned} \tilde{W}_n &= n\left(\hat{\pi}_{\setminus 0}-\pi(\hat{\beta})_{\setminus 0}\right)^\top \Sigma^{-1}_{\hat{\pi}_{\setminus 0}}\left(\hat{\pi}_{\setminus 0}-\pi(\hat{\beta})_{\setminus 0}\right) \\ &= \sum_{j=0}^k \frac{(Y_j-n\pi_j(\hat{\beta}))^2}{Y_j} = \sum_{j=0}^k \frac{(\mathrm{OBS}_j-\mathrm{EXP}_j)^2}{\mathrm{OBS}_j}. \end{aligned}$$

This differs from Pearson's χ^2 by using observed rather than expected counts in the denominator. However, this is rarely used; the computational effort is mostly associated with calculating EXP_j, and thus there is no essential computational saving compared to calculating G^2 or X^2.

8.1.3 Residuals

To check the validity of a model used, it is useful to supplement the numerical test statistics in (8.6) and (8.7) by inspecting appropriate residuals representing the difference between the observed and expected values, properly normalized. Inspection may be done graphically, to disclose systematic unexpected patterns, or numerically, by identifying residuals that are particularly large, possibly representing deviations from the model used that may then be further investigated. The *Pearson residuals* are the quantities

$$R_j = \frac{Y_j - n\pi_j(\hat{\beta})}{\sqrt{n\pi_j(\hat{\beta})}} = \frac{\mathrm{OBS}_j - \mathrm{EXP}_j}{\sqrt{\mathrm{EXP}_j}}$$

and Pearson's χ^2 statistic is then equal to $W = \sum_j R_j^2$.

The Pearson residuals are asymptotically normally distributed with expectation 0, but their asymptotic variance is smaller than 1 and equal to $1 - h_{jj}$, where h_{jj} are the diagonal elements of the *hat-matrix* of the model as given in (5.21), i.e. the matrix for the orthogonal projecton of y onto the space of tangents to the model with respect to the inner product determined by the Fisher information, see Lemma 5.32 for further details. It may therefore be more useful to consider the *standardized Pearson residuals* that correct for this reduced variance, i.e.

$$R_j^* = \frac{\mathrm{OBS}_j - \mathrm{EXP}_j}{\sqrt{\mathrm{EXP}_j(1 - \hat{h}_{jj})}}.$$

8.1.4 *Weldon's dice*

We illustrate the previous developments in a simple example, described in the classic text of Fisher (1934).

Example 8.2. [Weldon's dice data] *Walter Frank Raphael Weldon* (1860–1906) was a Professor of Zoology at Oxford University and worked on statistics with Frank Galton and Karl Pearson. Weldon rolled a set of 12 dice 26306 times (!) and recorded how many of the 12 dice were facing 5 or 6. The resulting data are displayed in Table 8.1. The total number Y with faces 5 or 6 when rolling 12 dice should follow a binomial distribution with length 12 and success parameter θ, yielding the expected numbers

$$\pi_j(\theta) = P_\theta(Y = j) = \binom{12}{j}\theta^j\,(1-\theta)^{12-j}.$$

If the dice were fair, we would have $\theta = 1/3$ and the expected numbers would be

$$\mathrm{EXP}_j = 26306 \times \binom{12}{j}\left(\frac{1}{3}\right)^j\left(\frac{2}{3}\right)^{12-j}$$

and these numbers are displayed in the third column of Table 8.1. It is apparent that the observed counts are too small for $j = 1, \ldots, 4$ and too large for $j > 4$, also reflected in the Pearson residuals, displayed in the fourth column.

Pearsons χ^2 for goodness-of-fit is equal to the sum of squares of these residuals and evaluates to $X^2 = 35.49$ when categories 10, 11, and 12 are combined to achieve reasonably large expected numbers. Compared to a χ^2-distribution with 10 degrees

Table 8.1 – Weldon's dice data: the number of dice among 12 which face 5 or 6 together with their expected number under the assumption that the dice are fair or not and the associated Pearson residuals. Residuals 1.96 or higher are indicated with bold face type. Categories 10, 11, and 12 are combined for the calculation of residuals. Source: Fisher (1934).

# 5 or 6	OBS	EXP fair	EXP biased	Residual fair	Residual biased
0	185	202.75	187.38	-1.25	-0.17
1	1149	1216.50	1146.51	-1.94	0.07
2	3265	3345.37	3215.24	-1.39	0.88
3	5475	5575.61	5464.70	-1.35	0.14
4	6114	6272.56	6269.35	**-2.00**	**-1.96**
5	5194	5018.05	5114.65	**2.48**	1.11
6	3067	2927.20	3042.54	**2.58**	0.44
7	1331	1254.51	1329.73	**2.16**	0.03
8	403	392.04	423.76	0.55	-1.01
9	105	87.12	96.03	1.92	0.92
10	14	13.07	14.69		
11	4	1.19	1.36	0.97	0.47
12	0	0.05	0.06		
	26306	26306	26306	$X^2 = 35.49$	$X^2 = 8.18$

of freedom—there are 11 categories when the last three are combined into one—yields a p-value of 0.00001 and we must reject the hypothesis that the dice are fair. The likelihood ratio statistic $G^2 = 35.10$ is not much different.

Alternatively, we could try to stick with the binomial distribution, but now estimate the success probability which here may be calculated to

$$\hat{\theta} = 106602/315672 = 0.3377$$

which is slightly larger than $1/3$. Expected values calculated for this value of θ are displayed in the fourth column of Table 8.1 and yield a much better description of the observed values, as reflected both in the Pearson residuals in the last column and in Pearson's χ^2 which now evaluate to $X^2 = 8.18$ yielding a p-value of 0.52 when compared to a χ^2-distribution with $10 - 1 = 9$ degrees of freedom, thus indicating that the data conform well with the binomial distribution. Here the likelihood ratio statistic is indistinguishably equal to $G^2 = 8.18$.

Interestingly, the 95% Wald confidence interval for θ evaluates to

$$\hat{\theta} \pm 1.96\sqrt{\hat{\theta}(1-\hat{\theta})/26306} = (0.3320, 0.3434)$$

which does include the value $1/3$ and hence from the frequency alone, we cannot conclude that the dice are unfair. Rather it is the systematic deviation observed in the binomial distribution that reveals the problem.

The data have been analyzed much by several authors over the years and different explanations have been suggested for the phenomenon, including the fact that some 7000 throws or so were actually made by an assistant. Since the opposite faces of 4, 5, and 6 are 3, 2, and 1, and the eyes of a die typically have been carved out, the carvings might just affect the center of gravity of the die to favour the larger number for the smaller. However—as far as I am aware—nobody has yet gone through the trouble and repeated Weldon's experiment.

We may also here emphasize that although the deviation from the binomial model with success parameter $1/3$ is *significant*, it is so small that it for most practical uses is *unimportant*, see also Example 7.1 for a similar situation. □

8.2 Genetic equilibrium models

Many models for multinomial observations originate from genetics, not least because R. A. Fisher was a geneticist and developed much statistical methodology with the purpose of solving scientific problems within this field. In this section we briefly describe two classical genetic models. It is amazing that these models were developed much before DNA was discovered and have survived the molecular revolution in genetics, just with a more precise understanding of what a gene is.

8.2.1 Hardy–Weinberg equilibrium

We consider a so-called *trait* that is determined by a single gene which might come in exactly two varieties, say a and A. These varieties are termed *alleles*. Humans are *diploid* individuals, meaning that chromosomes come in pairs, one inherited from the father, and one from the mother, so at a specific *locus* (position on the genome), the genetic composition might be any of aa, aA, and AA and this composition is referred to as the *genotype*.

The study of so-called SNPs (Single Nucleotide Polymorphisms) is an important part of modern genetics; then the allele A would represent a possible mutation on a specific location on the genome, whereas the allele a would represent the most common variant. For example, a could correspond to the typical and most common nucleobase pair in the population at that locus, say C-G, and A might correspond to a mutation to the other possible nucleobase pair A-T. Here A, C, G, and T are abbreviations of the names of the nucleobases *Adenine, Cytosine, Guanine,* and *Thymine* which are the main building blocks for the DNA strings in a chromosome.

It is technically difficult (although not impossible) to identify which of two alleles on a locus originate from the father and which from the mother, so only the genotype of an individual is easily observable.

A simple genetic model for the frequency of these genotypes is known as *Hardy–Weinberg equilibrium.* It can be shown that if members of a population mate at random and there are no selection or other adverse effects, then the distribution of the genotypes of a single individual will be as if the alleles a and A are allocated completely at random, so if we let $\beta \in B = (0, 1)$ denote the relative frequency of the allele A in the population, the probabilities of the genotype X of a random individual would be

$$P\{X = aa)\} = (1-\beta)^2, \quad P\{X = Aa\} = 2\beta(1-\beta), \quad P\{X = AA\} = \beta^2$$

or, in other words, the number of alleles of type A in an individual follows a binomial distribution of length 2 and probability of success β. Thus the genotype of an individual follows a multinomial distribution with parameters

$$\pi_0(\beta) = (1-\beta)^2, \quad \pi_1(\beta) = 2\beta(1-\beta), \quad \pi_2(\beta) = \beta^2.$$

This function is a smooth and injective homeomorphism and has Jacobian

$$D\pi(\beta)_{\setminus 0} = (2-4\beta, 2\beta)$$

which has full rank for $\beta \in (0,1)$; so the above specifies a curved multinomial model of dimension 1.

Observing genotypes $(Y_0 = y_0, Y_1 = y_1, Y_2 = y_2)$ of n individuals, the log-likelihood function becomes

$$\ell(\beta) = (2y_0 + y_1)\log(1-\beta) + (y_1 + 2y_2)\log(\beta) + y_1 \log 2 \tag{8.8}$$

yielding the score equation

$$\frac{(2y_0 + y_1)}{1-\beta} - \frac{y_1 + 2y_2}{\beta} = 0$$

with the unique solution

$$\hat{\beta} = \frac{y_1 + 2y_2}{2(y_0 + y_1 + y_2)} = \frac{y_1 + 2y_2}{2n}.$$

which is a valid solution if and only if $0 < y_1 + 2y_2 < 2n$. Since this is the only stationary point, and the log-likelihood function satisfies

$$\lim_{\beta \to 0} \ell(\beta) = \lim_{\beta \to 1} \ell(\beta) = -\infty,$$

the solution to the score equation must be a global maximum, and hence, we have identified the MLE. The MLE is asymptotically well-defined since

$$\lim_{n \to \infty} P_\beta\{0 < Y_1 + Y_2 < 2n\} = 1.$$

The Fisher information becomes

$$i_n(\beta) = \frac{\mathbf{E}_\beta((2Y_0 + Y_1))}{(1-\beta)^2} + \frac{\mathbf{E}_\beta((Y_1 + 2Y_2))}{\beta^2} = \frac{2n}{(1-\beta)} + \frac{2n}{\beta} = \frac{2n}{\beta(1-\beta)}$$

so we have $\hat{\beta} \overset{\text{as}}{\sim} \mathcal{N}(\beta, \beta(1-\beta)/(2n))$. This could also have been derived by realizing that $Y_1 + 2Y_2$ is the number of alleles of type A among $2n$ alleles, with β being the frequency of allele A.

If we now wish to test the hypothesis that a population is in Hardy–Weinberg equilibrium based on the observations, we just note that $\mathrm{OBS}_i = y_i$ and the expected numbers are

$$\mathrm{EXP}_0 = \frac{(2y_0 + y_1)^2}{4n}, \; \mathrm{EXP}_1 = \frac{(2y_0 + y_1)(y_1 + 2y_2)}{2n}, \; \mathrm{EXP}_2 = \frac{(y_1 + 2y_2)^2}{4n}.$$

These can now be used either to calculate the likelihood ratio statistic $G^2 = \Lambda$ using (8.6) or the Pearson χ^2 statistic X^2 using (8.7).

Example 8.3. [Cod in the Baltic Sea] We shall illustrate this with data from Sick (1965) where 86 cod were caught near the island of Bornholm and genotyped with respect to their haemoglobine type resulting in $y = (14, 20, 52)$. We get

$$\hat{\beta} = (20 + 2 \times 52)/(2 \times 86) = 124/172 = 0.721$$

and the corresponding expected numbers become $\mathrm{EXP} = (6.7, 34.6, 44.7)$ leading to $\Lambda = G^2 = 14.45$ and $X^2 = W = 15.31$. These numbers should be judged in an asymptotic $\chi^2(1)$ distribution, both leading to $p \sim 0.0001$ so we conclude that the cod population at this location was not in Hardy–Weinberg equilibrium. □

8.2.2 The ABO blood type system

A slightly more complicated genetic model is associated with the ABO blood type system. This trait is determined by a single gene on the 9th chromosome that has three types of allele: A, B or O. Here the A and B genes are co-dominant over O, so only persons with genotype OO has bloodtype O, whereas persons with genotypes OA or AA have bloodtype A, OB or BB have bloodtype B, and genotype AB yields bloodtype AB.

If the population is in Hardy–Weinberg equlibrium, genes can be assumed allocated at random to individuals. Thus if we let p denote the frequency of allele A and q the frequency of allele B, we get the following probabilities for the four observable bloodtypes

$$\pi_0 = (1-p-q)^2, \; \pi_1 = p^2 + 2p(1-p-q), \; \pi_2 = q^2 + 2q(1-p-q), \; \pi_3 = 2pq.$$

The Jacobian becomes

$$D\pi_{\setminus 0}(p,q) = \begin{pmatrix} 2(1-p-q) & -2p \\ -2q & 2(1-p-q) \\ 2q & 2p \end{pmatrix}$$

which has full rank 2 for $p, q > 0$ and $p + q < 1$ since if we have $\lambda_1 q + \lambda_2 p = 0$ we get $\lambda_2 = -\lambda_1 q/p$ and thus if

$$\lambda_1(1-p-q) - \lambda_2 p = \lambda_1(1-q) = 0$$

we must have $\lambda_1 = 0$ and therefore also $\lambda_2 = 0$. Here the maximum-likelihood estimator must be determined numerically.

Example 8.4. [Blood types of Danish individuals] Data on blood types of 1266 Danish individuals given in Hansen (2012, page 234) yield the counts

$$y = (535, 547, 140, 44),$$

leading to maximum-likelihood estimates $(\hat{p}, \hat{q}) = (0.270, 0.76)$ and thus expected values

$$\text{EXP} = (541.5, 539.4, 133.2, 52.0).$$

For the test statistics, we get $G^2 = 1.82$ and $X^2 = 1.76$ with corresponding p-values 0.185 and 0.178 when compared to a $\chi^2(3-2) = \chi^2(1)$ distribution. Thus there is no reason to doubt the equilibrium model. □

8.3 Contingency tables

In this section we shall discuss a variety of models for cross-classified data also known as *contingency tables*.

8.3.1 Comparing multinomial distributions

Here we consider related multinomial distributions and we wish to investigate whether they are identical.

8.3.1.1 Comparing two multinomial distributions

We first treat the case where we have mutually independent random variables $X_{01}, \ldots, X_{0n_0}$ and $X_{11}, \ldots, X_{1n_1}$ with values in $\mathcal{X} = (0, 1, \ldots, k)$ with

$$\pi_0 = (\pi_{0x}, x \in \mathcal{X}), \quad \pi_1 = (\pi_{1x}, x \in \mathcal{X})$$

where $\pi_0, \pi_1 \in \Delta^k$, the k-dimensional simplex as defined in (8.1). Each of the two families associated with π_1 and π_2 are regular exponential families, as established in Section 8.1, so the joint family is an example of an outer product as described in Section 3.4. The canonical parameter space is

$$\Theta = \Theta_0 \times \Theta_1 = \mathbb{R}^k \times \mathbb{R}^k,$$

the canonical parameter is

$$\theta = (\theta_0, \theta_1), \quad \theta_{ij} = \log \frac{\pi_{ij}}{\pi_{i0}}, \quad i = 0, 1; j = 1, \ldots, k$$

and the canonical statistic Y is

$$Y_{ij} = \sum_{m=1}^{n_i} t_j(X_{im}) = \sum_{m=1}^{n_i} \mathbf{1}_j(X_{im}) = \text{OBS}_{ij}, \quad i = 0, 1; j = 1, \ldots, k$$

so

$$\pi_{ij} = \frac{e^{\theta_{ij}}}{1 + \sum_{j=1}^{k} e^{\theta_{ij}}}, \quad i = 0, 1; j = 1, \ldots, k.$$

As before, the MLE of the unknown probabilities are the relative frequencies

$$\hat{\pi}_{ij} = \frac{Y_{ij}}{n_i} = \frac{\text{OBS}_{ij}}{n_i}, \quad i = 0, 1; j = 0, \dots, k$$

where

$$Y_{ij} = \sum_{m=1}^{n_i} t_j(X_{im}) = \sum_{m=1}^{n_i} \mathbf{1}_j(X_{im}) = \text{OBS}_{ij}, \quad i = 0, 1; j = 0, \dots, k$$

is the total number of observations in category j for group i. The MLE is well-defined if and only if $Y_{ij} > 0$ for all i, j.

We are interested in the *hypothesis of homogeneity*, i.e. that the distribution over categories is the same for the two groups $H_0 : \pi_0 = \pi_1$ or, equivalently, $H_0 : \theta_0 = \theta_1$, corresponding to the *direct product* of the multinomial families. This is a regular exponential family with canonical parameter space $\Xi = \mathbb{R}^k$ where ξ_j is the common value of the canonical parameters in the larger model

$$\xi_j = \theta_{0j} = \theta_{1j} = \log \frac{\pi_{ij}}{\pi_{i0}}, \qquad i = 0, 1; j = 1, \dots, k.$$

We may think of the smaller model as classifying $n = n_0 + n_1$ objects in the categories given by $\mathcal{X}$ and thus if we let $Y_{+j} = Y_{1j} + Y_{2j}$ be the total number of objects in category j, the MLE $\hat{\hat{\pi}}$ of the joint probability distribution under the hypothesis is

$$\hat{\hat{\pi}}_{ij} = \frac{Y_{+j}}{n}, \quad i = 0, 1; j = 0, \dots, k.$$

Using that log-likelihood functions for product models are additive as in (3.7) and (3.8), we now obtain the two maximized log-likelihood functions

$$\ell(\hat{\pi}) = \sum_{j=0}^{k} Y_{0j} \log \frac{Y_{0j}}{n_0} + \sum_{j=0}^{k} Y_{1j} \log \frac{Y_{1j}}{n_1}$$

and

$$\ell(\hat{\hat{\pi}}) = \sum_{j=0}^{k} Y_{0j} \log \frac{Y_{+j}}{n} + \sum_{j=0}^{k} Y_{1j} \log \frac{Y_{+j}}{n} = \sum_{j=0}^{k} Y_{+j} \log \frac{Y_{+j}}{n}$$

leading to the log-likelihood ratio statistic

$$\begin{aligned} G^2 &= \Lambda_n = 2\left(\ell(\hat{\pi}) - \ell(\hat{\hat{\pi}})\right) \\ &= 2\left(\sum_{i=0}^{1}\sum_{j=0}^{k} Y_{ij} \log \frac{Y_{ij}}{n_i} - \sum_{i=0}^{1}\sum_{j=0}^{k} Y_{ij} \log \frac{Y_{+j}}{n}\right) \\ &= 2\sum_{i=0}^{1}\sum_{j=0}^{k} Y_{ij} \log \frac{Y_{ij}}{n_i Y_{+j}/n} = 2\sum_{i=0}^{1}\sum_{j=0}^{k} \text{OBS}_{ij} \log \frac{\text{OBS}_{ij}}{\text{EXP}_{ij}} \end{aligned}$$

Table 8.2 – Weight of 300 trout caught in 1958 and 1960 in a Danish lake. Data from examination at Aarhus University June 1963.

	Weight of trout in grams					
Year	0–99	100–199	200–299	300-399	400+	Total
1958	12	36	74	19	9	150
1960	16	43	70	15	6	150
Total	28	79	144	34	15	300

where we have exploited that the *expected number* of objects in category j for group i under the hypothesis is

$$\text{EXP}_{ij} = n_{ij}\hat{\hat{\pi}}_j = n_i \frac{Y_{+j}}{n}.$$

The log-likelihood ratio statistic has an asymptotic $\chi^2(2k-k) = \chi^2(k)$ distribution under H_0 for $n \to \infty$ by Theorem 5.39; as does the corresponding Wald statistic (Corollary 5.38), leading also here to *Pearson's* χ^2 statistic

$$X^2 = W_n = \sum_{i=0}^{1}\sum_{j=0}^{k} \frac{(\text{OBS}_{ij} - \text{EXP}_{ij})^2}{\text{EXP}_{ij}}.$$

This follows from (8.7) and the fact that the independence of the two samples implies

$$W_n = W_{0n} + W_{1n}.$$

Example 8.5. [Weight distribution of trout] A factory was built in 1959 at a lake in Denmark, and its waste water after cleaning was released into the lake. To investigate the effect of the waste water on the lake environment, the weight of 150 trout caught in 1958 and 150 trout caught in 1960 were recorded. The results are displayed in Table 8.2. The likelihood ratio and Pearson test statistics are 2.38 and 2.37, respectively, and the associated asymptotic p-value is .67 when compared to a χ^2-distribution with 4 degrees of freedom, so there is no significant evidence of a change in the weight distribution. We note, however, that the distribution does seem to have shifted towards lower values but this type of systematic trend is not picked up by our analysis. We refer to the fact that the weight categories are ordered by saying that the weight is an *ordinal* variable. This suggests the use of alternative test statistics, see further in Section 8.3.6. □

8.3.1.2 Comparing two proportions

A special case of the above is when $|\mathcal{X}|$ is equal to two, so that $\theta = (\theta_0, \theta_1) \in \mathbb{R}^2$ with $Y_{01} \sim \text{binom}(n_0, \pi_0)$ and $Y_{11} \sim \text{binom}(n_1, \pi_1)$.

Then the hypothesis of homogeneity may be formulated as $H_0 : \pi_0 = \pi_1$, so we in effect are comparing two proportions. We shall then show that Pearson's χ^2

statistic simplifies. To see this, we calculate $\text{OBS}_{00} - \text{EXP}_{00}$ as

$$\text{OBS}_{00} - \text{EXP}_{00} = Y_{00} - n_0 \frac{Y_{+0}}{n} = \frac{Y_{00}Y_{11} - Y_{01}Y_{10}}{n} = \frac{\det Y}{n},$$

where

$$Y = \begin{pmatrix} Y_{00} & Y_{01} \\ Y_{10} & Y_{11} \end{pmatrix}$$

is the 2×2 matrix forming the contingency table. We have used that

$$n = Y_{00} + Y_{10} + Y_{01} + Y_{11}.$$

We also have

$$\text{OBS}_{00} - \text{EXP}_{00} = \text{EXP}_{01} - \text{OBS}_{01} = \text{OBS}_{11} - \text{EXP}_{11} = \text{EXP}_{10} - \text{OBS}_{10}$$

and hence Pearson's χ^2 simplifies to

$$X^2 = \frac{(\det Y)^2}{n^2} \left(\frac{1}{\text{EXP}_{00}} + \frac{1}{\text{EXP}_{01}} + \frac{1}{\text{EXP}_{10}} + \frac{1}{\text{EXP}_{11}} \right).$$

Note that this also means that there is only a single standardized residual to consider,

$$R^* = \frac{\text{OBS}_{00} - \text{EXP}_{00}}{n} \sqrt{\frac{1}{\text{EXP}_{00}} + \frac{1}{\text{EXP}_{01}} + \frac{1}{\text{EXP}_{10}} + \frac{1}{\text{EXP}_{11}}},$$

as all four residuals are equivalent.

When comparing proportions π_0 and π_1 that are not identical, we may wish to quantify exactly how and how much they differ. There are alternative ways of doing so by choosing different meaningful parameter functions known as *measures of association*. We shall briefly consider some of the most common of these, and show how to construct their confidence intervals.

Difference of proportions The simplest and most direct measure of association is the *difference of proportions* which is

$$\phi_{\text{diff}}(\pi_0, \pi_1) = \pi_1 - \pi_0 = -\phi_{\text{diff}}(1 - \pi_1, 1 - \pi_0)$$

and we note that comparing probabilities of failure is equivalent to probabilities of success, as this only modifies the sign. We always have

$$-1 \leq \phi_{\text{diff}} \leq 1$$

and $\phi_{\text{diff}} = 0$ if and only if $\pi_1 = \pi_0$. Based on data Y as above, we get a simple $1 - \alpha$ Wald based confidence interval for the difference of proportions as

$$C_{\text{diff}}^{1-\alpha}(Y) = \hat{\pi}_1 - \hat{\pi}_0 \pm z_{1-\alpha/2} \sqrt{\frac{\hat{\pi}_1(1 - \hat{\pi}_1)}{n_1} + \frac{\hat{\pi}_0(1 - \hat{\pi}_0)}{n_0}}$$

where $\hat{\pi}_i = Y_{i1}/n_i$, $n_i = Y_{i0} + Y_{i1}$, and $z_{1-\alpha/2}$ is the $1 - \alpha/2$ quantile in the standard normal distribution.

Relative risk A small difference of probabilities may still be important if both probabilities are small. If, say, the probabilities of dying from a certain disease is 0.01 when given one treatment and 0.001 when given another treatment, the difference may be considered neglible, but still one treatment must be considered strongly preferable. So it is of interest to consider the *relative risk*, which is the ratio of success probabilities rather than the difference

$$\phi_{\mathrm{RR}}(\pi_0, \pi_1) = \frac{\pi_1}{\pi_0}.$$

Note that here it matters whether we are considering successes or failures and the relation between the two is somewhat complex. In the disease example, the relative risk is 10 when considering probabilities of death, whereas the relative risk for survival is 0.99.

Since relative risks may be very small it is safest to construct a Wald interval for the logarithm and then exponentiate. The delta method yields that

$$\log \frac{\hat{\pi}_1}{\hat{\pi}_0} = \log \hat{\pi}_1 - \log \hat{\pi}_0 \overset{\text{as}}{\sim} \mathcal{N}\left(\log \frac{\pi_1}{\pi_0}, \frac{1-\pi_1}{\pi_1 n_1} + \frac{1-\pi_0}{\pi_0 n_0}\right)$$

and hence an approximate $1 - \alpha$ confidence interval for the relative risk is

$$C_{\mathrm{RR}}^{1-\alpha}(Y) = \left(\frac{\hat{\pi}_1}{\hat{\pi}_0} e^{-z_{1-\alpha/2}\tilde{s}}, \frac{\hat{\pi}_1}{\hat{\pi}_0} e^{z_{1-\alpha/2}\tilde{s}}\right)$$

where

$$\tilde{s}^2 = \frac{1-\hat{\pi}_1}{\hat{\pi}_1 n_1} + \frac{1-\hat{\pi}_0}{\hat{\pi}_0 n_0} = \frac{Y_{10}}{Y_{11} n_1} + \frac{Y_{00}}{Y_{01} n_0}.$$

Odds-ratio The relative risk is an asymmetric measure which may have its merits but also its difficulties, due to the lack of symmetry between success and failure. The odds for success is defined as the ratio of probabilities of success and failure, i.e. $\pi/(1-\pi)$ and this suggests comparing the probabilities by forming their *odds-ratio*

$$\phi_{\mathrm{OR}}(\pi_1, \pi_0) = \frac{\pi_1/(1-\pi_1)}{\pi_0/(1-\pi_0)} = \frac{\pi_1(1-\pi_0)}{(1-\pi_1)\pi_0} = \phi_{\mathrm{OR}}(1-\pi_1, 1-\pi_0)^{-1}.$$

The MLE for the odds-ratio is the *cross-product ratio*

$$\hat{\phi}_{\mathrm{OR}} = \frac{Y_{11} Y_{00}}{Y_{10} Y_{01}}.$$

Also here, the odds-ratio is best considered on the log-scale so we let

$$\theta_{10} = \log \phi_{\mathrm{OR}} = \log \frac{\pi_1/(1-\pi_1)}{\pi_0/(1-\pi_0)} = \log \frac{\pi_1}{(1-\pi_1)} - \log \frac{\pi_0}{(1-\pi_0)} = \theta_1 - \theta_0$$

where θ_1 and θ_2 are the canonical parameters in the two associated Bernoulli exponential families, see Example 3.19. Thus the asymptotic variance of the MLE of the

Table 8.3 – Myocardial infarctions after treatment with aspirin or placebo. Source: Agresti (2002) based on a study published in the *Lancet*, **338**, 1345–1349 (1991).

	Myocardial infarction		
Group	No	Yes	Total
Placebo	656	28	684
Aspirin	658	18	676
Total	1314	46	1360

log odds-ratio θ_{10} is

$$\overset{\text{a}}{\mathbf{V}}(\hat{\theta}_{10}) = \frac{1}{n_1\pi_1(1-\pi_1)} + \frac{1}{n_0\pi_0(1-\pi_0)}$$
$$= \frac{1}{n_1\pi_1} + \frac{1}{n_1(1-\pi_1)} + \frac{1}{n_0\pi_0} + \frac{1}{n_0(1-\pi_0)}$$

and the MLE for this variance is

$$\widehat{\overset{\text{a}}{\mathbf{V}}(\hat{\theta}_{10})} = \frac{1}{Y_{10}} + \frac{1}{Y_{11}} + \frac{1}{Y_{00}} + \frac{1}{Y_{01}}.$$

Thus we obtain an asymptotic $1-\alpha$ Wald confidence interval for the log-odds ratio θ_{10} as

$$C^{1-\alpha}_{\text{logOR}}(Y) = \log\frac{Y_{11}Y_{00}}{Y_{10}Y_{01}} \pm z_{1-\alpha/2}\sqrt{\frac{1}{Y_{10}} + \frac{1}{Y_{11}} + \frac{1}{Y_{00}} + \frac{1}{Y_{01}}}. \tag{8.9}$$

We conclude the section devoted to comparing binomial proportions with an example.

Example 8.6. [Swedish aspirin study] A study performed in Sweden was made to investigate the effect of aspirin use and myocardial infarction. A total of 1360 patients who all previously had suffered a stroke were randomly assigned to treatment with aspirin or placebo. Table 8.3 displays the number of patients who had or had not suffered myocardial infarction after a three-year follow-up period.

Pearson's χ^2 evaluates to $X^2 = 2.13$, yielding a p-value of 0.1444 when compared to a χ^2 distribution with one degree of freedom, so there is no strong evidence for the effect of aspirin with a sample of this size.

The 95% Wald confidence interval for the log-odds ratio θ_{10} as given in (8.9) evaluates to $(-.16, 1.05)$. Since 0 is included in this interval, we reach the same conclusion. The difference of proportions is not large:

$$\hat{\pi}_1 - \hat{\pi}_0 = \frac{18}{676} - \frac{28}{684} - = -0.014$$

as the probabilities of myocardial infarction are small themselves, whereas the relative risk

$$\frac{\hat{\pi}_1}{\hat{\pi}_0} = \frac{18}{676}\frac{684}{28} = 0.65$$

is considerably reduced in the aspirin group (although not significantly!). The reader may calculate a confidence interval for the relative risk as an exercise. □

8.3.1.3 Comparing several multinomial distributions

The considerations above generalize readily to the case of comparing r multinomial distributions. Indeed, if we let

$$X_{01}, \ldots, X_{0n_0}; \; \ldots \; X_{(r-1)1}, \ldots, X_{(r-1)n_{r-1}}$$

denote random variables with values in $\mathcal{X}$,

$$Y_{ij} = \sum_{m=1}^{n_i} t_j(X_{im}) = \sum_{m=1}^{n_i} \mathbf{1}_j(X_{im}) = \text{OBS}_{ij}, \quad i = 0, \ldots, r-1; j = 0, \ldots, s-1$$

be the total number of observations in category j for group i, and

$$Y_{+j} = \sum_{i=0}^{r-1} Y_{ij}$$

the total number of observations in category j, we obtain the likelihood ratio statistic for the hypothesis of homogeneity

$$H_0 : \pi_0 = \cdots = \pi_{r-1}$$

to be

$$\Lambda_n = 2 \sum_{i=0}^{r-1} \sum_{j=0}^{s-1} Y_{ij} \log \frac{Y_{ij}}{n_i Y_{+j}/n} = 2 \sum_{i=0}^{r-1} \sum_{j=0}^{s-1} \text{OBS}_{ij} \log \frac{\text{OBS}_{ij}}{\text{EXP}_{ij}},$$

where as before the *expected number* of objects in category j for group i under the hypothesis is

$$\text{EXP}_{ij} = n_{ij} \hat{\hat{\pi}}_j = n_i \frac{Y_{+j}}{n}.$$

The log-likelihood ratio statistic now has an asymptotic χ^2 distribution with degrees of freedom

$$\text{degrees of freedom} = r(s-1) - (s-1) = (r-1)(s-1)$$

under H_0 for $n \to \infty$ by Theorem 5.39; as does the corresponding Wald statistic (Corollary 5.38), leading also here to *Pearson's* χ^2 statistic

$$W_n = X^2 = \sum_{i=0}^{r-1} \sum_{j=0}^{s-1} \frac{(\text{OBS}_{ij} - \text{EXP}_{ij})^2}{\text{EXP}_{ij}}.$$

Traditionally, such data are presented in an $r \times s$ *contingency table* which has the form in Table 8.4. The *cells* of the table are the entries i, j and the *cell counts* are the random variables Y_{ij} .

Table 8.4 – An $r \times s$ contingency table for the comparison of r multinomial distributions with s categories. Random variables Y_{ij} are the number among n_i objects with response j in group i.

	Response				
Group	0	1	$\cdots$	$s-1$	Total
0	Y_{00}	Y_{01}	$\cdots$	$Y_{0,s-1}$	$n_{0,+}$
1	Y_{10}	Y_{11}	$\cdots$	$Y_{1,s-1}$	$n_{1,+}$
$\vdots$	$\vdots$	$\vdots$	$\ddots$	$\vdots$	$\vdots$
$r-1$	$Y_{r-1,0}$	$Y_{r-1,1}$	$\cdots$	$Y_{r-1,s-1}$	$n_{r-1,+}$
Total	$Y_{+,0}$	$Y_{+,1}$	$\cdots$	$Y_{+,s-1}$	n

8.3.2 *Independence of classification criteria*

Next, let us consider objects $X_1, \ldots, X_n$ classified according to two criteria , i.e. with values in $\mathcal{X} = \mathcal{I} \times \mathcal{J}$ where $\mathcal{I} = (0, \ldots, r-1)$ and $\mathcal{J} = (0, \ldots, s-1)$, resulting in a contingency table as before, just that now also the row-totals $n_{i+} = Y_{i+}$ are random.

Without further restrictions, this corresponds to an unrestricted multinomial exponential family as considered in Section 8.1.1, just with the state space coded differently so that the category $(i, j) = (0, 0)$ is coded as 0; in other words, the model corresponds to an exponential family with

$$\theta_{ij} = \log \frac{\pi_{ij}}{\pi_{00}}; \; \pi_{ij} = \frac{e^{\theta_{ij}}}{\sum_{u=0}^{r-1} \sum_{v=0}^{s-1} e^{\theta_{uv}}}, \qquad i, j \in \mathcal{X}_0 = \mathcal{X} \setminus ((0,0)),$$

with the canonical parameters satisfying $\theta_{00} = 0$ and $\theta_{ij} \in \mathbb{R}$ for $(i, j) \neq (0, 0)$.

We are interested in investigating whether the categories in the cross-classification are independent or, in other words, whether the coordinate projections $I : \mathcal{X} \mapsto \mathcal{I}$ and $J : \mathcal{X} \mapsto \mathcal{J}$ are independent random variables. In other words, our *hypothesis of independence* expressed in terms of π is

$$H_0 : \pi_{ij} = \pi_{i+}\pi_{+j},$$

or, expressed in terms of the canonical parameters above:

$$H_0 : \theta_{ij} = \alpha_i + \beta_j \tag{8.10}$$

where $\alpha_0 = \beta_0 = 0$ and $\alpha_i, \beta_j \in \mathbb{R}$ for $i = 1, \ldots, r-1$ and $j = 1, \ldots, s-1$ since under the hypothesis of independence we have

$$\theta_{ij} = \log \frac{\pi_{ij}}{\pi_{00}} = \log \frac{\pi_{i+}\pi_{+j}}{\pi_{0+}\pi_{+0}} = \alpha_i + \beta_j$$

where

$$\alpha_i = \log \frac{\pi_{i+}}{\pi_{0+}}, \quad \beta_j = \log \frac{\pi_{+j}}{\pi_{+0}}.$$

The hypothesis in (8.10) defines a linear subspace of the $rs - 1$ dimensional canonical parameter space of the unrestricted model with dimension $r + s - 2$; and hence, by Theorem 3.14, this is again a regular exponential family and—see Example 3.27—thus a curved multinomial model as discussed in Section 8.1.2. Hence, the associated likelihood ratio and Pearson's χ^2 statistics are those given in (8.6) and (8.7) and become

$$G^2 = 2\sum_{i=0}^{r-1}\sum_{j=0}^{s-1} \text{OBS}_{ij} \log \frac{\text{OBS}_{ij}}{\text{EXP}_{ij}}, \quad X^2 = \sum_{i=0}^{r-1}\sum_{j=1}^{s-1} \frac{(\text{OBS}_{ij} - \text{EXP}_{ij})^2}{\text{EXP}_{ij}}.$$

Both of these statistics are asymptotically χ^2 distributed with degrees of freedom determined by the difference of the dimensions of the models

$$\text{degrees of freedom} \quad = \quad rs - 1 - (r + s - 2) = (r-1)(s-1).$$

We need to find the expected cell counts EXP_{ij} under the hypothesis of independence. The log-likelihood function under the hypothesis is

$$\begin{aligned} \ell(\pi) &= \sum_{i=0}^{r-1}\sum_{j=0}^{s-1} Y_{ij} \log \pi_{ij} = \sum_{i=0}^{r-1}\sum_{j=0}^{s-1} Y_{ij} \log(\pi_{i+}\pi_{+j}) \\ &= \sum_{i=0}^{r-1}\sum_{j=0}^{s-1} Y_{ij} \log \pi_{i+} + \sum_{i=1}^{r-1}\sum_{j=0}^{s-1} Y_{ij} \log \pi_{+j} \\ &= \sum_{i=0}^{r-1} Y_{i+} \log \pi_{i+} + \sum_{j=0}^{s-1} Y_{+j} \log \pi_{+j}. \end{aligned}$$

Thus the log-likelihood function is a sum of two log-likelihood functions, each depending on their own separate parameters and each of the same form as the standard multinomial likelihood for classifying objects separately into the groups $\mathcal{I}$ and $\mathcal{J}$. It follows that the MLE under the hypothesis is given as

$$\hat{\hat{\pi}}_{i+} = \frac{Y_{i+}}{n}, \quad \hat{\hat{\pi}}_{+j} = \frac{Y_{+j}}{n}$$

and thus the expected cell counts are

$$\text{EXP}_{ij} = n\hat{\hat{\pi}}_{i+}\hat{\hat{\pi}}_{+j} = n\frac{Y_{i+}}{n}\frac{Y_{+j}}{n} = \frac{Y_{i+}Y_{+j}}{n}.$$

Note that under the hypothesis of homogeneity, the expected cell counts were equal to $\text{EXP}_{ij} = n_{i+}Y_{+j}/n$; but since in that case we have $Y_{i+} = n_{i+}$, *the estimated cell counts are identical to those under the hypothesis of independence.* It follows that also the test statistics are identical, and we note that so is their asymptotic χ^2 distributions. The exact distribution of the test statistics differ, as Y_{i+} are random in the independence model, but fixed in the model of homogeneity.

Table 8.5 – A 4 × 4 contingency table for the cross-classification of 592 students at the University of Delaware according to the colour of their eyes and their hair, collected by Snee (1974).

	Eye colour				
Hair colour	Brown	Blue	Hazel	Green	Total
Black	68	20	15	5	108
Brown	119	84	54	29	286
Red	26	17	14	14	71
Blond	7	94	10	16	127
Total	220	215	93	64	592

Table 8.6 – Standardized Pearson's residuals for the cross-classification according to hair and eye colour in Table 8.5. Residuals larger than 3 are highlighted by bold type.

	Eye colour			
Hair colour	Brown	Blue	Hazel	Green
Black	**6.14**	**-4.25**	-0.58	-2.29
Brown	2.16	**-3.40**	2.05	-0.51
Red	-0.10	-2.31	0.99	2.58
Blond	**-8.33**	**9.97**	-2.74	0.73

Example 8.7. [Hair and eye colour] To illustrate the developments above, we consider a cross-classification of 592 students at the University of Delaware according to their hair colour and eye colour. The data are displayed in Table 8.5.

We are interested in investigating whether these characteristics of individuals are independent or related. The Pearson χ^2 statistic evaluates to $X^2 = 138.3$ which yields an asymptotic p-value of 0 when evaluated in a χ^2 distribution with degrees of freedom being $(4-1)(4-1) = 9$. Thus the characteristics are *not* independent. The likelihood ratio statistic becomes $G^2 = 146.4$ leading to the same conclusion.

To understand more about the nature of deviations from independence, we investigate the standardized residuals, displayed in Table 8.6. It is apparent that a main feature of these data is that very few students with black or brown hair have blue eyes, whereas blue eyes are very common for students with blond hair. □

8.3.3 Poisson models for contingency tables

For certain types of cross-classified data, it is not reasonable to assume that the total number n of classified objects is fixed. This holds, for example when the entries in the table represent the number of events (errors, accidents, deaths, diseased individuals) in a specified period or region, classified according to characteristics $\mathcal{I} = (1, \ldots, |I|)$ and $\mathcal{J} = (1, \ldots, |J|)$. In such cases, the total number of events n is not controlled.

As in the previous subsection, data may still usefully be presented in a contingency table, just with the modification that also the entry in the lower right corner of that table, containing the total number of objects now usefully may be denoted Y_{++}, to emphasize its random nature.

A generic model for data of this type is that the entries in the contingency table are independent and Poisson distributed with

$$P_\lambda(Y_{ij} = y_{ij}) = \frac{\lambda_{ij}^{y_{ij}}}{y_{ij}!}e^{-\lambda_{ij}}, \quad i = 0, \ldots, r-1; j = 0, \ldots, s-1.$$

This corresponds to the outer product construction—described in Section 3.4—of individual simple Poisson models, and hence, this family of distributions is a regular and minimally represented exponential family with canonical parameter $\theta \in \Theta = \mathbb{R}^{rs}$ where

$$\theta_{ij} = \log \lambda_{ij}, \quad i = 0, \ldots, r-1; j = 0, \ldots, s-1$$

and the table of counts as the canonical statistic. We shall refer to this model as the *unrestricted* Poisson model. In this exponential family, the MLE is simply equal to the observed table of counts

$$\hat{\lambda}_{ij} = y_{ij} \quad i = 0, \ldots, r-1; j = 0, \ldots, s-1. \tag{8.11}$$

In the following we shall investigate hypotheses and models obtained by specifying various restrictions on the parameters.

8.3.3.1 The simple multiplicative Poisson model

The first submodel we consider is the *simple multiplicative Poisson model*, determined by restricting the expectation to have the multiplicative form

$$\lambda_{ij} = \mu\rho_i\eta_j, \quad i = 0, \ldots, r-1; j = 0, \ldots, s-1 \tag{8.12}$$

for some $\mu \in \mathbb{R}_+$, $\rho \in \mathbb{R}_+^r$ and $\eta \in \mathbb{R}_+^s$ with $\rho_0 = \eta_0 = 1$ or, equivalently, restricting the canonical parameter to have the additive form

$$\theta_{ij} = \log \lambda_{ij} = \gamma + \alpha_i + \beta_j, \quad i = 0, \ldots, r-1; j = 0, \ldots, s-1$$

for some $\gamma \in \mathbb{R}$, $\alpha \in \mathbb{R}^r$, and $\beta \in \mathbb{R}^s$ with $\alpha_0 = \beta_0 = 0$.

This model was developed and investigated by the Danish statistician *Georg Rasch* (1901 – 1980) in Rasch (1960). It specifies that $\theta = \log \lambda \in L$ where L is a linear subspace of $\mathbb{R}^{rs}$ of dimension $r + s - 1$ and is known as a *log-linear model*.

By Theorem 3.14, the multiplicative Poisson model is a regular exponential model and a curved (linear) submodel of the unrestricted Poisson model. Indeed we may write the joint density of the model as

$$\begin{aligned} f_{(\gamma,\alpha,\beta)}(Y = y) &= \prod_{i=0}^{r-1}\prod_{j=0}^{s-1} e^{(\gamma+\alpha_i+\beta_j)y_{ij} - e^{\gamma+\alpha_i+\beta_j}} \\ &= e^{\gamma y_{++} + \sum_{i=1}^{r-1}\alpha_i y_{i+} + \sum_{j=1}^{s-1}\beta_j y_{+j} - \psi(\mu,\alpha,\beta)} \end{aligned}$$

with respect to the discrete base measure ν

$$\nu(y) = \prod_{i=0}^{r-1} \prod_{j=0}^{s-1} \frac{1}{y_{ij}!}$$

identifying the canonical statistic as

$$t_0(y)^\top = (y_{++}, y_{1+}, \ldots, y_{r-1,+}, y_{+1}, \ldots, y_{+,s-1})$$

and the cumulant function

$$\psi(\gamma, \alpha, \beta) = \sum_{i=0}^{r-1} \sum_{j=0}^{s-1} e^{\gamma + \alpha_i + \beta_j} = \sum_{i=0}^{r-1} \sum_{j=0}^{s-1} \lambda_{ij}.$$

We note that the canonical statistic is in one-to-one linear correspondence with the *marginal totals* of the contingency table

$$\tilde{t}_0(y) = (y_{0+}, \ldots, y_{r-1,+}, y_{+0}, \ldots, y_{+,s-1})$$

since we have

$$y_{++} = y_{1+} + \cdots + y_{r-1,+} = y_{+1} + \ldots + y_{+,s-1}$$

and

$$y_{0+} = y_{++} - y_{1+} - \cdots - y_{r-1,+}, \quad y_{+0} = y_{++} - y_{+1} - \cdots - y_{+,s-1}.$$

We thus obtain the MLE in this model by equating the observed canonical statistic to its expectation or, equivalently, the MLE is the unique point in the model that satisfies the equations

$$\hat{\hat{\lambda}}_{i+} = y_{i+}, i = 0, \ldots, r-1; \quad \hat{\hat{\lambda}}_{+j} = y_{+j}, j = 0, \ldots, s-1$$

where we have exploited the linear relationship between the canonical statistic t and the marginal totals $\tilde{t}$. We next note that if we let $\hat{\hat{\lambda}}$ be

$$\hat{\hat{\lambda}}_{ij} = \frac{y_{i+}y_{+j}}{y_{++}},$$

these equations are indeed satisfied since adding the numerators over either of the indices yields the relevant marginal totals. Since we have

$$\sum_{i=0}^{r-1} \sum_{j=0}^{s-1} \hat{\lambda}_{ij} = \sum_{i=0}^{r-1} \sum_{j=0}^{s-1} y_{ij} = y_{++}$$

and also

$$\sum_{i=0}^{r-1} \sum_{j=0}^{s-1} \hat{\hat{\lambda}}_{ij} = \sum_{i=0}^{r-1} \sum_{j=0}^{s-1} \frac{y_{i+}y_{+j}}{y_{++}} = y_{++},$$

the log-likelihood ratio statistic becomes

$$\begin{aligned} G^2 &= 2\sum_{i=0}^{r-1}\sum_{j=0}^{s-1} y_{ij}\log\frac{\hat{\lambda}_{ij}}{\hat{\hat{\lambda}}_{ij}} - 2\sum_{i=0}^{r-1}\sum_{j=0}^{s-1}(\hat{\lambda}_{ij} - \hat{\hat{\lambda}}_{ij}) \\ &= 2\sum_{i=0}^{r-1}\sum_{j=0}^{s-1} y_{ij}\log\frac{\hat{\lambda}_{ij}}{\hat{\hat{\lambda}}_{ij}} = 2\sum_{i=0}^{r-1}\sum_{j=0}^{s-1}\mathrm{OBS}_{ij}\log\frac{\mathrm{OBS}_{ij}}{\mathrm{EXP}_{ij}} \end{aligned}$$

where again OBS_{ij} and EXP_{ij} are the observed and expected cell counts

$$\mathrm{OBS}_{ij} = y_{ij}, \quad \mathrm{EXP}_{ij} = \hat{\hat{\lambda}}_{ij} = \frac{y_{i+}y_{+j}}{y_{++}}.$$

Since in this model, individual cell counts are independent and the Fisher information in the simple Poisson model is $i(\lambda) = \lambda^{-1}$—derived in Example 1.26—we get the Wald statistic

$$X^2 = W = \sum_{i=0}^{r-1}\sum_{j=0}^{s-1} i(\hat{\hat{\lambda}}_{ij})(\hat{\lambda}_{ij} - \hat{\hat{\lambda}}_{ij})^2 = \sum_{i=0}^{r-1}\sum_{j=0}^{s-1}\frac{(\mathrm{OBS}_{ij} - \mathrm{EXP}_{ij})^2}{\mathrm{EXP}_{ij}}.$$

We note again that both of the log-likelihood ratio and Wald test statistics in this model are *the same function of the cell counts* as in the case of testing for homogeneity of multinomial distributions or independence of cross-classifications. And—as argued in Example 5.41—we may use the standard asymptotic results if just all of λ_{ij} are large, so both of these statistics are asymptotically χ^2 distributed with degrees of freedom determined by the difference of the model dimensions:

$$\text{degrees of freedom} = rs - (r + s - 1) = (r-1)(s-1)$$

which also is exactly the same as we found in the homogeneity and independence cases. Again we note that the asymptotic distributions of the test statistics are identical in contrast to the exact distributions, as these involve different distributional assumptions for the sample marginals.

Example 8.8. [Incidence of Covid 19] Table 8.7 shows the number of confirmed cases of Covid 19 infection in four local Danish communities in the period 4–10 October 2020. Copenhagen West combines data from Ishøj, Brøndby, and Rødovre municipalities and Copenhagen City combines data from Copenhagen and Frederiksberg municipalities. We shall investigate whether this table may be well-described by a multiplicative Poisson model.

The log-likelihood ratio statistic can be calculated to $G^2 = 151.5$ and Pearson's χ^2 statistic evaluates to $X^2 = 139.7$. Since the degrees of freedom for the asymptotic χ^2 distribution is $(4-1)(8-1) = 21$, this yields an associated p-value which is numerically 0, thus in effect ruling out the multiplicative Poisson model as a reasonable description of data, even though some of the entries in the table are quite small, thus casting some doubt on the validity of the χ^2-approximation. A Monte Carlo-based method—to be discussed in Section 8.3.4.3 below—yields $p = 0.0005$, also effectively ruling out the multiplicative Poisson model as a good description of the table.

Table 8.7 – Number of confirmed infections in the period 4–10 October 2020 with Covid-19 in four Danish local communities. Source: Statens Serum Institut.

	Age group								
Area	0–9	10-19	20-29	30-39	40-49	50-59	60–69	70+	Total
Copenhagen West	7	42	6	11	16	18	14	9	123
Aarhus	11	74	103	33	28	26	13	7	295
Copenhagen City	36	51	153	70	64	36	16	11	437
Slagelse	3	14	2	2	13	14	3	1	52
Total	57	181	264	116	121	94	46	28	907

Table 8.8 – Standardized Pearson residuals for a multiplicative Poisson model applied to the Covid 19 data in Table 8.7. Residuals larger than 3 are in bold type.

	Age group							
Area	0–9	10–19	20–29	30–39	40–49	50–59	60–69	70+
Copenhagen West	-0.29	**4.24**	**-6.36**	-1.37	-0.12	1.67	**3.43**	2.92
Aarhus	-2.20	2.68	2.67	-1.00	-2.37	-1.06	-0.63	-0.86
Copenhagen City	2.34	**-6.02**	**3.77**	2.81	1.11	-2.03	-1.87	-0.96
Slagelse	-0.16	1.29	**-4.13**	-1.99	2.55	**4.04**	0.24	-0.50

To investigate the deviations further, we consider the standardized Pearson residuals, as displayed in Table 8.8. There are several of these residuals that are far too big; for example, there are far too many incidences in age groups 10–19 and 60–69 in Copenhagen west (42 and 14) compared to what would be expected (24.55 and 6.24) and far too few in the age group 20–29 in Copenhagen West (six observed cases and 35.8 expected). Also, the number of observed cases in Copenhagen City in the age group 20–29 is far too large (observed 153 and 87.21 expected).

There are many possible explanations of these deviations from the multiplicative Poisson model, but it is clear that the phenomenon of Covid-19 is more complex than the multiplicative Poisson model is able to accommodate. One notable omission in the analysis is that the age distributions in the four areas could be quite different, potentially explaining some of the inhomogeneities in the Covid-19 incidence. The population sizes in the various age groups are displayed in Table 8.9 and we note

Table 8.9 – Population size in local areas in Denmark by age group. Source: Danmarks Statistik.

	Age group							
Area	0–9	10–19	20–29	30–39	40–49	50–59	60–69	70+
Copenhagen West	12,025	11,459	12,126	13,239	13,179	13,226	10,532	13,192
Aarhus	37,238	34,360	86,348	46,231	38,009	38,180	32,370	36,697
Copenhagen City	80,772	61,161	172,299	135,122	96,076	78,365	54,249	59,109
Slagelse	7,642	9,132	10,069	8,342	10,042	11,482	9,765	12,681

that whereas the distribution over age groups in Copenhagen West and Slagelse are almost uniform, Copenhagen City and Aarhus both have a much larger population in the age group 20–29 than Slagelse and Copenhagen West, which might explain the relatively large number of infected individuals in these groups. Below we shall discuss models that are able to incorporate structural information of this type. □

8.3.3.2 The shifted multiplicative Poisson model

We shall consider a modification of the multiplicative Poisson model that takes a *background risk* into account. More precisely, we assume that the mean number of events in category ij has the form

$$\lambda_{ij} = \mu\rho_i\eta_j B_{ij}, \quad i = 0, \ldots, r-1; j = 0, \ldots, s-1$$

where

$$B_{ij} > 0, \quad i = 0, \ldots, r-1; j = 0, \ldots, s-1$$

is a table of known positive numbers, meant to indicate the magnitude of known factors that affect the number of events in groups i and j with $\mu, \rho_i, \eta_j \in \mathbb{R}_+$ and $\rho_0 = \eta_0 = 1$ as before. Models of this type are commonly used for pricing in non-life insurance, where B_{ij}, for example, is the number of customers in risk group ij and Y_{ij} the number of claims from customers in that risk group in a specific time period.

Rewriting the model in terms of canonical parameters $\theta_{ij} = \log \lambda_{ij}$, we get

$$\theta_{ij} = \gamma + \alpha_i + \beta_j + \log B_{ij}$$

so this is an *affine* subfamily in contrast to the simple multiplicative model considered above which defined a *linear* subfamily. We shall refer to this model as the *shifted multiplicative Poisson model*. Theorem 3.14 implies that this is a regular exponential model and an affine submodel of the unrestricted Poisson model.

As for the simple multiplicative model, we may write the joint density as

$$\begin{aligned} f_{(\mu,\alpha,\beta)}(Y = y) &= \prod_{i=0}^{r-1}\prod_{j=0}^{s-1} e^{(\gamma+\alpha_i+\beta_j+\log B_{ij})y_{ij} - e^{\gamma+\alpha_i+\beta_j+\log B_{ij}}} \\ &= \exp(\gamma y_{++} + \sum_{i=1}^{r-1}\alpha_i y_{i+} + \sum_{j=1}^{s-1}\beta_j y_{+j} - \psi^*(\mu,\alpha,\beta)) \end{aligned}$$

with respect to the discrete base measure $\tilde{\nu}$

$$\tilde{\nu}(y) = \prod_{i=0}^{r-1}\prod_{j=0}^{s-1}\frac{B_{ij}^{y_{ij}}}{y_{ij}!}$$

identifying the canonical statistic as

$$t(y)^\top = (y_{++}, y_{1+}, \ldots, y_{r-1+}, y_{+1}, \ldots, y_{+,s-1})$$

and the cumulant function

$$\tilde{\psi}(\gamma, \alpha, \beta) = \sum_{i=0}^{r-1}\sum_{j=0}^{s-1} e^{\gamma+\alpha_i+\beta_j+\log B_{ij}} = \sum_{i=0}^{r-1}\sum_{j=0}^{s-1} \lambda_{ij}.$$

Note that the affine component $\log B_{ij}$ changes the base measure and the cumulant function, but not the sufficient statistics which also here are

$$t(y)^\top = (y_{++}, y_{1+}, \ldots, y_{r-1+}, y_{+1}, \ldots, y_{+,s-1})$$

We thus obtain the MLE in the shifted Poisson model by equating the observed canonical statistic to its expectation or, equivalently, the MLE is the unique solution to the equations

$$\hat{\hat{\lambda}}_{i+} = y_{i+}, i = 0, \ldots, r-1; \quad \hat{\hat{\lambda}}_{+j} = y_{+j}, j = 0, \ldots, s-1, \tag{8.13}$$

provided it is well defined. We have again exploited the linear relationship between the canonical statistic t and the marginal totals $\tilde{t}$.

$$\begin{aligned}\sum_{j=0}^{s-1} \mu\rho_i\eta_j B_{ij} &= y_{i+}, i = 0, \ldots, r-1 \\ \sum_{i=0}^{r-1} \mu\rho_i\eta_j B_{ij} &= y_{+j}, j = 0, \ldots, s-1.\end{aligned}$$

In contrast to the simple multiplicative model, the likelihood equations cannot be solved explicitly since λ_{ij} has a more complex relation to the parameters of the model. An exception is the case when $r = s = 2$, see Exercise 8.8.

However, the form of the equations suggests a simple iterative procedure for solving them. The procedure is initiated at $\mu = y_{++}/B_{++}$, and $\rho_i = \eta_j = 1$ and now updates the parameters as

$$\mu \leftarrow \frac{y_{0+}}{\sum_{j=0}^{s-1} \eta_j B_{0j}}, \quad \rho_i \leftarrow \frac{y_{i+}}{\sum_{j=0}^{s-1} \mu\eta_j B_{ij}}, i = 1, \ldots, r-1 \tag{8.14}$$

$$\mu \leftarrow \frac{y_{+0}}{\sum_{i=0}^{r-1} \rho_i B_{i0}}, \quad \eta_j \leftarrow \frac{y_{+j}}{\sum_{i=0}^{r-1} \mu\rho_i B_{ij}}, j = 1, \ldots, s-1. \tag{8.15}$$

using previously calculated values of the parameters on the right-hand side to update the values on the left-hand side. The iteration may become a bit clearer when expressed in terms of repeatedly updating the estimates for the means λ_{ij} as:

$$\lambda_{ij} \leftarrow \lambda_{ij}\frac{y_{i+}}{\lambda_{i+}}, i = 0, \ldots, r-1, \quad \lambda_{ij} \leftarrow \lambda_{ij}\frac{y_{+j}}{\lambda_{+j}}, j = 0, \ldots, s-1. \tag{8.16}$$

Indeed this procedure is known as *iterative proportional scaling* (IPS), since at each step the table of expected values is scaled proportionally by the ratio of observed to expected marginals. The algorithm can be shown to be convergent and converge to the MLE which is well defined as long as all marginal totals are positive, see Section B.3 for details.

Table 8.10 – Standardized Pearson residuals for a shifted multiplicative Poisson model applied to the Covid-19 data in Table 8.7. Residuals larger than 3 are in bold type.

	Age group							
Area	0–9	10–19	20–29	30–39	40–49	50–59	60–69	70+
Copenhagen West	-0.74	2.40	**-3.82**	-0.95	-0.67	0.69	2.45	1.85
Aarhus	-2.00	2.41	0.94	0.29	-1.33	-0.71	-0.85	-0.97
Copenhagen City	2.41	**-4.18**	2.44	0.93	0.61	-1.32	-0.77	0.13
Slagelse	-0.02	0.21	-2.73	-1.36	2.24	2.80	-0.51	-1.12

To see that the updates in (8.16) are equivalent to those in (8.14) and (8.15), we express the marginal scalings in terms of the model parameters so that, for example, the row updates are

$$\mu\rho_i\eta_j B_{ij} \leftarrow \mu\rho_i\eta_j B_{ij}\frac{y_{i+}}{\sum_k \mu\rho_i\eta_k B_{ik}} = \eta_j B_{ij}\frac{y_{i+}}{\sum_k \eta_k B_{ik}}.$$

Thus we see that the factor $\mu\rho_i$ is updated by the ratio of the row total to a weighted sum of the base risks B_{ij}, whereas other parameters are unchanged. For $i = 0$ we get the first update in (8.14) and for $i > 0$ we get the second by division with μ. The calculations are analogous for the column updates.

Example 8.9. [Incidence of Covid-19] This is a continuation of Example 8.8, where we had to abandon the multiplicative Poisson model with a suspicion that the different age distributions in the areas under investigation could give an inadequate description of the situation.

To investigate whether this is a possible explanation of the deviation from the Poisson model, we consider a shifted multiplicative model where the number Y_{ij} of confirmed Covid-19 infections in area i and age group j is assumed to be independent and Poisson distributed with expectation λ_{ij}, where now

$$\lambda_{ij} = \mu\rho_i\eta_j B_{ij}, \quad i = 0,\ldots,3; j = 0,\ldots,7$$

with $\mu, \rho_i, \eta_j \in \mathbb{R}_+$ and $\rho_0 = \eta_0 = 1$ and B_{ij} is the total number of individuals living in area i in the age group j.

This model is clearly an improvement since we find a likelihood ratio statistic of $G^2 = 71.3$ and Pearson's χ^2 evaluates to $X^2 = 65.8$. However, both of these give asymptotic p-values around 10^{-6} when compared to the asymptotic χ^2 distribution with 21 degrees of freedom as before, so we conclude that the shifted model is still not an acceptable description of the data.

The standardized residuals are displayed in Table 8.10, and we note that the most extreme residuals from the simple multiplicative Poisson analysis have disappeared or have become less extreme. There are still extreme residuals in the age group 20–29 in Copenhagen West, where the number of infections is smaller than expected, and the same holds for the age group 10–19 in Copenhagen City. The expected numbers from the shifted multiplicative model are displayed in Table 8.11 and comparing with the observed incidences in Table 8.7 we note that only six cases were observed

Table 8.11 – Expected number of infected individuals based on a shifted multiplicative Poisson model applied to the COVID-19 data in Table 8.7.

	Age group							
Area	0–9	10–19	20–29	30–39	40–49	50–59	60–69	70+
Copenhagen West	8.95	31.39	21.21	14.15	18.45	15.67	7.93	5.26
Aarhus	17.79	60.39	96.92	31.70	34.14	29.03	15.64	9.38
Copenhagen City	27.22	75.85	136.46	65.38	60.89	42.04	18.50	10.66
Slagelse	3.04	13.37	9.41	4.76	7.51	7.27	3.93	2.70

among 20–29 year olds in Copenhagen West, whereas 21 were expected, and only 51 cases were observed in Copenhagen City among 10–19 year olds, whereas 76 were expected.

Still, there are too many residuals that are moderately large, as the formal test also indicates. This may be due to the Poisson distribution not being adequate for events of these type, as infections among individuals are not singular, random events, but tend to appear in clusters of larger outbreaks as individuals infect each other. This would typically lead to the variance being higher than predicted by the Poisson distribution—a phenomenon known as *over-dispersion*—and thus the Pearson χ^2 statistic becomes large. We would then estimate an over-dispersion factor σ^2 as $\tilde{\sigma}^2 = X^2/21 = 65.8/21 = 3.13$.

For illustration we may calculate confidence intervals for the parameters in the model, but since the Poisson variance is too small, we inflate the confidence intervals by multiplying the estimated standard deviations of the canonical parameter estimates by the square root of the over-dispersion factor, i.e. with $\sqrt{3.13} = 1.76$.

The inflated confidence intervals are displayed in Table 8.12. We note that all areas have infection rates below those in Copenhagen West since 1 is outside the confidence intervals for the rates associated with areas. Also, the factors related to age groups 10–19 and 20–29 stand out as the highest among the age groups, but caution should be taken when interpreting the confidence intervals in the light of the shortcomings of the model that we have established. □

Table 8.12 – Inflated 95% confidence intervals for parameters μ, ρ, η in the shifted multiplicative Poisson model as applied to the Covid-19 incidence data. The inflated confidence interval for the base infection rate, i.e. the infection rate μ in age group 0–9 in Copenhagen West μ is 4.3–12.7 per thousand.

Aarhus	Copenhagen City	Slagelse
0.44-0.93	0.32-0.65	0.30-0.95

10–19	20–29	30–39	40–49	50–59	60–69	70+
2.2-6.2	1.4-3.9	0.82-2.5	1.1-3.3	0.89-2.8	0.51-2	0.24-1.19

8.3.4 Sampling models for tables of counts

We have previously noted a striking similarity between the models for contingency tables associated with homogeneity, with independence of cross-classifications, and multiplicative Poisson models. Their main difference were associated with assumptions on whether various totals were considered fixed or random.

In the multiplicative Poisson model, all entries were random and hence all totals as well; the model for studying independence of cross-classifications had the grand total $Y_{++} = n$ fixed, but other entries random; when considering homogeneity, the row totals $Y_{i+} = n_i$ and hence the grand total were fixed.

Clearly, these models represent different ways in which the data have been collected, but it also represents the choice of representation space used in the statistical modelling. For example, in the study concerning myocardial infarction, the allocation to aspirin vs. placebo was in fact made by randomization, but as the randomization itself was not relevant for the problem, we chose to consider the marginal totals fixed, rather than binomially distributed. This reflects that we consider this additional variation irrelevant, so we prefer to avoid this complication.

Luckily we noticed that all test statistics and all asymptotic distributions of these did not depend on which of the three situations we consider and, as we shall see, this is not a coincidence as the models are directly related in a specific and simple way as we shall demonstrate below.

8.3.4.1 From multiplicative Poisson to independence

Recall that the Poisson model has density

$$P_\lambda(Y = y) = \prod_{i=0}^{r-1}\prod_{j=0}^{s-1} \frac{\lambda_{ij}^{y_{ij}}}{y_{ij}!} e^{-\lambda_{ij}}.$$

The marginal total Y_{++} is a sum of independent Poisson distributed random variables and is therefore Poisson distributed with parameter $\lambda_{++} = \sum_{ij} \lambda_{ij}$ and has therefore density

$$P_\lambda(Y_{++} = y_{++}) = \frac{\lambda_{++}^{y_{++}}}{y_{++}!} e^{-\lambda_{++}}.$$

If we wish to consider the observations in a representation space where $Y_{++} = n$ is fixed and thus ignore the possibly irrelevant information in the grand total, we may calculate the conditional distribution which for any y with $y_{++} = n$ is

$$\begin{aligned}
P_\lambda(Y = y \mid Y_{++} = n) &= \frac{P_\lambda(Y = y \text{ and } Y_{++} = n)}{P_\lambda(Y_{++} = n)} \\
&= \frac{\prod_{i=0}^{r-1}\prod_{j=0}^{s-1} \frac{\lambda_{ij}^{y_{ij}}}{y_{ij}!} e^{-\lambda_{ij}}}{\frac{\lambda_{++}^{y_{++}}}{y_{++}!} e^{-\lambda_{++}}} \\
&= \frac{n!}{\prod_{ij} y_{ij}!} \prod_{i=0}^{r-1}\prod_{j=0}^{s-1} \frac{\lambda_{ij}^{y_{ij}}}{\lambda_{++}^{y_{ij}}} \\
&= \binom{n}{y_{00}, \ldots, y_{r-1,s-1}} \prod_{i=0}^{r-1}\prod_{j=0}^{s-1} \pi_{ij}^{y_{ij}},
\end{aligned}$$

where now $\pi_{ij} = \lambda_{ij}/\lambda_{++}$. Thus the unrestricted Poisson model becomes the unrestricted multinomial model when considering the grand total fixed.

In addition, if we assume that the parameter has multiplicative structure as in (8.12), i.e. if $\lambda_{ij} = \mu\rho_i\eta_j$, we have

$$\pi_{ij} = \frac{\mu\rho_i\eta_j}{\mu\rho_+\eta_+} = \frac{\rho_i}{\rho_+}\frac{\eta_i}{\eta_+} = \pi_{i+}\pi_{+j}.$$

Hence, the multiplicative Poisson model becomes the model for independence of cross-classifications and *vice versa* since if

$$\pi_{ij} = \frac{\lambda_{ij}}{\lambda_{++}} = \pi_{i+}\pi_{+j} = \frac{\lambda_{i+}}{\lambda_{++}}\frac{\lambda_{+j}}{\lambda_{++}}$$

we also have

$$\lambda_{ij} = \lambda_{++}\frac{\lambda_{i+}}{\lambda_{++}}\frac{\lambda_{+j}}{\lambda_{++}} = \mu\rho_i\eta_j$$

which has the multiplicative form desired.

8.3.4.2 *From independence to homogeneity*

If we now proceed yet one step further and also consider the marginal row totals $Y_{i+} = n_i$ fixed, we get similarly

$$\begin{aligned} P_\lambda(Y = y \mid Y_{i+} = n_i, i = 0, \ldots, r-1) &= \frac{P_\lambda(Y = y \text{ and } Y_{i+} = n_i, i = 0, \ldots, r-1)}{P_\lambda(Y_{i+} = n_i, i = 0, \ldots, r-1)} \\ &= \frac{\prod_{i=0}^{r-1}\prod_{j=0}^{s-1}\frac{\lambda_{ij}^{y_{ij}}}{y_{ij}!}e^{-\lambda_{ij}}}{\prod_{i=0}^{r-1}\frac{\lambda_{i+}^{y_{i+}}}{y_{i+}!}e^{-\lambda_{i+}}} \\ &= \prod_{i=0}^{r-1}\frac{n_i!}{\prod_j y_{ij}!}\prod_{j=0}^{s-1}\frac{\lambda_{ij}^{y_{ij}}}{\lambda_{i+}^{y_{ij}}} \\ &= \prod_{i=0}^{r-1}\binom{n_i}{y_{i0}, \ldots, y_{i,s-1}}\prod_{j=0}^{s-1}\pi_{ij}^{y_{ij}} \end{aligned}$$

where now $\pi_{ij} = \lambda_{ij}/\lambda_{+j}$. Thus, again, as we condition with the row totals, the unrestricted multinomial model becomes the model of independent and unrelated multinomial distributions for the rows of the table. And, if now λ has multiplicative structure, we get

$$\pi_{ij} = \frac{\lambda_{ij}}{\lambda_{+j}} = \frac{\mu\rho_i\eta_j}{\mu\rho_+\eta_j} = \frac{\rho_i}{\rho_+}$$

so the multiplicative Poisson model becomes the model of homogeneity of multinomial distributions.

8.3.4.3 *Exact conditional tests*

If we are focusing our interest on aspects of the distribution that are unrelated to the actual size of the marginals, for example if we are interested whether or not the hypothesis of multiplicativity, independence, or homogeneity is valid, we may wish to go one step further and consider all marginal totals fixed. Using calculations similar to those above it may be shown that—*under any of these null hypotheses*—the conditional distribution of the entries of the table given the marginal totals is

$$P(Y = y \mid Y_{i+} = n_{i+}, i = 0, \ldots, r-1; Y_{+j} = n_{+j}, j = 0, \ldots, s-1)$$
$$= \frac{\prod_{i=0}^{r-1} n_{i+}! \prod_{j=0}^{s-1} n_{+j}!}{n! \prod_{i=0}^{r-1} \prod_{j=0}^{s-1} y_{ij}!} \tag{8.17}$$

which is known as the *multiple hypergeometric distribution.*

An important feature of this distribution is that it is *free of unknown parameters* as long as either of the hypotheses of multiplicativity, independence of classification criteria, and homogeneity of multinomials is fulfilled. Although the distribution appears complicated for large tables, there is a simple Monte Carlo algorithm for sampling from this distribution due to Patefield (1981). Hence, the method provides a generic basis for calculating Monte Carlo p-values for essentially any test statistic of interest and thus an interesting alternative to using asymptotic results, in particular when some of the entries in a table are small. Although Monte Carlo methods must often be used to calculate the p-value, we refer to tests using the multiple hypergeometric distribution as *exact conditional tests.*

In the special case of a 2×2 contingency table, the conditional distribution in (8.17) simplifies to an ordinary hypergeometric distribution as all entries Y_{ij} in the table are deterministic functions of Y_{00} and the marginal totals. More precisely, we have

$$P(Y_{00} = y \mid Y_{0+} = n_{0+}, Y_{1+} = n_{1+}, Y_{+0} = n_{+0}, Y_{+1} = n_{+1}) = \frac{\binom{n_{0+}}{y}\binom{n_{1+}}{n_{+0}-y}}{\binom{n_{++}}{n_{+0}}}.$$

Example 8.10. [Smoking and myocardial infarction] We shall illustrate the use of exact conditional test applied to data from a study of association between smoking and myocardial infarction, displayed in Table 8.13. The category 'Heavy smoking' corresponds to regular smoking of more than 25 cigarettes a day. The numbers in the table are small and one may fear that the asymptotic p-values are far from the actual ones. In fact, the Pearson χ^2 statistic evaluates to $W = 6.96$ which compared to a

Table 8.13 – Number of patients suffering myocardial infarction compared to smoking habits and controls. Source: Agresti (2002), based on a study published in the *Lancet*, **313**, 743–747 (1979).

	Smoking level			
	No smoking	Moderate smoking	Heavy smoking	Total
Control	25	25	12	62
Myocardial infarction	0	1	3	4
Total	25	26	15	66

χ^2-distribution with 2 degrees of freedom yields an asymptotic p-value of $p = 0.03$, whereas the exact p-value may be calculated to be $p = 0.052$. The likelihood ratio test evaluates to $G^2 = 6.69$ with an asymptotic p-value of $p = 0.035$, whereas a Monte Carlo approximation based on $N = 5000$ samples yields $p = 0.074$. These tests ignore the fact that the variable indicating smoking habits has an ordinal nature, and may not reveal a tendency of the probability of infarction increasing with the amount of smoking. □

Note that if we reject the hypothesis when the conditional p-value is less that α, the test also maintains the overall level α. For if we let $\mathcal{K}(N_{+1})$ denote the critical region for the conditional test, we then have the conditional level $H(\mathcal{K}(N_{+1})) \leq \alpha$ and thus

$$\begin{aligned} P_\theta(\mathcal{K}(N_{+1})) &= \sum_y P_\theta(\mathcal{K}(N_{+1}) \mid N_{+1} = y) P_\theta(N_{+1} = y) \\ &= \sum_y H(\mathcal{K}(y)) P_\theta(N_{+1} = y) \\ &\leq \sum_y \alpha P_\theta(N_{+1} = y) = \alpha. \end{aligned}$$

8.3.5 *Fisher's exact test*

We shall here focus on the case where $r = s = 2$ and illustrate the concepts with data from Example 8.7 in Hansen (2012).

Here $n = 130$ individuals were classified according to whether or not they used an old or a new type of computer screen in their work, and whether they had problems with reflections from the screen or not. This resulted in the following table:

	Reflection problems		
Screen type	No	Yes	Total
Old	15	50	65
New	27	38	65
Total	42	88	130

Several different models are possible for these data, as it is not directly specified how the data were collected. Were they simply collecting data from all individuals at the department involved, did they deliberately choose a sample size of $n = 130$, or were they ensuring that exactly 65 of each screen type were asked about reflection problems?

Since we do not know how the marginals were collected, we may use a sampling model with all marginals fixed, thus performing an analysis based on the exact conditional distribution. More precisely, we consider $X_1, \ldots, X_{42}$ to be binary variables corresponding to those that have reflection problems with their screens and indicating whether the screen is old or new, s. If there is no difference between old and new screens, we may consider these to be sampled *at random and without replacement* from the set of all screens, having 65 of each type. Now we let $X = N_{11} = \sum_{i=1}^{42} X_i$

we have that X follows a *hypergeometric distribution*

$$P(N_{11} = x) = h(x \mid 42, 65, 65) = \frac{\binom{65}{x}\binom{65}{42-x}}{\binom{130}{42}}.$$

Fisher's exact test, described in Fisher (1934) uses the size of this probability itself as test statistic:

$$d(x) = 1 - h(x \mid 42, 65, 65)$$

so values of x are considered extreme if they have very small probability in the hypergeometric distribution. Thus, in this hypergeometric model, the p-value becomes

$$p(x) = \sum_{(x:d(x) \geq h(15 \mid 42,65,65))} h(x \mid 42, 65, 65) = 0.03847.$$

The sum in the expression ranges over $x \in (0, \ldots, 15, 27, \ldots 42)$, as these are the values with smaller probability than the observed. The procedure is illustrated in Figure 8.1.

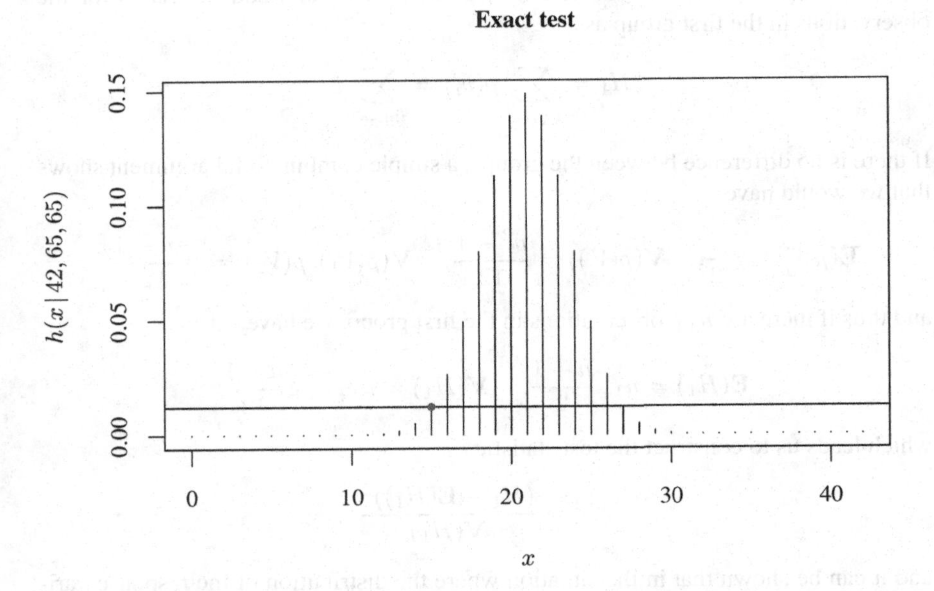

Figure 8.1 – The hypergeometric distribution used for the exact test in the screen glare example. The probability of the observation is indicated by a dot. The horizontal line has height $h(15 \mid 42, 65, 65)$ and the p-value is the sum of the sizes of the point probabilities that are below or on that line.

8.3.6 Rank statistics for ordinal data

We shall illustrate the flexibility of exact conditional tests by comparing multinomial distributions with *ordinal* categories, i.e. in situations such as in Example 8.5 concerned with the weight of trout or Example 8.10, where there is a special emphasis of discovering deviations from the hypothesis in the form of a translational shift in the distribution: is there a tendency of the response distribution shifting to the right or left when changing category. In the trout example, are the fish generally smaller? In the smoking example, do the infarction group generally smoke more than the controls?

The standard test statistics G^2 and X^2 do not take the ordinal structure into account, as a permutation of the response categories would have no influence on the values of either. So we would wish to consider alternatives. One such alternative is the *Mann–Whitney* or *Wilcoxon* rank statistic M^2 which we shall describe below.

Consider a table of the form in Table 8.4, but with only $r = 2$ categories to be compared. Imagine first that there is at most one observation in any response category and the data therefore may be completely sorted as

$$(v_1, g_1), (v_2, g_2), \ldots, (v_n, g_n)$$

with $g_i \in (1, 2)$ denoting the group and v_i the value of the ith observation where now

$$v_1 < v_2 < \cdots < v_n.$$

We define the *rank* of an observation v_i as $\rho(v_i) = i$ and add the ranks for the observations in the first group as

$$H_1 = \sum_{i:g_i=1} \rho(v_i) = \sum_{i:g_i=1} i$$

If there is no difference between the groups, a simple combinatorial argument shows that we would have

$$\mathbf{E}(\rho(V)) = \frac{n}{2}, \quad \mathbf{V}(\rho(V)) = \frac{n^2-1}{12}, \quad \mathbf{V}(\rho(V_1), \rho(V_2)) = \frac{-n-1}{12}$$

and thus if there are n_{1+} observations in the first group, we have

$$\mathbf{E}(H_1) = n_{1+}\frac{n+1}{2}, \quad \mathbf{V}(H_1) = n_{1+}n_{2+}\frac{n+1}{12}$$

which leads us to construct the test statistic

$$M^2 = \frac{(H_1 - \mathbf{E}(H_1))^2}{\mathbf{V}(H_1)}$$

and it can be shown that in the situation where the distribution of the response variable V is continuous so there are no ties among the observations, M^2 is approximately χ^2-distributed with one degree of freedom when both groups are large.

However, when applying this to ordinal data, the ranks are not well-defined, as many observations have the same value. We then instead allocate the *mid-rank* to all data in the same response category. For example, in category 0 we have observations $Y_{+0} = Y_{10} + Y_{20}$ and these observations all receive the rank $(1 + Y_{+0})/2$. Similarly, the data in group j are given ranks $\tau_j + (1 + Y_{+j})/2$, where $\tau_0 = 0$ and $\tau_j = Y_{+0} + \cdots Y_{+,j-1}$ for $j > 0$. Thus we define the *tied rank sum* H_1^* for group 1 as

$$H_1^* = \sum_{j=0}^{s-1} \left(\tau_j + \frac{1 + Y_{+j}}{2} \right) Y_{1j}.$$

However, now it is less easy to determine the variance of the statistic. If we normalize it as before, we get

$$M^2 = \frac{(H_1^* - \mathbf{E}(H_1))^2}{\mathbf{V}(H_1)}$$

the asymptotic χ^2 distribution for M^2 may now be less adequate, and we may therefore evaluate the p-value in the exact conditional distribution with all marginals fixed, using, say, a Monte Carlo algorithm.

Example 8.11. If we consider the trout data in Example 8.5, the value of M^2 becomes 2.02, yielding an asymptotic p-value (ignoring ties) of $p = 0.15$. If we instead calculate the Monte Carlo p-value based on $N = 5000$ simulated tables, we get $p = 0.13$; none of these are significant, so we are still not able to document an effect of the waste water release on the weight of trout in the lake although the p-value is much smaller than for the standard G^2 or X^2 test statistics, where we found $p = 0.67$.

For the data on smoking and myocardial infarction, we get $M^2 = 4.87$ which yields an asymptotic p-value of $p = 0.027$, whereas the Monte Carlo p-value based on 5000 simulated tables is $p = 0.035$. Now this should be compared to the results for G^2 with a Monte Carlo p-value of $p = 0.074$. Thus the ordinal test statistic will just reject the hypothesis of equality between the infarction and control group on a 5% level, whereas G^2 will not. □

8.4 Exercises

Exercise 8.1. 150 Petri dishes prepared with streptomycin were each subjected to one million E-coli bacteria. If a bacterium mutates to become resistant, it will grow a colony of surviving bacteria. Otherwise the streptomycin will kill the bacteria. After growth, the following number of surviving colonies was observed:

Number of colonies per dish	0	1	2	3	4
Number of Petri dishes	98	40	8	3	1

a) Argue why it would be reasonable to expect that the number of colonies in a Petri dish would be Poisson distributed;

b) Estimate the parameter under the assumption of a Poisson distributed number of colonies.

c) Investigate whether the observations conform with the Poisson model.

Note that as the Poisson distribution has infinite support, a little modification is needed to transform the data into a multinomial distribution to answer question c). This may, for example be done by merging the categories $3, 4, \ldots$ into a category ≥ 3 with a count of 4, or even merging to ≥ 2 with a count of 12. This now becomes a curved multinomial family, and the analysis may proceed along the same lines as in Example 8.2.

Exercise 8.2. The *logarithmic* distribution is a distribution on integers with density

$$f_\theta(x) = \frac{\theta^x}{x(-\log(1-\theta))}, \quad x = 1, 2, \ldots$$

for $\theta \in (0, 1)$. Williams (1943) investigated the distribution of the number of publications per author in a 1913 volume of *Review of Applied Entomology* with the following result:

Number of articles per author	1	2	3	4	5	6	7	8	9	10
Number of authors	285	70	32	10	4	3	3	1	2	1

There are a total of 411 authors and 656 articles. Does the logarithmic distribution describe these data sufficiently well?

Again, as in the previous exercise, some categories need to be merged, for example introducing a category of ≥ 7 with a count of 7.

Exercise 8.3. The haptoglobin type for humans are determined by a single diallelic gene with alleles Hb1 and Hb2. The genotypes are denoted Hb1, Hb1Hb2, and Hb2.

Haptoglobin types were determined for 607 men and 1439 women, selected at random from the Danish population, resulting in the following table:

	Haptoglobin type			
	Hb1	Hb1Hb2	Hb2	Total
Men	83	289	235	607
Women	245	678	516	1439

a) Is the distribution of haptoglobin type the same for men and women?
b) Is the population in Hardy–Weinberg equilibrium?

Exercise 8.4. Colour blindness is more common among males than females. This could be due to colour being inherited via a gene on the X-chromosome, where males have only one, whereas females have two. Waaler (1927) collected the following data on colour blindness of 18121 school children in Oslo:

	Boys	Girls	Total
Colour blind	725	40	765
Normal	8324	9032	17356
Total	9049	9072	18121

If this were the case, and μ denotes the probability that a boy is colour blind, one would expect that the probability that a girl is colour blind should be equal to μ^2 corresponding to having two copies of the colour blindness gene. Investigate whether the data support such a hypothesis.

Exercise 8.5. Calculate a 95% confidence interval for the relative risk of getting myocardial infarction when using aspirin as compared to placebo in Example 8.6.

Exercise 8.6. A teaching assistant at the University of Copenhagen consumed 384 hard-boiled eggs in the academic year 1968/69 and noted every day how many of the eggs broke during cooking so that the egg-white flowed into the boiling water, and how many cracked without the egg-white leaving the shell. For 130 of these eggs, he used a so-called egg-piercer that pierced a small hole at the bottom of the egg, to prevent the egg from breaking. The results of his experiment are summarized below:

	Broken	Cracked	Whole	Total
Pierced	5	16	109	130
Unpierced	19	36	199	254
Total	24	52	308	384

Is the egg-piercer effective?

Exercise 8.7. The data below are concerned with 1545 fraternal twins of opposite sex and their criminal behaviour. Each twin pair is classified after their criminal status, resulting in the following table.

	Female criminal	Female not criminal	Total
Male criminal	16	286	302
Male not criminal	24	1219	1243
Total	40	1505	1545

a) Are the criminal status of the male and female twin independent?

b) Is there a difference between the sexes with respect to being criminal?

Exercise 8.8. Consider the shifted multiplicative Poisson model in the case where $r = s = 2$. Show that the likelihood equations are equivalent to the following equation in ρ:

$$\frac{(Y_{00} + \rho)(Y_{11} + \rho)}{(Y_{10} - \rho)(Y_{01} - \rho)} = \frac{B_{00}B_{11}}{B_{10}B_{01}}$$

with

$$\lambda_{00} = Y_{00} + \rho, \ \lambda_{01} = Y_{01} - \rho, \ \lambda_{10} = Y_{10} - \rho, \ \lambda_{11} = Y_{11} + \rho,$$

expressing that the estimated cross-product ratio is equal to the cross-product ratio of the background risk B. Use this to give an explicit formula for the maximum likelihood estimator $\hat{\lambda}$ of λ in this special case.

Exercise 8.9. In 1974 there was a suspicion that the number of lung cancer cases was extraordinarily high in the city of Fredericia (Clemmensen et al., 1974). To investigate whether this suspicion were justified, cancer incidence in Fredericia was compared to cancer incidence in Vejle in different age groups. The tables below display the number of cancer cases in two age groups in the two cities, and the sizes of the population in the same age groups.

	Cancer cases			Population	
	55–64	70+	Total	55–59	60–64
Vejle	12	15	27	3398	1158
Fredericia	22	18	40	3859	1114
Total	34	33	67	7257	2272

Now use the results from the previous exercise to answer the following questions:

a) Is a shifted multiplicative Poisson model adequate for these data?

b) Does the suspicion concerning a high incidence of lung cancer in Fredericia seem justified?

Exercise 8.10. Compare the p-value for the exact conditional test in the screen glare example in Section 8.3.4.3 with asymptotic p-values using the G^2 and X^2 statistics.

Appendix A

Auxiliary Results

A.1 Euclidean vector spaces

Definition A.1. A *Euclidean vector space* $(V, \langle \cdot, \rangle)$ is a finite-dimensional vector space over $\mathbb{R}$ with an inner product $\langle \cdot, \cdot \rangle$, satisfying

i) For all $u, v \in V$: $\langle u, v \rangle = \langle v, u \rangle$;

ii) For all $u, v, w \in V$ and all $a, b \in \mathbb{R}$: $\langle u, av + bw \rangle = a\langle u, v \rangle + b\langle v, w \rangle$;

iii) For all $v \in V \setminus \{0\}$: $\langle v, v \rangle > 0$.

Note that i) and ii) combined imply that $\langle \cdot, \cdot \rangle$ is *bilinear*, i.e. linear in both of its arguments. From the inner product, we may now define the *Euclidean norm* $\|v\| = \sqrt{\langle v, v \rangle}$ and *Euclidean distance* $d(u, v) = \|u - v\|$. We note the following:

Proposition A.2. *Let V be a vector space over $\mathbb{R}$. Then any symmetric bilinear form $\langle \cdot, \cdot \rangle$ on V is determined by its values $\langle v, v \rangle, v \in V$ on the diagonal.*

Proof. This follows from the relation

$$\langle u, v \rangle = \frac{1}{4} \left(\langle u + v, u + v \rangle - \langle u - v, u - v \rangle \right)$$

which is easily established using i) and ii) above. □

Observe that $\mathbb{R}^d$ is an example of such a space, with the standard inner product given by $\langle u, v \rangle = u^\top v$ and any Euclidean space is isomorphic to $\mathbb{R}^d$; an isomorphism is constructed by choosing an *orthonormal basis*, i.e. a system $(e_1, \ldots, e_d)$ of elements of V satisfying $\langle e_i, e_j \rangle = 0$ if $i \neq j$ and $\|e_i\| = 1$. Then any element $u \in V$ has a unique representation as

$$u = \langle u, e_1 \rangle e_1 + \cdots \langle u, e_d \rangle e_d = \alpha_1 e_1 + \cdots + \alpha_d e_d$$

where $\alpha = \alpha(u) = (\alpha_1, \ldots, \alpha_d)^\top$ is the vector of *coordinates* of u with respect to the chosen basis. The correspondence $u \leftrightarrow \alpha$ is an isomorphism so that, in particular, if v has coordinates β with respect to the chosen basis, we have

$$\langle u, v \rangle = \alpha^\top \beta \text{ and hence } \|u\|^2 = \alpha_1^2 + \cdots + \alpha_d^2.$$

The Borel σ-algebra $\mathbb{B}(V)$ on a Euclidean space V is the σ-algebra generated by open sets of V and any isomorphism as above is also bimeasurable so we may identify the

DOI: 10.1201/9781003272359-A

Borel sets of V with those of $\mathbb{R}^d$, the particular identification depending on the choice of basis. Similarly, the *standard Lebesgue measure* on V is obtained by identifying any $v \in V$ with its coordinates in an orthonormal basis; it can be assured that this specification does not depend on the particular orthonormal basis chosen; we refrain from giving the details.

If W is another vector space we let $\mathcal{L}(V, W)$ denote all linear maps from V to W and recall that all linear forms $f \in \mathcal{L}(V, \mathbb{R})$ may be uniquely be represented via the inner product as

$$f(v) = \langle u_f, v\rangle.$$

The *adjoint* A^* of a linear map $A \in \mathcal{L}(V, W)$ is the unique linear map $A^* \in \mathcal{L}(W, V)$ satisfying for all $v, w \in V \times W$ that

$$\langle v, A^*w\rangle_V = \langle Av, w\rangle_W$$

where $\langle \cdot, \cdot\rangle_V$ and $\langle \cdot, \cdot\rangle_W$ are the inner products on V and W. Clearly, we have $(A^*)^* = A$.

A map $A \in \mathcal{L}(V, V)$ is *self-adjoint* if $A^* = A$. Thus for $V = \mathbb{R}^d$ and $W = \mathbb{R}^q$ with standard inner product, the transpose $A^\top \in \mathbb{R}^{d\times q}$ is the matrix for the adjoint of the map with matrix $A \in \mathbb{R}^{q\times d}$. Then a map is self-adjoint if and only if its matrix is symmetric.

Let now $L \subseteq V$ be a linear subspace of V. The *orthogonal projection* Π_L onto L of $u \in V$ is determined as the unique point $\Pi_L u \in L$ satisfying for all $w \in L$

$$\langle u - \Pi_L u, w\rangle = 0 \tag{A.1}$$

or, in other words, the unique point $\Pi_L u \in L$ satisfying that $u - \Pi_L u$ is orthogonal to L with respect to the inner product. Here and elsewhere we shall omit the qualifier 'orthogonal' as all projections we consider are orthogonal. The projection $\Pi_L u$ is also the point in L that is closest to u in Euclidean distance:

$$\Pi_L u = \operatorname{argmin}_{v\in L} \|u - v\|^2.$$

If $v_1, \dots, v_m$ is a set of mutually orthogonal vectors that span L, the projection may be expressed in terms of these as

$$\Pi_L u = \frac{\langle u, v_1\rangle}{\|v_1\|^2} v_1 + \cdots + \frac{\langle u, v_m\rangle}{\|v_m\|^2} v_m \tag{A.2}$$

which in particular simplifies when $v_1, \dots, v_m$ is an orthonormal basis as then $\|v_i\|^2 = 1$. The relation (A.2) follows since then for any $w = \alpha_1 v_1 + \cdots \alpha_m v_m \in L$ we have

$$\langle \Pi_L u, w\rangle = \sum_{i=1}^{m}\sum_{j=1}^{m} \alpha_j \frac{\langle u, v_i\rangle}{\|v_i\|^2}\langle v_i, v_j\rangle = \sum_{j=1}^{m} \alpha_j \langle u, v_j\rangle = \langle u, w\rangle$$

since $\langle v_i, v_j\rangle = 0$ if $i \neq j$.

Theorem A.3. *A projection is linear, i.e.* $\Pi \in \mathcal{L}(V, V)$. *A linear map* P *is a projection if and only if* P *is idempotent and self-adjoint. Then* P *is the projection onto its range* $L = \mathrm{rg}(P)$.

Proof. Linearity of Π follows directly from (A.1) and linearity of the inner product since if $y = \alpha u + \beta v$ we have

$$\langle \alpha u + \beta v - (\alpha \Pi u + \beta \Pi v), w\rangle = \alpha\langle u - \Pi u, w\rangle + \beta\langle v - \Pi v, w\rangle = 0.$$

Assume that P is self-adjoint and idempotent. Then for all $u, v \in V$ we have

$$\langle u - Pu, Pv\rangle = \langle u, Pv\rangle - \langle Pu, Pv\rangle = \langle u, Pv\rangle - \langle u, P^2 v\rangle = 0$$

and hence Pu is the projection onto $\mathrm{rg}(P)$.

Conversely, if $P = \Pi u$ is the projection of u onto $\mathrm{rg}(P)$, we must have $P^2 u = Pu$ as Pu is the closest point to Pu in $\mathrm{rg}(P)$. Also, from (A.1)

$$\langle u, Pv\rangle = \langle Pu, Pv\rangle = \langle Pu, v\rangle$$

showing that P is self-adjoint. □

If $L \subseteq V$ is a subspace of V, its *orthogonal complement* $L^\perp$ is the subspace of vectors orthogonal to L:

$$L^\perp = \{u \in V \mid \langle u, v\rangle = 0 \text{ for all } v \in L\}.$$

If Π_L is the projection onto L, $I - \Pi_L$ is the projecton onto $L^\perp$ as $I - \Pi_L$ is idempotent and self-adjoint if and only if Π_L is. Then any vector $v \in V$ has a unique decomposition as $v = u + w$ where $u \in L$ and $w \in L^\perp$

$$v = \Pi_L v + (I - \Pi_L)v = u + w$$

so $V = L \oplus L^\perp$. We note the following relation between the image and null-space of a linear map:

Proposition A.4. *Let* $A \in \mathcal{L}(V, W)$ *be a linear map between Euclidean spaces* $(V, \langle\cdot,\cdot\rangle_1)$ *and* $(W, \langle\cdot,\cdot\rangle_2)$ *and* $A^* \in \mathcal{L}(W, V)$ *its adjoint. Then*

$$\mathrm{rg}(A)^\perp = \ker(A^*).$$

Hence, if A *is surjective,* A^* *is injective and vice versa. Also, we have orthogonal decompositions*

$$V = \mathrm{rg}(A) \oplus \ker(A^*) = \ker(A) \oplus \mathrm{rg}(A^*).$$

Proof. Assume $w \in \mathrm{rg}(A)$, i.e. $w = Av$ and $z \in \ker(A^*)$. Then

$$\langle w, z\rangle_2 = \langle Av, z\rangle_2 = \langle v, A^* z\rangle_1 = \langle v, 0\rangle_1 = 0$$

and hence $w \perp \ker(A^*)$. Conversely, if $z \in \ker(A^*)$, reading this relation from right to left yields that $z \perp \mathrm{rg}(A)$. if A is surjective we thus have

$$\{0\} = \mathrm{rg}(A)^\perp = \ker(A^*)$$

whence A^* is injective. The *vice versa* follows since $A = (A^*)^*$. □

The case where $A = A^*$ is self-adjoint yields the following corollary:

Corollary A.5. *Let $(V, \langle \cdot, \cdot \rangle)$ be a Euclidean space and $\Sigma \in \mathcal{L}(V, V)$ self-adjoint. Then $V = \mathrm{rg}(\Sigma) \oplus \ker(\Sigma)$.*

Further we have the following useful lemma:

Lemma A.6. *If $\Sigma = AA^*$ we have $\mathrm{rg}(A) = \mathrm{rg}(\Sigma)$.*

Proof. The inclusion $\mathrm{rg}(\Sigma) \subseteq \mathrm{rg}(A)$ follows since

$$u = \Sigma v = AA^* v = A(A^* v).$$

For the reverse inclusion, assume $u \in \mathrm{rg}(A)$ and thus $u = Av$ for some $v \in V$. By Proposition A.4 we may write $v = v_1 + v_2$ where $v_1 = A^* w \in \mathrm{rg}(A^*)$ and $v_2 \in \ker(A)$ implying

$$u = Av = AA^* w + Av_2 = AA^* w = \Sigma w$$

and hence $u \in \mathrm{rg}(\Sigma)$, as required. □

A.2 Convergence of random variables

This section collects some standard results from probability theory; consult, for example Jacod and Protter (2004) for details.

A.2.1 *Convergence in probability*

Recall the following from Jacod and Protter (2004, p. 143 ff.):

Definition A.7. [Convergence in probability] A sequence $X_1, \ldots, X_n, \ldots$ of random variables with values in $\mathbb{R}^k$ is said to *converge in probability* to a random variable X if for all $\delta > 0$

$$\lim_{n \to \infty} P\{\|X_n - X\| > \delta\} = 0$$

where $\|x\|$ is any of the equivalent norms on $\mathbb{R}^k$. Symbolically we write $X_n \xrightarrow{P} X$ or $\mathrm{plim}_{n \to \infty} X_n = X$.

We note that

Proposition A.8. *If $f : \mathbb{R}^k \to \mathbb{R}^m$ is continuous and $X_n \xrightarrow{P} X$, then it holds that $f(X_n) \xrightarrow{P} f(X)$.*

The (weak) *law of large numbers* says that

Theorem A.9 (Law of Large Numbers). *If $X_1, \ldots, X_n, \ldots$ are independent and identically distributed with values in $\mathbb{R}^k$ and $\mathbf{E}(\|X_1\|) < \infty$, then the average converges in probability to the mean $\xi = \mathbf{E}(X_i)$.*

$$\bar{X}_n = \frac{X_1 + \cdots + X_n}{n} \xrightarrow{P} \xi \quad \textit{for } n \to \infty.$$

See Remark 20.1 in Jacod and Protter (2004) for this. There is a similar strong law, saying that under the same conditions, the set of $\omega \in \Omega$ where the average $\bar{X}_n(\omega)$ does not converge, is a null set. However, it is the weak version above that plays a role in Statistics.

A.2.2 Convergence in distribution

Let X be a random variable with values in $\mathbb{R}^k$ and F its distribution function, i.e.

$$F(x) = F(x_1, \dots, x_k) = P\{X_1 \leq x_1, \dots X_k \leq x_k\}.$$

We then define

Definition A.10. [Convergence in distribution] A sequence $X_1, \dots, X_n, \dots$ of random variables with values in $\mathbb{R}^k$ and distribution functions F_n is said to *converge in distribution* to X if it holds for all continuity points x of F that

$$\lim_{n \to \infty} F_n(x) = F(x).$$

Symbolically we write $X_n \xrightarrow{\mathcal{D}} X$ or $\operatorname{dlim}_{n \to \infty} X_n = X$, or even $X_n \xrightarrow{\mathcal{D}} F$, and we also say that $(X_n)_{n \in \mathbb{N}}$ converges in distribution to F. Convergence in distribution is preserved for continuous transformations, a result that is known as the *continuous mapping theorem*:

Theorem A.11. *If $f : \mathbb{R}^k \to \mathbb{R}^m$ is continuous and $X_n \xrightarrow{\mathcal{D}} X$, then it holds that $f(X_n) \xrightarrow{\mathcal{D}} f(X)$.*

Note that if $X = \xi$ is a constant random variable, convergence in distribution and in probability coincide:

$$X_n \xrightarrow{\mathcal{D}} X \iff X_n \xrightarrow{P} \xi.$$

We shall often use the following result, which is an easy consequence of the continuous mapping theorem above and the fact that if $X_n \xrightarrow{\mathcal{D}} X$ and $Y_n \xrightarrow{P} \eta$ then $(X_n, Y_n) \xrightarrow{\mathcal{D}} (X, c)$; see for example van der Vaart (2012, p. 11) for details.

Corollary A.12. *Assume $(X_1, Y_1), \dots, (X_n, Y_n), \dots$ is a sequence of random variables with values in $\mathbb{R}^k \times \mathbb{R}^m$ and $f : \mathbb{R}^k \times \mathbb{R}^m \to \mathbb{R}^l$ is continuous. If $X_n \xrightarrow{\mathcal{D}} X$ and $Y_n \xrightarrow{P} \eta$ then $f(X_n, Y_n) \xrightarrow{\mathcal{D}} f(X, \eta)$.*

Definition A.13. Consider two sequences of random variables $X_1, \dots, X_n, \dots$ and $Y_1, \dots, Y_n, \dots$. We say that the sequences are *asymptotically equivalent* and write

$$X_n \stackrel{\text{as}}{=} Y_n$$

if $Y_n - X_n \xrightarrow{P} 0$ for $n \to \infty$.

Corollary A.12 is also known as Slutsky's lemma. It follows that asymptotically equivalent sequences have the same limiting distribution:

Proposition A.14. *If $X_n \stackrel{\text{as}}{=} Y_n$ then $X_n \xrightarrow{\mathcal{D}} X \iff Y_n \xrightarrow{\mathcal{D}} X$; if $f : \mathbb{R}^k \to \mathbb{R}^m$ is continuous and $X_n \xrightarrow{\mathcal{D}} X$ then $X_n \stackrel{\text{as}}{=} Y_n \implies f(X_n) \stackrel{\text{as}}{=} f(Y_n)$.*

Proof. Let $Z_n = Y_n - X_n$. Then if $X_n \xrightarrow{\mathcal{D}} X$, Corollary A.12 (Slutsky's lemma) yields

$$Y_n = X_n + Z_n \xrightarrow{\mathcal{D}} X$$

and the converse follows by symmetry. Further, if f is continuous, we have

$$f(Y_n) - f(X_n) = f(X_n + Z_n) - f(X_n)$$

so if $X_n \xrightarrow{\mathcal{D}} X$, Slutsky's lemma yields that

$$f(Y_n) - f(X_n) \xrightarrow{\mathcal{D}} f(X + 0) - f(X) = 0$$

and hence $f(X_n) \overset{\text{as}}{=} f(Y_n)$. □

An important instance of convergence in distribution is associated with the normal distribution. More precisely we define:

Definition A.15. [Asymptotic normality] A sequence $X_1, \dots, X_n$ of random variables with values in $\mathbb{R}^k$ is said to be *asymptotically normally distributed* with *asymptotic mean* ξ and *asymptotic variance* Σ/n if

$$\sqrt{n}(X_n - \xi) \xrightarrow{\mathcal{D}} Z$$

where $Z \sim \mathcal{N}_k(0, \Sigma)$. We then write $X_n \overset{\text{as}}{\sim} \mathcal{N}_k(\xi, \Sigma/n)$.
Note that, formally, this means that for all $x \in \mathbb{R}^k$

$$\lim_{n\to\infty} P\{\sqrt{n}(X_n - \xi) \leq x\} = \mathbf{\Phi}_\Sigma(x),$$

where the inequality sign is interpreted coordinatewise. Here $\mathbf{\Phi}_\Sigma$ is the distribution function of the multivariate normal distribution

$$\mathbf{\Phi}_\Sigma(x) = \int_{-\infty}^{x_1} \cdots \int_{-\infty}^{x_k} \frac{e^{-y^\top \Sigma^{-1} y/2}}{\sqrt{(2\pi)^k \det \Sigma}} \, dy_k \cdots dy_1.$$

We now recall the following fundamental result:

Theorem A.16 (Central Limit Theorem). *If $X_1, \dots, X_n, \dots$ are independent and identically distributed with values in $\mathbb{R}^k$ and $\mathbf{E}||X_1||^2 < \infty$, then their average is asymptotically normally distributed as*

$$\bar{X}_n = \frac{X_1 + \cdots + X_n}{n} \overset{\text{as}}{\sim} \mathcal{N}_k\left(\xi, \frac{1}{n}\Sigma\right)$$

where $\xi = \mathbf{E}(X_i)$ and $\Sigma = \mathbf{V}(X_i)$.
In other words, the displayed expression in the theorem can be written

$$\sqrt{n}(\bar{X}_n - \xi) \xrightarrow{\mathcal{D}} Z, \quad Z \sim \mathcal{N}_k(0, \Sigma).$$

We note in passing that a sequence of asymptotically normal random variables always converges in probability to its asymptotic mean:

Corollary A.17. *Let $X_1, \dots, X_n, \dots$ be a sequence of random variables with values in $\mathbb{R}^k$ such that $X_n \overset{\text{as}}{\sim} \mathcal{N}_k(\xi, \Sigma/n)$. Then* $\operatorname{plim}_{n\to\infty} X_n = \xi$.

Proof. We have

$$\lim_{n\to\infty} P\{\|X_n - \xi\| > \delta\} \leq \lim_{n\to\infty} \sum_{i=1}^{k} P\{\sqrt{n}|X_{ni} - \xi_i| > \delta\sqrt{n}\}.$$

But for any fixed $K > 0$ we have that

$$\begin{aligned} \lim_{n\to\infty} \sum_{i=1}^{k} P\{\sqrt{n}|X_{ni} - \xi_i| > \delta\sqrt{n}\} &\leq \lim_{n\to\infty} \sum_{i=1}^{k} P\{\sqrt{n}|X_{ni} - \xi_i| > K\} \\ &= \sum_{i=1}^{k} 2(1 - \Phi(K/\sigma_{ii})). \end{aligned}$$

Letting now $K \to \infty$ makes the last expression tend to zero and the desired conclusion follows. □

We shall also be interested in convergence in distribution and asymptotic equivalence of quadratic forms. So let $Z_n, n \in \mathbb{N}$ be a sequence of random variables with values in $\mathbb{R}^d$ and and $K_n, n \in \mathbb{N}$ a sequence of random $d \times d$ matrices.

Lemma A.18. *Assume that the sequence $Z_n, n \in \mathbb{N}$ converges in distribution to Z and the sequence $K_n, n \in \mathbb{N}$ converges in probability to K. Then it holds that*

$$Z_n^\top K_n Z_n \overset{\text{as}}{=} Z_n^\top K Z_n \overset{\mathcal{D}}{\to} Z^\top K Z.$$

Proof. Write

$$Z_n^\top K_n Z_n - Z_n^\top K Z_n = Z_n^\top (K_n - K) Z_n.$$

Since $K_n - K \overset{P}{\to} 0$, Slutsky's lemma yields that $Z_n^\top K_n Z_n \overset{\text{as}}{=} Z_n^\top K Z_n$. The convergence in distribution follows from the continuous mapping theorem applied to the continuous map $z \mapsto z^\top K z$. □

A.2.3 The delta method

From the definition of asymptotic normality it follows that if A is an $m \times k$ matrix representing a linear map from $\mathbb{R}^k$ to $\mathbb{R}^m$ and $X_n \overset{\text{as}}{\sim} \mathcal{N}_k(\xi, \Sigma/n)$, then

$$Y_n = AX_n \overset{\text{as}}{\sim} \mathcal{N}(A\xi, A\Sigma A^\top/n).$$

This follows from Theorem A.11 since $x \to Ax$ is continuous and

$$\sqrt{n}(Y_n - A\xi) = A\sqrt{n}(X_n - \xi) \overset{\mathcal{D}}{\to} AZ$$

and if $Z \sim \mathcal{N}_k(0, \Sigma)$ then $AZ \sim \mathcal{N}_m(0, A\Sigma A^\top)$.

The following important result says essentially that differentiable functions behave like linear functions in this respect.

Theorem A.19 (The delta method). *Let $X_1, \ldots, X_n, \ldots$ be a sequence of random variables with values in $\mathbb{R}^k$ such that $X_n \overset{\text{as}}{\sim} \mathcal{N}_k(\xi, \Sigma/n)$. Let U be an open and convex neighbourhood of ξ and assume that $f : U \to \mathbb{R}^m$ is differentiable. It then holds that*

$$\sqrt{n}(f(X_n) - f(\xi)) \overset{\text{as}}{=} \sqrt{n}\, Df(\xi)(X_n - \xi) \xrightarrow{\mathcal{D}} Df(\xi)Z$$

where $Z \sim \mathcal{N}_k(0, \Sigma)$ and $Df(\xi)$ is the Jacobian matrix of partial derivatives

$$Df(\xi) = \{Df(\xi)_{ij}\} = \left\{ \left. \frac{\partial f_i(x)}{\partial x_j} \right|_{x=\xi} \right\}.$$

We thus have

$$Y_n = f(X_n) \overset{\text{as}}{\sim} \mathcal{N}_m(f(\xi), Df(\xi)\,\Sigma\, Df(\xi)^\top/n)$$

Proof. Differentiability of f implies

$$f(X_n) - f(\xi) = Df(\xi)(X_n - \xi) + \epsilon(X_n - \xi)\|X_n - \xi\|$$

where $\epsilon(x) \to 0$ for $x \to 0$. Now multiply both sides with $\sqrt{n}$ to get

$$\sqrt{n}(f(X_n) - f(\xi)) = \sqrt{n}\, Df(\xi)(X_n - \xi) + \epsilon(X_n - \xi)\|\sqrt{n}(X_n - \xi)\|$$

But $X_n \xrightarrow{P} \xi$ by Corollary A.17 so $\epsilon(X_n - \xi) \xrightarrow{P} 0$ and thus

$$\sqrt{n}(f(X_n) - f(\xi)) \overset{\text{as}}{=} \sqrt{n}\, Df(\xi)(X_n - \xi).$$

Since $\sqrt{n}(X_n - \xi) \xrightarrow{\mathcal{D}} \mathcal{N}_k(0, \Sigma)$, Corollary A.12 implies

$$\sqrt{n}(f(X_n) - f(\xi)) \xrightarrow{\mathcal{D}} \mathcal{N}_m(0, Df(\xi)\Sigma Df(\xi)^\top)$$

as also required. □

Note that we could without loss of generality assume that $U = \mathbb{R}^k$ as f always can be extended to $\mathbb{R}^k$ in a measurable way. Note also, that if $f(x) = Ax$, then $Df(\xi) = A$ so this indeed extends the linear case.

A.3 Results from real analysis

A.3.1 Inverse and implicit functions

We shall need the following fundamental theorems about inverse functions and functions which are implicitly defined through equations, see, for example (Rudin, 1976, p. 221–227) or many other books on real analysis. We recall that a function f is *smooth* if f is infinitely often differentiable.

Theorem A.20 (Inverse function theorem). *Let $f : U \to \mathbb{R}^k$ be a smooth function where $U \subseteq \mathbb{R}^k$ is open. Suppose that f is injective and $\det(Df(x)) \neq 0$ for all $x \in U$. Then $f(U)$ is open and f is a diffeomorphism of U onto $f(U)$. Further, if $h = f^{-1}$ then h is smooth and*

$$Dh(f(x)) = (Df(x))^{-1} \tag{A.3}$$

or, in words, the derivative of the inverse function is the inverse of the derivative.

The formula in (A.3) is obtained by composite differentiation (Rudin, 1976, Theorem 9.25) in the equation $h(f(x)) = x$, which yields

$$Dh\, Df = I_k$$

and thus implies (A.3).

Theorem A.21 (Implicit function theorem). *Let $f : \Omega \to \mathbb{R}^m$ be a smooth function where $\Omega \subseteq \mathbb{R}^{k+m}$ is open and let $z \in \mathbb{R}^m$ be fixed. Let further*

$$\Omega^0 = \{(x, y) \in \Omega : f(x, y) = z\}$$

and let $\omega^0 = (x^0, y^0) \in \Omega^0$. Assume that the determinant $\det A(\omega^0)$ *of the last m columns of the Jacobian matrix*

$$A(x, y) = \left\{\frac{\partial f_i(x, y)}{\partial y_j}\right\}_{i=1,\ldots,m;j=k+1,\ldots,k+m}$$

is non-zero. Then there exist open intervals $I \subseteq \mathbb{R}^k$ and $J \subseteq \mathbb{R}^m$ with

$$\omega^0 \in W = I \times J \subseteq \Omega$$

and a smooth map $g : I \to J$ such that

$$\Omega_0 \cap W = \{(x, g(x)) : x \in I\}.$$

Moreover, the partial derivatives of g are determined as

$$\frac{\partial g_i(x)}{\partial x_j} = -\sum_u A^{-1}(x, g(x))_{iu} \frac{\partial f_u(x, g(x))}{\partial x_j}. \tag{A.4}$$

To obtain the formula for the partial derivatives of g we have used a calculation known as *implicit differentiation*. If we have shown that in some neighbourhood, y is a smooth function $y = g(x)$ of x we simply use composite differentiation in the equation $f(x, g(x)) = z$. Differentiating on both sides of the equation with respect to x, we get

$$D_1 f(x, g(x)) I_k + D_2 f(x, g(x)) Dg(x) = 0,$$

where $D_1 f$ is the matrix of partial derivatives of $f(x, y)$ with respect to x and $D_2 f$ with respect to y. Now $D_2 f$ is what has been termed A in the theorem above, so if we solve the equation with respect to $Dg(x)$ we get

$$Dg = -A^{-1} D_1 f$$

which exactly is equation (A.4), written in compact form.

We note that these theorems also have versions where f is only assumed to be r times continuously differentiable for $r \geq 1$. The conclusions then warrants the implicit or inverse functions to be continuously differentiable of the same order.

A.3.2 Taylor approximation

Taylor approximation is an important and often used technique in mathematical statistics. It exists in several versions, with the remainder term expressed by an integral, with the remainder term represented by an intermediate value (Lagrange's version), or just with an estimate of the size of the remainder term (Peano's version). We only use Taylor approximation of the second-order version so these are the only versions given here.

Theorem A.22 (Taylor's formula with remainder term in integral form). *Let $\Omega \subseteq \mathbb{R}^k$ be an open and convex set and let $f : \Omega \to \mathbb{R}$ be a real-valued and smooth function. Then for any $\omega_0, \omega \in \Omega$ it holds that*

$$\begin{aligned} f(\omega) &= f(\omega_0) + Df(\omega_0)(\omega - \omega_0) + \\ &\quad (\omega - \omega_0)^\top \left(\int_0^1 (1-t) D^2 f\left(\omega_0 + t(\omega - \omega_0)\right) dt \right) (\omega - \omega_0). \quad \text{(A.5)} \end{aligned}$$

Here Df is the derivative and $D^2 f$ the Hessian of f.

Proof. We have for any real-valued and smooth function g

$$g(u) - g(0) = \int_0^u g'(t)\, dt.$$

For u fixed we integrate by parts to obtain

$$\begin{aligned} g(u) &= g(0) + [(t-u)g'(t)]_0^u - \int_0^u (t-u) g''(t)\, dt \\ &= g(0) + ug'(0) + \int_0^u (u-t) g''(t)\, dt. \end{aligned} \qquad \text{(A.6)}$$

Now use this for $g(u) = f(\omega_0 + u(\omega - \omega_0))$. Then we have

$$g(0) = f(\omega_0), \; g'(0) = Df(\omega_0)(\omega - \omega_0)$$

and

$$g''(u) = (\omega - \omega_0)^\top D^2 f\left(\omega_0 + u(\omega - \omega_0)\right)(\omega - \omega_0).$$

Since $g(1) = f(\omega)$ we get (A.5) when letting $u = 1$ in (A.6). □

For completeness we also state the more standard version with Lagrange's remainder term, again only in the second-order version.

Theorem A.23 (Taylor's formula with Lagrange's remainder term). *Let $\Omega \subseteq \mathbb{R}^k$ be open and convex set and $f : \Omega \to \mathbb{R}$ a real-valued and smooth function. Then for any $\omega_0, \omega \in \Omega$, there is an ω^* and ω^{**} on the line segment $\{\omega_0 + t(\omega - \omega_0) : t \in [0,1]\}$ with*

$$f(\omega) = f(\omega_0) + Df(\omega^*)(\omega - \omega_0),$$

and

$$f(\omega) = f(\omega_0) + Df(\omega_0)(\omega - \omega_0) + \frac{1}{2}(\omega - \omega_0)^\top D^2 f(\omega^{**})(\omega - \omega_0).$$

Here Df is the derivative and $D^2 f$ the Hessian of f.

Proof. We define the function $g(t) = f(\omega_0 + t(\omega - \omega_0))$ and use the one-dimensional result (Rudin, 1976, Theorem 5.15) for this function, yielding that

$$f(\omega) = g(1) = g(0) + g'(t^*) = g(0) + g'(0) + g''(t^{**})/2 \qquad \text{(A.7)}$$

for some $t^*, t^{**} \in (0, 1)$. By composite differentiation, we get

$$g'(t) = Df(\omega_0 + t(\omega - \omega_0))(\omega - \omega_0)$$

and

$$g''(t) = (\omega - \omega_0)^\top D^2 f(\omega_0 + t(\omega - \omega_0))(\omega - \omega_0).$$

Letting $\omega^* = \omega_0 + t^*(\omega - \omega_0)$ and $\omega^{**} = \omega_0 + t^{**}(\omega - \omega_0)$ yields the result. □

Finally we wish to use a form that works even for functions with values in $\mathbb{R}^m$ rather than just real-valued functions.

Theorem A.24 (Taylor's formula with Peano's remainder term). *Let $\Omega \subseteq \mathbb{R}^k$ be an open and convex set and let $f : \Omega \to \mathbb{R}^m$ be a smooth function. Then for any $\omega_0, \omega \in \Omega$ it holds that*

$$\begin{aligned} f(\omega) &= f(\omega_0) + Df(\omega_0)(\omega - \omega_0) + \frac{1}{2}(\omega - \omega_0)^\top D^2 f(\omega_0)(\omega - \omega_0) \\ &\quad + \epsilon(\omega - \omega_0)\|\omega - \omega_0\|^2 \end{aligned}$$

where

$$\lim_{x \to 0} \epsilon(x) = 0.$$

Proof. See, for example Exercise 9.30 in Rudin (1976) and use the Lagrange version, exploiting that the difference $D^2 f(\omega^{**}) - D^2 f(\omega_0) \to 0$ when $\omega \to \omega_0$. This argument provides such an epsilon function ϵ_i for each of the functions $f_i, i = 1, \ldots, m$. Letting $\epsilon(x) = (\epsilon_1(x), \ldots, \epsilon_m(x))^\top$ yields the result. □

A.4 The information inequality

We shall establish a useful inequality. First recall that for $x > 0$

$$\log x \leq x - 1 \qquad \text{(A.8)}$$

and (A.8) is strict unless $x = 1$. This follows from Taylor's formula with the Lagrange remainder term, expanding around $x = 1$:

$$\log x = x - 1 - \frac{(x-1)^2}{2y^2}$$

for some y between 1 and x. We then have

Lemma A.25 (Information inequality). *Let $a, b \in \mathbb{R}^k_+$ be vectors of positive numbers with $\sum_i a_i = \sum_i b_i$. Then*

$$\sum_i b_i \log a_i \leq \sum_i b_i \log b_i \tag{A.9}$$

and the inequality is strict unless $a_i = b_i$ for all i.

Proof. From (A.8) we get

$$\begin{aligned}\sum_i b_i \log a_i - \sum_i b_i \log b_i &= \sum_i b_i \log \frac{a_i}{b_i}\\ &\leq \sum_i b_i \left(\frac{a_i}{b_i} - 1\right) = \sum_i a_i - \sum_i b_i = 0\end{aligned}$$

which gives the result since the inequality is strict unless $a_i = b_i$ for all i. □

A.5 Trace of a matrix

The *trace* $\mathrm{tr}(A)$ of a square matrix $A = \{a_{ij}\}$ is the sum of its diagonal elements $\mathrm{tr}(A) = \sum_i a_{ii}$ and has, for example the following properties

1. $\mathrm{tr}(\gamma A + \mu B) = \gamma\,\mathrm{tr}(A) + \mu\,\mathrm{tr}(B)$ for $\gamma, \mu \in \mathbb{R}$;
2. $\mathrm{tr}(A) = \mathrm{tr}(A^\top)$;
3. $\mathrm{tr}(AB) = \mathrm{tr}(BA)$ for any $r \times s$-matrix A and an $s \times r$-matrix B.
4. $\mathbf{E}(\mathrm{tr}(X)) = \mathrm{tr}(\mathbf{E}(X))$ for any random square matrix X with entries having finite expectation.

Here expectations are taken coordinatewise in the last expression. These facts follow by direct calculation from the definitions of the trace. For example

$$\mathrm{tr}(AB) = \sum_i (AB)_{ii} = \sum_i \sum_j a_{ij} b_{ji} = \sum_j \sum_i b_{ji} a_{ij} = \sum_j (BA)_{jj} = \mathrm{tr}(BA).$$

Appendix B

Technical Proofs

In this chapter, we collect some proofs that may appear more technical than informative.

B.1 Analytic properties of exponential families

We need a couple of important technical lemmas associated with the normalizing constant in regular exponential families.

B.1.1 Integrability of derivatives

Lemma B.1. *Consider a regular exponential family with canonical parameter space Θ, canonical statistic t, and representation space $\mathcal{X}$. Let further and θ, h satisfy that $\theta \pm h \in \Theta$. It then holds for any $n \in \mathbb{N}$ that*

$$\mathbf{E}_\theta\left(|h^\top t(X)|^n\right) < \infty \tag{B.1}$$

Proof. Consider the function

$$g(x,n) = \frac{1}{n!}(h^\top t(x))^n e^{\theta^\top x}.$$

We first show that this function is integrable with respect to $\mu \otimes m$ where m is counting measure on $\mathbb{N}$. Indeed since $e^{|u|} \leq e^u + e^{-u}$ for all $u \in \mathbb{R}$ we have

$$\begin{aligned}
\int \sum_{n=0}^{\infty} |g(n,x)|\,\mu(dx) &= \int \sum_{n=0}^{\infty} e^{\theta^\top t(x)} \frac{1}{n!} |h^\top t(x)|^n\,\mu(dx) \\
&= \int e^{\theta^\top t(x)} e^{|h^\top t(x)|}\,\mu(dx) \\
&\leq \int e^{\theta^\top t(x)} e^{h^\top t(x)}\,\mu(dx) + \int e^{\theta^\top t(x)} e^{-h^\top t(x)}\,\mu(dx) \\
&= c(\theta + h) + c(\theta - h) < \infty
\end{aligned}$$

since we have assumed that $\theta \pm h \in \Theta$. Fubini's theorem now yields that $g(n,\cdot)$ is integrable for all n which yields (B.1). □

DOI: 10.1201/9781003272359-B

In addition we have the following:

Lemma B.2. *Consider a regular exponential family with canonical parameter space Θ, canonical statistic t, representation space $\mathcal{X}$, and base measure μ. Then for every $\theta \in \Theta$ and every $m_1, \ldots, m_k \in \mathbb{N}_0$ there is a neighbourhood U_θ and a μ-integrable function h_θ such that for $\eta \in U_\theta$ it holds that*

$$\left| \frac{\partial^{m_1+\cdots+m_k}}{\partial \eta_1^{m_1} \cdots \partial \eta_k^{m_k}} e^{\eta^\top t(x)} \right| = \left| \prod_{i=1}^k t_i(x)^{m_i} e^{\eta^\top t(x)} \right| \leq h_\theta(x).$$

Proof. Let $a_j, j = 1, \ldots 2^k$ denote the 2^k vectors with coordinates $a_{ji} = \pm 1$. Let further C_ε denote the cube with $\varepsilon a_j, j = 1, \ldots, 2^k$ as corners and let

$$U_\theta = \{\eta \mid \eta - \theta \in C_\varepsilon\}.$$

We have for $\eta = \theta + h \in U_\theta$

$$\begin{aligned}\left| \prod_{i=1}^k t_i(x)^{m_i} e^{\eta^\top t(x)} \right| &= e^{\theta^\top t(x)} \prod_{i=1}^k |t_i(x)|^{m_i} e^{h^\top t(x)} \\ &\leq e^{\theta^\top t(x)} \prod_{i=1}^k |t_i(x)|^{m_i} \sum_{j=1}^{2^k} e^{\varepsilon a_j^\top t(x)}\end{aligned}$$

If ε is chosen sufficiently small, it holds that $\theta + a_j \in \Theta$ for $j = 1 \ldots, 2^k$ and hence we may let

$$h_\theta(x) = \sum_{j=1}^{2^k} \prod_{i=1}^k |t_i(x)|^{m_i} e^{(\theta + \varepsilon a_j)^\top t(x)}$$

as each term in the sum is integrable by Theorem 3.8. □

B.1.2 Quadratic regularity of exponential families

Proof of Proposition 4.5. Curved exponential families are smooth and locally stable, so we only have to establish the additional regularity condition (4.3). We have

$$f_\beta(x) = e^{\phi(\beta)^\top t(x) - \psi(\phi(\beta))}.$$

Taylor's formula with Lagrange's remainder term (Theorem A.23) yields

$$f_\gamma(x) - f_\beta(x) = \left.\frac{\partial f_\beta(x)}{\partial \beta}\right|_{\beta=\gamma^*} (\gamma - \beta) = f_{\gamma^*}(x) J(\gamma^*)^\top \left(t(x) - \tau(\phi(\gamma^*)\right) (\gamma - \beta)$$

where γ^* is between γ and β. Thus if we let $\theta^* = \phi(\gamma^*)$, $\theta = \phi(\beta)$, and $\delta = \theta^* - \theta$ we have

$$|f_\gamma(x) - f_\beta(x)| \leq f_\beta(x) |\gamma - \beta| J(\gamma^*)^\top (t(x) - \tau(\gamma^*)) |e^{\delta^\top t(x) - \psi(\gamma^*) + \psi(\beta)}.$$

Using the same argument as in the proof of Lemma B.2, we have for $|\gamma - \beta|$ sufficiently small so $\delta \in C_\varepsilon$ that

$$|J(\gamma^*)^\top(t(x) - \tau(\gamma^*))|e^{\delta^\top t(x) - \psi(\gamma^*) + \psi(\beta)} \leq K_\theta \|t(x)\| \sum_{i=1}^{2^k} \left(e^{\varepsilon a_i t_i(x)}\right) = H_\theta(X)$$

for some constant K_θ and if ε is chosen sufficiently small, we have $\theta \pm 2\varepsilon a_i \in \Theta$ and therefore

$$\mathbf{E}_\theta\{H_\theta(X)^2\} < \infty,$$

as desired. □

B.2 Asymptotic existence of the MLE

Here we shall provide a proof of Lemma 5.25, stating that the MLE in a curved exponential family is asymptotically well-defined and given as a smooth function of the average canonical statistic.

Proof of Lemma 5.25. We fix an element $\beta_0 \in B$, let $\theta_0 = \phi(\beta_0)$, and $\eta_0 = \tau(\phi(\beta_0)) \in \tau(\phi(B))$. We also introduce the function

$$\tilde{\lambda}(\eta, \theta) = \theta^\top \eta - \psi(\theta)$$

and note that then $\lambda(\eta, \beta) = \tilde{\lambda}(\eta, \phi(\beta))$.

For $\eta = \eta_0$ we know from the regular case that $\tilde{\lambda}(\eta_0, \theta)$ has a unique maximum over $\theta \in \Theta$ for $\theta = \tau^{-1}(\eta_0) = \phi(\beta_0)$ implying that also $\lambda(\eta_0, \beta)$ has a unique maximum in β for $\beta = \beta_0$; thus we have $g(\eta_0) = g(\tau(\phi(\beta_0)))$ which is (5.14).

Next choose an ϵ so that $C_\epsilon = \{\beta \mid \|\beta - \beta_0\| \leq \epsilon\} \subseteq B$; this is possible because B is an open set. Because ϕ is a homeomorphism, ϕ^{-1} is continuous and therefore there is a δ such that

$$\|\beta - \beta_0\| > \epsilon \implies \|\phi(\beta) - \phi(\beta_0)\| > \delta. \tag{B.2}$$

Since also Θ is open, we may assume that $\{\theta \mid \|\theta - \theta_0\| \leq \delta\} \subseteq \Theta$ by choosing δ sufficiently small.

We note that if $\eta \to \eta_0$, then the functions $\tilde{\lambda}(\eta, \cdot)$ converge uniformly to $\tilde{\lambda}(\eta_0, \cdot)$ on bounded sets since

$$|\tilde{\lambda}(\eta, \theta) - \tilde{\lambda}(\eta_0, \theta)| = |\theta^\top(\eta - \eta_0)| \leq \|\theta\| \|\eta - \eta_0\|.$$

This holds in particular on the sphere $S_\delta = \{\theta \mid \|\theta - \theta_0\| = \delta\}$ and $S_\delta \cup \{\theta_0\}$. Since

$$\sup_{\theta \in S_\delta} \tilde{\lambda}(\eta_0, \theta) < \tilde{\lambda}(\eta_0, \theta_0) = \lambda(\eta_0, \beta_0),$$

there is therefore a neighbourhood U_{η_0} of η_0 so that for all $\eta \in U_{\eta_0}$ it also holds that

$$\sup_{\theta \in S_\delta} \tilde{\lambda}(\eta, \theta) < \tilde{\lambda}(\eta, \theta_0) = \lambda(\eta, \beta_0).$$

As ψ is strictly convex, $\tilde{\lambda}(\eta, \cdot)$ is strictly concave for any η, we may streghten the conclusion also to hold for all values outside that sphere:

$$\sup_{\{\theta \mid \|\theta-\theta_0\| \geq \delta\}} \tilde{\lambda}(\eta, \theta) < \lambda(\eta, \beta_0); \tag{B.3}$$

for if there were a θ outside this sphere with $\tilde{\lambda}(\eta, \theta) \geq \lambda(\eta, \beta_0)$ and $\theta^* \in S_\delta$ denotes the point on the line segment between θ and θ_0, we would have

$$\tilde{\lambda}(\eta, \theta^*) < \tilde{\lambda}(\eta, \theta) \text{ and } \tilde{\lambda}(\eta, \theta^*) < \tilde{\lambda}(\eta, \theta_0)$$

contradicting that $\tilde{\lambda}(\eta, \cdot)$ is concave.

The continuous function $\lambda(\eta, \cdot)$ must attain its maximum over the compact set C_ϵ. From (B.2) and (B.3) we conclude that $\lambda(\eta, \beta) < \lambda(\eta, \beta_0)$ for all $\beta \notin C_\epsilon$ and thus the maximum over C_ϵ is actually a global maximum. Since the maximizer is an interior point in B, it must be a stable point and is thus implicitly defined by the equation $S(\eta, \beta) = 0$ where

$$S(\eta, \beta) = \frac{\partial}{\partial \beta} \lambda(\eta, \beta) = (\eta - \tau(\phi(\beta))^\top J(\beta).$$

We now wish to use the implicit function theorem (Theorem A.21) on this equation and get for the partial derivatives with respect to β

$$\begin{aligned} A(\eta, \beta)_{ij} &= \frac{\partial S_i(\eta, \beta)}{\partial \beta_j} \\ &= \sum_{u=1}^{k} \frac{\partial^2 \phi_u(\beta)}{\partial \beta_i \partial \beta_j} (\eta_u - \tau_u(\phi(\beta))) - \sum_{u=1}^{k} \frac{\partial \phi_u(\beta)}{\partial \beta_i} \kappa(\phi(\beta))_{uv} \frac{\partial \phi_v(\beta)}{\partial \beta_j}, \end{aligned}$$

or in a more compact form

$$A(\eta, \beta) = \sum_u D^2 \phi(\beta)_u (\eta_u - \tau_u(\phi(\beta))) - J(\beta)^\top \kappa(\phi(\beta)) J(\beta). \tag{B.4}$$

For $\eta = \tau(\phi(\beta))$ the first term in (B.4) is zero and thus if $\eta_0 = \tau(\phi(\beta_0))$ we get

$$A(\eta_0, \beta_0) = -J(\beta_0)^\top \kappa(\phi(\beta_0)) J(\beta_0) = -i(\beta_0). \tag{B.5}$$

Since κ is positive definite and $J(\beta)$ has full rank we have that if

$$0 = -\lambda^\top A(\eta_0, \beta_0) \lambda = -(J(\beta_0)\lambda)^\top \kappa(\phi(\beta_0)) J(\beta_0) \lambda$$

we must have $J(\beta_0)\lambda = 0$ and hence $-A(\eta_0, \beta_0)$ is positive definite. Therefore, $\det A(\eta_0, \beta_0) \neq 0$ and Theorem A.21 applies, We conclude that there is a neighbourhood $U'_{\eta_0} \subseteq U_{\eta_0}$ where g is well-defined as a smooth function of η. Further, we have

$$\frac{\partial S(\eta, \beta)^\top}{\partial \eta} = J(\beta)^\top. \tag{B.6}$$

Now implicit differentiation combination with (B.5) and (B.6) yields that the function g has Jacobian equal to

$$Dg(\eta) = i(g(\eta))^{-1}J(g(\eta))^{\top}.$$

We have now established the conclusions of the theorem in a neighbourhood $\eta \in U'_{\eta_0}$ of an arbitrary $\eta_0 \in \tau(\phi(B))$. We finally let $O = \cup_{\eta_0 \in \tau(\phi(B))} U'_{\eta_0}$ and the proof is complete. □

B.3 Iterative proportional scaling

In this section we establish convergence of the IPS algorithm for two-way classifications in contingency tables, as described in Section 8.3.3.2.

Let $\mathcal{I} = \{0, \ldots, r-1\}$, $\mathcal{J} = \{0, \ldots, s-1\}$, and let $B = \{B_{ij}, i \in \mathcal{I}, j \in \mathcal{J}\}$ be an array of positive numbers $B_{ij} > 0$. Further, let $\mathcal{M} = \mathcal{M}(B)$ be the set of arrays with positive entries that have the same cross-product ratios as B, i.e. that for all $i, i^* \in \mathcal{I}, j, j^* \in \mathcal{J}$ satisfy

$$\frac{m_{ij}m_{i^*j^*}}{m_{ij^*}m_{i^*j}} = \frac{B_{ij}B_{i^*j^*}}{B_{ij^*}B_{i^*j}}. \tag{B.7}$$

Lemma B.3. *A table of means $\lambda = \{\lambda_{ij}\}$ is in the shifted multiplicative Poisson family determined by B if and only if $\lambda \in \mathcal{M}(B)$.*

Proof. If $\lambda_{ij} = \mu\rho_i\eta_j B_{ij}$ we get

$$\frac{\lambda_{ij}\lambda_{i^*j^*}}{\lambda_{ij^*}\lambda_{i^*j}} = \frac{\mu\rho_i\rho_j B_{ij}\,\mu\rho_{i^*}\eta_{j^*}B_{i^*j^*}}{\mu\rho_i\eta_{j^*}B_{ij^*}\,\mu\rho_{i^*}\eta_j B_{i^*j}} = \frac{B_{ij}B_{i^*j^*}}{B_{ij^*}B_{i^*j}}$$

and hence $\lambda \in \mathcal{M}(B)$.

Conversely, if $\lambda \in \mathcal{M}(B)$, we may choose $i^* = j^* = 0$ and thus

$$\lambda_{ij} = \frac{\lambda_{i0}\lambda_{0j}}{\lambda_{00}}\frac{B_{ij}B_{00}}{B_{i0}B_{0j}} = \frac{\lambda_{00}}{B_{00}}\frac{\lambda_{i0}B_{00}}{\lambda_{00}B_{i0}}\frac{\lambda_{0j}B_{00}}{\lambda_{00}B_{0j}}B_{ij} = \mu\rho_i\eta_j B_{ij}$$

where

$$\mu = \frac{\lambda_{00}}{B_{00}}, \quad \rho_i = \frac{\lambda_{i0}B_{00}}{\lambda_{00}B_{i0}}, \quad \eta_j = \frac{\lambda_{0j}B_{00}}{\lambda_{00}B_{0j}}$$

are all positive and $\rho_0 = \eta_0 = 1$. This completes the proof. □

Thus we may parametrize the shifted multiplicative Poisson model by $\mathcal{M}(B)$. Consider now the marginal totals from a two-way contingency table

$$y_{1+} + \cdots + y_{r-1,+} = y_{+1} + \cdots + y_{+,s-1} = y_{++}$$

and the log-likelihood for $m \in \mathcal{M}(B)$:

$$\ell_y(m) = \sum_{i=0}^{r-1}\sum_{j=0}^{s-1}(y_{ij}\log m_{ij} - m_{ij}). \tag{B.8}$$

We note that if m satisfies the likelihood equations (8.13), we have $m_{++} = y_{++}$ so if we define

$$\mathcal{M}_y(B) = \{m \in \mathcal{M}(B) : m_{++} = y_{++}\},$$

we may as well maximize ℓ_y over $\mathcal{M}_y(B)$ rather than $\mathcal{M}(B)$. If we let

$$\log m_{ij} = \gamma + \alpha_i + \beta_j + \log B_{ij}$$

for $m \in \mathcal{M}_y(B)$, (B.8) simplifies to

$$\ell_y(m) = \sum_{i=0}^{r-1} y_{i+}\alpha_i + \sum_{j=0}^{s-1} y_{+j}\beta_j + \sum_{i=0}^{r-1}\sum_{j=0}^{s-1} y_{ij} \log B_{ij} + y_{++}(\gamma - 1).$$

We then have an important lemma.

Lemma B.4. *If all marginal totals of y are positive and $m^n, n \in \mathbb{N}$ is a sequence of elements in $\mathcal{M}_y(B)$ so that $m^n_{ij} \to 0$ for $n \to \infty$, then $\ell_y(m^n) \to -\infty$ for $n \to \infty$.*

Proof. Consider the array x given as $x_{ij} = y_{i+}y_{+j}/y_{++}$. Then all entries of x are positive and has the same marginal totals as y.

$$x_{i+} = y_{i+}, i \in \mathcal{I}, \quad x_{+j} = y_{+j}, j \in \mathcal{J}, \quad x_{++} = y_{++}.$$

Thus the formal Poisson likelihood based on x is

$$\begin{aligned}\ell_x(m) &= \sum_{i=0}^{r-1} x_{i+}\alpha_i + \sum_{j=0}^{s-1} x_{+j}\beta_j + \sum_{i=0}^{r-1}\sum_{j=0}^{s-1} x_{ij} \log B_{ij} + x_{++}(\gamma - 1) \\ &= \ell_y(m) + \sum_{i=0}^{r-1}\sum_{j=0}^{s-1} (x_{ij} - y_{ij}) \log B_{ij} = \ell_y(m) + \text{constant}.\end{aligned}$$

Thus, apart from a constant, we have for $m \in \mathcal{M}_y(B)$

$$\ell_y(m^n) = \sum_{i=0}^{r-1}\sum_{j=0}^{s-1} y_{ij} \log m^n_{ij} + c_1 = \sum_{i=0}^{r-1}\sum_{j=0}^{s-1} x_{ij} \log m^n_{ij} + c_2$$

where c_1 and c_2 are constants. But since $x_{ij} > 0$ for all ij and entries of m are bounded above in $\mathcal{M}_y(B)$, the last expression clearly tends to $-\infty$ for $m^n_{ij} \to 0$, as desired. □

Next, we let $m^0 \in \mathcal{M}_y(B)$ be determined as

$$m^0_{ij} = \frac{y_{++}}{B_{++}} B_{ij}$$

and define

$$\mathcal{M}^0_y(B) = \{m \in \mathcal{M}(B) : m_{++} = y_{++} \text{ and } \ell_y(m) \geq \ell_y(m^0)\}.$$

We may then show

Corollary B.5. *If y has all marginal totals positive, the MLE of m based on y exists.*

Proof. Any maximizer of ℓ_y over $\mathcal{M}_y(B)$ is also a maximizer over $\mathcal{M}^0_y(B)$ and vice versa. By Lemma B.4, the latter is closed and bounded, hence compact and since ℓ_y is continuous in m, it attains its maximum over $\mathcal{M}^0_y(B)$, hence over $\mathcal{M}_y(B)$ and $\mathcal{M}(B)$. □

We now define the *proportional scaling* operations T_R and T_C on $\mathcal{M}(B)$ as

$$T_R(m)_{ij} = m_{ij}\frac{y_{i+}}{m_{i+}},\ i \in \mathcal{I}, j \in \mathcal{J};\quad T_C(m)_{ij} = m_{ij}\frac{y_{+j}}{m_{+j}},\ i \in \mathcal{I}, j \in \mathcal{J}$$

and note the following

Lemma B.6. *The scaling operations are continuous operations on $\mathcal{M}^0_y(B)$ and satisfy*

$$T_R(\mathcal{M}^0_y(B)) \subseteq \mathcal{M}^0_y(B), \quad T_C(\mathcal{M}^0_y(B)) \subseteq \mathcal{M}^0_y(B).$$

Further, they adjust the marginals to satisfy

$$T_R(m)_{i+} = y_i,\ i = 1 \in \mathcal{I};\quad T_C(m)_{+j} = s_j,\ j = 1 \in \mathcal{J}.$$

In addition, it holds for all $m \in \mathcal{M}^0_y(B)$ that $\ell_y(T_R(m)) \geq \ell_y(m)$ with equality if and only if $m_{i+} = y_{i+}, i \in \mathcal{I}$ and similarly $\ell_y(T_C(m)) \geq \ell_y(m)$ with equality if and only if $m_{+_j} = y_{+j}, j \in \mathcal{J}$

Proof. The operations are obviously continuous. To see that T_R preserves cross-product ratios, we let $\gamma_i = y_{i+}/m_{i+}$ and get

$$\frac{T_R(m)_{ij}T_R(m)_{i^*j^*}}{T_R(m)_{ij^*}T_R(m)_{i^*j}} = \frac{m_{ij}\gamma_i m_{i^*j^*}\gamma_{i^*}}{m_{ij^*}\gamma_i m_{i^*j}\gamma_{i^*}} = \frac{m_{ij}m_{i^*j^*}}{m_{ij^*}m_{i^*j}} = \frac{B_{ij}B_{i^*j^*}}{B_{ij^*}B_{i^*j}}$$

and similarly for T_C.

Next we have

$$T_R(m)_{i+} = \sum_{j=0}^{s-1} m_{ij}\frac{y_{i+}}{m_{i+}} = \frac{y_{i+}}{m_{i+}}m_{i+} = y_{i+}$$

and similarly for T_C, showing that marginals are adjusted, implying also that $T_R(m)_{++} = T_C(m)_{++} = y_{++}$ so we have shown that $T_R(\mathcal{M}_y(B)) \subseteq \mathcal{M}_y(B)$ and similarly for T_C.

Now consider the change in likelihood after updating the rows, say. We get

$$\begin{aligned}\ell_y(T_R(m)) &= \sum_{i=0}^{r-1}\sum_{j=0}^{s-1} y_{ij}(\log m_{ij} + \log y_{i+} - \log m_{i+}) \\ &= \ell_y(m) + \sum_{i=0}^{r-1} y_{i+}(\log y_{i+} - \log m_{i+}).\end{aligned}$$

The information inequality in Lemma A.25 therefore implies

$$\ell_y(T_R(m)) \geq \ell_y(m) \tag{B.9}$$

with equality if and only if $y_{i+} = m_{i+}$ for all $i \in \mathcal{I}$. Similarly we get for all $m \in \mathcal{M}(B)$ with $m_{++} = n$ that

$$\ell_y(T_C(m)) \geq \ell_y(m) \tag{B.10}$$

and the inequality is strict unless $y_{+j} = m_{+j}$ for all $j \in \mathcal{J}$. and since the likelihood increases whenever we scale by (B.9) and (B.10) it follows from Lemma B.6 that we also have $T_R(\mathcal{M}_y^0(B)) \subseteq \mathcal{M}_y^0(B)$ and $T_C(\mathcal{M}_y^0(B)) \subseteq \mathcal{M}_y^0(B)$ and the proof is complete. □

We next define the iteration

$$m^{n+1} = T_R(T_C(m^n)), n = 0, 1, 2 \ldots$$

and may now show:

Theorem B.7. *The sequence m^n of arrays obtained by iterative proportional scaling converges to the maximum likelihood estimate $\hat{m}$ of the mean in the shifted multiplicative Poisson model:*

$$\lim_{n\to\infty} m^n = \hat{m}.$$

Proof. The set $\mathcal{M}_y^0(B)$ is compact so we just need to show that every convergent subsequence of the sequence above has $\hat{m}$ as limit. So let $m^{n_k}, k = 1, 2, \ldots$ denote a convergent subsequence and let m^* be its limit. Since the log-likelihood increases at every scaling, we have that $\ell_y(m^n)$ is non-decreasing in n and thus

$$\ell_y(T_R(T_C(m^*))) = \lim_{k\to\infty} \ell_y(T_R(T_C(m^{n_k}))) \leq \ell_y(m^{n_{k+1}})) = \ell_y(m^*).$$

But since also

$$\ell_y(T_R(T_C(m^*))) \geq \ell_y(T_C(m^*)) \geq \ell_y(m^*)$$

we must have equality and thus

$$T_R(T_C(m^*)) = T_C(m^*), \quad T_C(m^*) = m^*, \quad T_R(m^*) = m^*,$$

implying that m^* satisfies the likelihood equations

$$m^*_{i+} = y_{i+}, i \in \mathcal{I}, \quad m^*_{+j} = y_{+j}, j \in \mathcal{J}$$

and hence is the unique MLE. □

Bibliography

Agresti, A. (2002). *Categorical Data Analysis* (2nd ed.). New York: John Wiley & Sons.

Andersen, A. H. (1969). Asymptotic results for exponential families. *Bulletin of the International Statistical Institute 43*, 241–242.

Berk, R. H. (1972). Consistency and asymptotic normality of maximum likelihood estimates for exponential models. *The Annals of Mathematical Statistics 43*, 193–204.

Borel, E. (1943). *Les Probabilités et la Vie*. Paris: Presses Universitaires de France.

Boyd, S. and L. Vandenberghe (2004). *Convex Optimization*. Cambridge, UK: Cambridge University Press.

Clemmensen, J., G. Hansen, A. Nielsen, J. Røjel, J. Steensberg, S. Sørensen, and J. Toustrup (1974). Lung cancer and air pollution in Fredericia. *Ugeskrift for Læger 136*, 2260–2268. In Danish.

Cramér, H. (1946). *Mathematical Methods of Statistics*. Princeton, NJ: Princeton University Press.

Fisher, R. A. (1922). On the mathematical foundations of theoretical statistics. *Philosophical Transactions of the Royal Society, Series A 222*, 309–368.

Fisher, R. A. (1934). *Statistical Methods for Research Workers* (5th ed.). Edinburgh: Oliver and Boyd.

Fisher, R. A. (1947). The analysis of covariance method for the relation between a part and the whole. *Biometrics 3*, 65–68.

Hald, A. (1952). *Statistical Theory with Engineering Applications*. New York: John Wiley & Sons.

Hald, A. (1990). *A History of Probability and Statistics and Their Applications before 1750*. New York: John Wiley & Sons.

Hald, A. (1998). *A History of Mathematical Statistics from 1750 to 1930*. New York: John Wiley & Sons.

Hansen, E. (2012). *Introduktion til Matematisk Statistik* (3rd ed.). København: Institut for Matematiske Fag, Københavns Universitet. In Danish.

Horn, R. A. and C. R. Johnson (2013). *Matrix Analysis* (2nd ed.). Cambridge: Cambridge University Press.

Jacod, J. and P. Protter (2004). *Probability Essentials* (2nd ed.). Berlin Heidelberg: Springer-Verlag.

Patefield, W. M. (1981). Algorithm AS 159. An efficient method of generating random $r \times c$ tables with given row and column totals. *Applied Statistics 30*, 91–97.

Popper, K. (1959). *The Logic of Scientific Discovery*. Abingdon-on-Thames, UK: Routledge.

Rasch, G. (1960). *Probabilistic Models for Some Intelligence and Attainment Tests*, Volume 1 of *Studies in Mathematical Psychology*. Copenhagen: Danmarks Pædagogiske Institut.

Rényi, A. (1953). On the theory of order statistics. *Acta Mathematica Academiae Scientarum Hungarica 4*, 191–231.

Rudin, W. (1976). *Principles of Mathematical Analysis* (3rd ed.). Kogakusha, Japan: McGraw Hill.

Schilling, R. L. (2017). *Measures, Integrals and Martingales* (2nd ed.). Cambridge, UK: Cambridge University Press.

Sick, K. (1965). Haemoglobin polymorphism of cod in the Baltic and the Danish Belt Sea. *Hereditas 54*, 19–48.

Snee, R. D. (1974). Graphical display of two-way contingency tables. *The American Statistician 28*, 9–12.

van der Vaart, A. W. (2012). *Asymptotic Statistics*. Cambridge, UK: Cambridge University Press.

von Scheele, C., G. Svensson, and J. Rasmusson (1935). Om Bestämning av Potatisens Stärkelse och Torrsubstanshalt med Tillhjälp av dess specifika Vikt. *Nordisk Jordbrugsforskning*, 22, 17.

Waaler, G. H. M. (1927). Über die Erblichkeitsverhältnisse der verschiedenen Arten von angeborener Rotgrünblindheit. *Acta Ophthalmologica 5*, 309–345.

Wald, A. (1949). Note on the consistency of the maximum likelihood estimate. *The Annals of Mathematical Statistics 20*, 595–601.

Wilks, S. S. (1938). The large sample distribution of the likelihood ratio for testing composite hypotheses. *The Annals of Mathematical Statistics 9*, 60–62.

Williams, C. B. (1943). The number of publications written by biologists. *Annals of Eugenics 12*, 143–146.

Index

Printed in the United States
by Baker & Taylor Publisher Services

Printed in the United States
by Baker & Taylor Publisher Services